Remote Sensing and GIS Integration

Theories, Methods, and Applications

Qihao Weng, Ph.D.

New York Chicago San Francisco
Lisbon London Madrid Mexico City
Milan New Delhi San Juan
Seoul Singapore Sydney Toronto

The McGraw-Hill Companies

Cataloging-in-Publication Data is on file with the Library of Congress

1 2 3 4 5 6 7 8 9 0 DOC/DOC 0 1 5 4 3 2 1 0 9

ISBN 978-0-07-160653-0
MHID 0-07-160653-X

Sponsoring Editor
Taisuke Soda

Editing Supervisor
Stephen M. Smith

Production Supervisor
Richard C. Ruzycka

Acquisitions Coordinator
Michael Mulcahy

Project Manager
Somya Rustagi,
Glyph International

Copy Editor
James K. Madru

Proofreader
Sonal Chandel,
Glyph International

Indexer
Robert Swanson

Art Director, Cover
Jeff Weeks

Composition
Glyph International

Printed and bound by RR Donnelley.

McGraw-Hill books are available at special quantity discounts to use as premiums and sales promotions, or for use in corporate training programs. To contact a representative, please e-mail us at bulksales@mcgraw-hill.com.

This book is printed on acid-free paper.

Contents

Foreword

When Qihao Weng asked me to write a foreword to his book, I had two immediate reactions. I was, of course, at first flattered and honored by his invitation but when I read further in his letter I shockingly realized that 20 years had gone by since Geoffrey Edwards, Yvan Bédard, and I published our paper on the integration of remote sensing and GIS in *Photogrammetric Engineering & Remote Sensing* (*PE&RS*). Twenty years is a long time in a fast-moving field such as ours that is concerned with geospatial data collection, management, analysis, and dissemination. I am very excited that Qihao had the enthusiasm, the stamina, and, last but not the least, the time to compile a comprehensive summary of the status of GIS/remote sensing integration today.

When Geoff, Yvan, and I wrote our paper it was not only the first partially theoretical article on the integration of the two very separate technologies at that time, but it was also meant to be a statement for the forthcoming National Center for Geographic Information and Analysis (NCGIA) Initiative 12: Integration of Remote Sensing and GIS. The leading scientists for this initiative—Jack Estes, Dave Simonett, Jeff Star, and Frank Davis—were all from the University of California at Santa Barbara NCGIA site, so I thought that we had to do something to prove our value to this group of principal scientists. To my delight, we achieved the desired result.

Actually, the making of this paper started to some degree by accident. Geoff Edwards discovered that he and I had both submitted papers with very similar titles and content to the GIS National Conference in Ottawa and asked me if we could combine our efforts. I immediately agreed and saw the chance to publish a research article in the upcoming special *PE&RS* issue on GIS. Geoff and Yvan worked at Laval University in Quebec, I was at the University of Maine in Orono, and, at this very important time, we all worked with Macintoshes and sent our files back and forth through the Internet without being concerned with data conversion issues.

When I look back upon those times, I ponder the research questions that we thought were the most pressing ones 20 years ago. How many of them have been solved by now, how many of them still exist, and how many new ones have appeared in the meantime? Is there still a dichotomy between GIS and remote sensing/image processing? Are the scientific communities that are concerned with the development of GIS and remote sensing still separated? Are data formats, conversion, and the lack of standards still the most pressing research questions? Is it not that we are used to switch from map view to satellite picture to bird's eye view or street view by a simple click in our geobrowser? Has not Google Earth taught us a lesson that technology can produce seamless geospatial databases from diverse datasets including, and relying on, remote sensing images that act as the backbone for geographic orientation? Do we not expect to be linked to geospatial databases through UMTS, wireless LAN, or hotspots wherever we are? Have we not seen a sharp increase in the use of remotely sensed data with the advent of very high resolution satellites and digital aerial cameras? In one sentence: Have we solved all problems that are associated with the integration of remote sensing and GIS?

It is here that Qihao Weng's book takes up this issue at a scientific level. His book presents the progress that we have made with respect to theories, methods, and applications. He also points out the shortcomings and new research questions that have arisen from new technologies and developments. Twenty years ago, we did not mention GPS, LiDAR, or the Internet as driving forces for geospatial progress. Now, we have to rethink our research questions, which often stem from new technologies and applications that always seem to be ahead of theories and thorough methodological analyses. Especially, the application part of this book looks at case studies that are methodically arranged into certain areas. It reveals how many applications are nowadays based on the cooperation of remote sensing with other geospatial data. As a matter of fact, it is hard to see any geospatial analysis field that does not benefit from incorporating remotely sensed data. On the other hand, it is also true that the results of automated interpretation of remotely sensed images have greatly been improved by an integrated analysis with diverse geospatial and attribute data managed in a GIS.

In 1989, when Geoff Edwards, Yvan Bédard, and I wrote our paper on the integration of remote sensing and GIS, these two technologies were predominantly separated from, or even antagonistic to, each other. Today, this dichotomy no longer exists. GISs incorporate remotely sensed images as an integral part of their geospatial databases and image processing systems incorporate GIS analysis capabilities in their processing software. I even doubt that the terms GIS (for data processing) and remote sensing (for data collection) hold the same importance now as they did 20 years ago. We have seen over the last 10 to 15 years the emergence of a new scientific discipline that encompasses these two technologies. Whether we refer to this field as geospatial science, geographic information science, geomatics, or geoinformatics, one thing is consistent: remote sensing, image analysis, and GIS are part of this discipline.

I congratulate Qihao Weng on accomplishing the immense task that he undertook in putting this book together. We now have the definitive state-of-the-art book on remote sensing/GIS integration. Twenty years from now, it will probably serve as the reference point from which to start the next scientific progress report. I will certainly use his book in my remote sensing and GIS classes.

Manfred Ehlers
University of Osnabrück
Osnabrück, Germany

Preface

Over the past three to four decades, there has been an explosive increase in the use of remotely sensed data for various types of resource, environmental, and urban studies. The evolving capability of geographic information systems (GIS) makes it possible for computer systems to handle geospatial data in a more efficient and effective way. The attempt to take advantage of these data and modern geospatial technologies to investigate natural and human systems and to model and predict their behaviors over time has resulted in voluminous publications with the label *integration.* Indeed, since the 1990s, the remote sensing and GIS literature witnessed a great deal of research efforts from both the remote sensing and GIS communities to push the integration of these two related technologies into a new frontier of scientific inquiry.

Briefly, the integration of remote sensing and GIS is mutually beneficial for the following two reasons: First, there has been a tremendous increase in demand for the use of remotely sensed data combined with cartographic data and other data gathered by GIS, including environmental and socioeconomic data. Products derived from remote sensing are attractive to GIS database development because they can provide cost-effective large-coverage data in a raster data format that are ready for input into a GIS and convertible to a suitable data format for subsequent analysis and modeling applications. Moreover, remote sensing systems usually collect data on multiple dates, making it possible to monitor changes over time for earth-surface features and processes. Remote sensing also can provide information about certain biophysical parameters, such as object temperature, biomass, and height, that is valuable in assessing and modeling environmental and resource systems. GIS as a modeling tool needs to integrate remote sensing data with other types of geospatial data. This is particularly true when considering that cartographic data produced in GIS are usually static in nature, with most being collected on a single occasion and then archived. Remotely sensed data can be used to correct, update, and maintain GIS databases. Second, it is still true that GIS is a predominantly data-handling technology, whereas remote sensing is primarily a data-collection technology.

Many tasks that are quite difficult to do in remote sensing image processing systems are relatively easy in a GIS, and vice versa. In a word, the need for the combined use of remotely sensed data and GIS data and for the joint use of remote sensing (including digital image processing) and GIS functionalities for managing, analyzing, and displaying such data leads to their integration.

This year marks the twentieth anniversary of the publishing of the seminal paper on integration by Ehlers and colleagues (1989), in which the perspective of an evolutionary integration of three stages was presented. In December 1990, the National Center for Geographic Information and Analysis (NCGIA) launched a new research initiative, namely, Initiative 12: Integration of Remote Sensing and GIS. The initiative was led by Drs. John Estes, Frank Davis, and Jeffrey Star and was closed in 1993. The objectives of the initiative were to identify impediments to the fuller integration of remote sensing and GIS, to develop a prioritized research agenda to remove those impediments, and to conduct or facilitate research on the topics of highest priority. Discussions were concentrated around five issues: institutional issues, data structures and access, data processing flow, error analysis, and future computing environments. (See www.ncgia.ucsb.edu/research/initiatives.html.) The results of the discussions were published in a special issue of *Photogrammetric Engineering & Remote Sensing* in 1991 (volume 57, issue 6).

In nearly two decades, we witnessed many new opportunities for combining ever-increasing computational power, modern telecommunications technologies, more plentiful and capable digital data, and more advanced analytical algorithms, which may have generated impacts on the integration of remote sensing and GIS for environmental, resource, and urban studies. It would be interesting to examine the progress being made by, problems still existing for, and future directions taken by the current technologies of computers, communications, data, and analysis. I decided to put together such a book to reflect part of my work over the past 10 years and found it challenging, at the beginning, to determine what, how, and why materials should or should not be engaged.

This book addresses three interconnected issues: theories, methods, and applications for the integration of remote sensing and GIS. First, different theoretical approaches to integration are examined. Specifically, this book looks at such issues as the levels, methodological approaches, and models of integration. The review then goes on to investigate practical methods for the integrated use of remote sensing and GIS data and technologies. Based on theoretical and methodological issues, this book next examines the current impediments, both conceptually and technically, to integration and their possible solutions. Extensive discussions are directed toward the impact of computers, networks, and telecommunications technologies; the impact of the availability of high-resolution satellite images and light detection and ranging (LiDAR) data; and, finally, the impact of new image-analysis algorithms on integration. The theoretical discussions end with my perspective on future developments. A large portion of this book is dedicated to showcasing a series of application areas involving the integration of remote sensing and GIS. Each application area starts with an analysis of state-of-the-art methodology followed by a detailed presentation of a case study. The application areas include urban land-use and land-cover mapping, landscape characterization and analysis, urban feature extraction, building extraction with LiDAR data, urban heat island and local climate analysis, surface runoff modeling and analysis, the relationship between air quality and land-use patterns, population estimation, quality-of-life assessment, urban and regional development, and public health.

Qihao Weng, Ph.D.

Acknowledgments

My interest in the topic of the integration of remote sensing and GIS can be traced back to the 1990s when I studied at the University of Georgia under the supervision of the late Dr. Chor-Pang Lo. He strongly encouraged me to take this research direction for my dissertation. I am grateful for his encouragement and continued support until he passed away in December 2007. In the spring of 2008, I was granted a sabbatical leave. A long-time collaborator, Dr. Dale Quattrochi, invited me to come to work with him, but the NASA fellowship did not come in time for my leave. Just at the moment of relaxation, a friend at McGraw-Hill, Mr. Taisuke Soda, sent me an invitation to write a book on the integration of remote sensing and GIS.

I wish to extend my most sincere appreciation to several recent Indiana State University graduates who have contributed to this book. Listed in alphabetical order, they are: Ms. Jing Han, Dr. Xuefei Hu, Dr. Guiying Li, Dr. Bingqing Liang, Dr. Hua Liu, and Dr. Dengsheng Lu. I thank them for data collection and analysis and for drafting some of the chapters. My collaborator, Dr. Xiaohua Tong of Tongji University at Shanghai, contributed to the writing of Chapters 2 and 6. Drs. Paul Mausel, Brain Ceh, Robert Larson, James Speer, Cheng Zhao, and Michael Angilletta, who are or were on the faculty of Indiana State University, reviewed earlier versions of some of the chapters.

My gratitude further goes to Professor Manfred Ehlers, University of Osnabrück, Germany, who was kind enough to write the Foreword for this book. His seminal works on the integration of remote sensing and GIS have always inspired me to pursue this evolving topic.

Finally, I am indebted to my family, to whom this book is dedicated, for their enduring love and support.

It is my hope that the publication of this book will provide stimulation to students and researchers to conduct more in-depth work and analysis on the integration of remote sensing and GIS. In the course of writing this book, I felt more and more like a student again, wanting to focus my future study on this very interesting topic.

About the Author

Qihao Weng is a professor of geography and the director of the Center for Urban and Environmental Change at Indiana State University. He is also a guest/adjunct professor at Wuhan University and Beijing Normal University, and a guest research scientist at the Beijing Meteorological Bureau. From 2008 to 2009, he visited NASA as a senior research fellow. He earned a Ph.D. in geography from the University of Georgia. At Indiana State, Dr. Weng teaches courses on remote sensing, digital image processing, remote sensing–GIS integration, and GIS and environmental modeling. His research focuses on remote sensing and GIS analysis of urban ecological and environmental systems, land-use and land-cover change, urbanization impacts, and human-environment interactions. In 2006 he received the Theodore Dreiser Distinguished Research Award, Indiana State's highest faculty research honor. Dr. Weng is the author of more than 100 peer-reviewed journal articles and other publications.

CHAPTER 1

Principles of Remote Sensing and Geographic Information Systems (GIS)

This chapter introduces to the principles of remote sensing and geographic information systems (GIS). Because there are many textbooks of remote sensing and GIS, the readers of this book may take a closer look at any topic discussed in this chapter if interested. It is my intention that only the most recent pertinent literature is included. The purpose for these discussions on remote sensing and GIS principles is to facilitate the discussion on the integration of remote sensing and GIS set forth in Chap. 2.

1.1 Principles of Remote Sensing

1.1.1 Concept of Remote Sensing

Remote sensing refers to the activities of recording, observing, and perceiving (sensing) objects or events in far-away (remote) places. In remote sensing, the sensors are not in direct contact with the objects or events being observed. Electromagnetic radiation normally is used as the information carrier in remote sensing. The output of a remote sensing system is usually an image representing the scene being observed. A further step of image analysis and interpretation is required to extract useful information from the image. In a more restricted sense, *remote sensing* refers to the science and technology of acquiring information about the earth's surface (i.e., land and ocean) and atmosphere using sensors onboard airborne (e.g., aircraft or balloons) or spaceborne (e.g., satellites and space shuttles) platforms. Depending on the scope, remote sensing may be broken down into (1) satellite remote sensing (when satellite platforms are used), (2) photography and photogrammetry (when photographs are used to capture visible light), (3) thermal remote sensing (when the thermal infrared portion of the spectrum is used), (4) radar remote sensing (when microwave wavelengths are used), and (5) LiDAR remote sensing (when laser pulses are transmitted toward the ground and the distance between the sensor and the ground is measured based on the return time of each pulse).

The technology of remote sensing evolved gradually into a scientific subject after World War II. Its early development was driven mainly by military uses. Later, remotely sensed data became widely applied for civil applications. The range of remote sensing applications includes archaeology, agriculture, cartography, civil engineering, meteorology and climatology, coastal studies, emergency response, forestry, geology, geographic information systems, hazards, land use and land cover, natural disasters, oceanography, water resources, and so on. Most recently, with the advent of high spatial-resolution imagery and more capable techniques, urban and related applications of remote sensing have been rapidly gaining interest in the remote sensing community and beyond.

1.1.2 Principles of Electromagnetic Radiation

Remote sensing takes one of the two forms depending on how the energy is used and detected. Passive remote sensing systems record the reflected energy of electromagnetic radiation or the emitted energy from the earth, such as cameras and thermal infrared detectors. Active remote sensing systems send out their own energy and record the reflected portion of that energy from the earth's surface, such as radar imaging systems.

Electromagnetic radiation is a form of energy with the properties of a wave, and its major source is the sun. Solar energy traveling in the form of waves at the speed of light (denoted as c and equals to 3×10^8 ms^{-1}) is known as the *electromagnetic spectrum*. The waves propagate through time and space in a manner rather like water waves, but they also oscillate in all directions perpendicular to their direction of travel. Electromagnetic waves may be characterized by two principal measures: wavelength and frequency. The wavelength λ is the distance between successive crests of the waves. The frequency μ is the number of oscillations completed per second. Wavelength and frequency are related by the following equation:

$$C = \lambda \times \mu \tag{1.1}$$

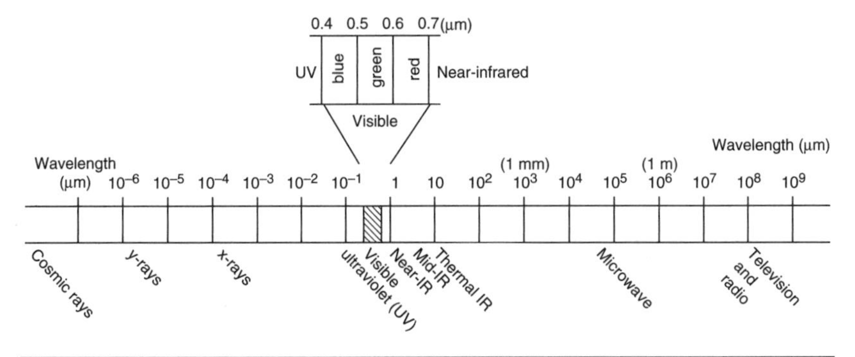

FIGURE 1.1 Major divisions of the electromagnetic spectrum.

The electromagnetic spectrum, despite being seen as a continuum of wavelengths and frequencies, is divided into different portions by scientific convention (Fig. 1.1). Major divisions of the electromagnetic spectrum, ranging from short-wavelength, high-frequency waves to long-wavelength, low-frequency waves, include gamma rays, x-rays, ultraviolet (UV) radiation, visible light, infrared (IR) radiation, microwave radiation, and radiowaves.

The visible spectrum, commonly known as the *rainbow of colors* we see as visible light (sunlight), is the portion of the electromagnetic spectrum with wavelengths between 400 and 700 billionths of a meter (0.4–0.7 μm). Although it is a narrow spectrum, the visible spectrum has a great utility in satellite remote sensing and for the identification of different objects by their visible colors in photography.

The IR spectrum is the region of electromagnetic radiation that extends from the visible region to about 1 mm (in wavelength). Infrared waves can be further partitioned into the near-IR, mid-IR, and far-IR spectrum, which includes thermal radiation. IR radiation can be measured by using electronic detectors. IR images obtained by sensors can yield important information on the health of crops and can help in visualizing forest fires even when they are enveloped in an opaque curtain of smoke.

Microwave radiation has a wavelength ranging from approximately 1 mm to 30 cm. Microwaves are emitted from the earth, from objects such as cars and planes, and from the atmosphere. These microwaves can be detected to provide information, such as the temperature of the object that emitted the microwave. Because their wavelengths are so long, the energy available is quite small compared with visible and IR wavelengths. Therefore, the fields of view must be large enough to detect sufficient energy to record a signal. Most passive microwave sensors thus are characterized by low spatial resolution. Active microwave sensing systems (e.g., radar) provide their own source of microwave radiation to illuminate the targets on the ground.

A major advantage of radar is the ability of the radiation to penetrate through cloud cover and most weather conditions owing to its long wavelength. In addition, because radar is an active sensor, it also can be used to image the ground at any time during the day or night. These two primary advantages of radar, all-weather and day or night imaging, make radar a unique sensing system.

The electromagnetic radiation reaching the earth's surface is partitioned into three types by interacting with features on the earth's surface. *Transmission* refers to the movement of energy through a surface. The amount of transmitted energy depends on the wavelength and is measured as the ratio of transmitted radiation to the incident radiation, known as *transmittance.* Remote sensing systems can detect and record both reflected and emitted energy from the earth's surface. *Reflectance* is the term used to define the ratio of the amount of electromagnetic radiation reflected from a surface to the amount originally striking the surface. When a surface is smooth, we get *specular* reflection, where all (or almost all) of the energy is directed away from the surface in a single direction. When the surface is rough and the energy is reflected almost uniformly in all directions, *diffuse* reflection occurs. Most features of the earth's surface lie somewhere between perfectly specular or perfectly diffuse reflectors. Whether a particular target reflects specularly or diffusely or somewhere in between depends on the surface roughness of the feature in comparison with the wavelength of the incoming radiation. If the wavelengths are much smaller than the surface variations or the particle sizes that make up the surface, diffuse reflection will dominate. Some electromagnetic radiation is absorbed through electron or molecular reactions within the medium. A portion of this energy then is reemitted, as *emittance,* usually at longer wavelengths, and some of it remains and heats the target.

For any given material, the amount of solar radiation that reflects, absorbs, or transmits varies with wavelength. This important property of matter makes it possible to identify different substances or features and separate them by their spectral signatures (spectral curves). Figure 1.2 illustrates the typical spectral curves for three major terrestrial features: vegetation, water, and soil. Using their reflectance differences, we can distinguish these common earth-surface materials. When using more than two wavelengths, the plots in multidimensional space tend to show more separation among the materials. This improved ability to distinguish materials owing to extra wavelengths is the basis for multispectral remote sensing.

Before reaching a remote sensor, the electromagnetic radiation has to make at least one journey through the earth's atmosphere and two journeys in the case of active (i.e., radar) systems or passive systems that detect naturally occurring reflected radiation. Each time a ray passes through the atmosphere, it undergoes absorption and scattering. Absorption is mostly caused by three types of atmospheric gasses, that is, ozone, carbon dioxide, and water vapor. The electromagnetic

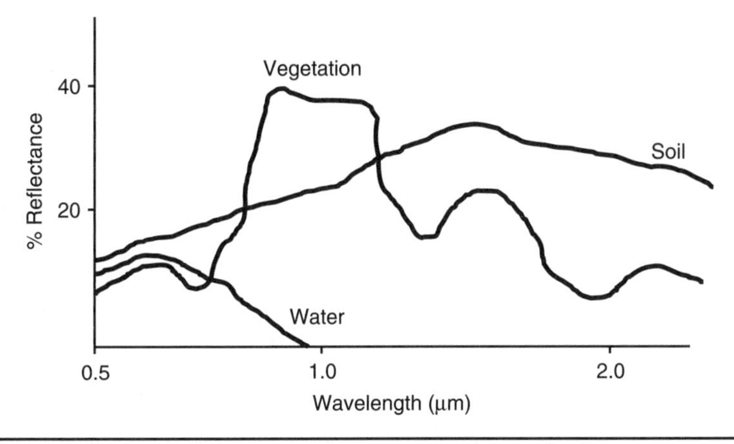

FIGURE 1.2 Spectral signatures of water, vegetation, and soil.

radiation is absorbed and reradiated again in all directions and probably over a different range of wavelengths. Scattering is mainly caused by N_2 and O_2 molecules, aerosols, fog particles, cloud droplets, and raindrops. The electromagnetic radiation is lost by being redirected out of the beam of radiation, but wavelength does not change. In either case, the energy is attenuated in the original direction of the radiation's propagation. However, there are certain portions of the electromagnetic spectrum that can pass through the atmosphere with little or no attenuation. The four principal windows (by wavelength interval) open to effective remote sensing from above the atmosphere include (1) visible–near-IR (0.4–2.5 μm), (2) mid-IR (3–5 μm), (3) thermal IR (8–14 μm), and (4) microwave (1–30 cm).

1.1.3 Characteristics of Remotely Sensed Data

Regardless of passive or active remote sensing systems, all sensing systems detect and record energy "signals" from earth surface features and/or from the atmosphere. Familiar examples of remote-sensing systems include aerial cameras and video recorders. More complex sensing systems include electronic scanners, linear/area arrays, laser scanning systems, etc. Data collected by these remote sensing systems can be in either analog format (e.g., hardcopy aerial photography or video data) or digital format (e.g., a matrix of "brightness values" corresponding to the average radiance measured within an image pixel). Digital remote sensing images may be input directly into a GIS for use; analog data also can be used in GIS through an analog-to-digital conversion or by scanning. More often, remote sensing data are first interpreted and analyzed through various methods of information extraction in order to provide needed data layers for GIS. The success of data collection from remotely sensed imagery requires an understanding of four basic resolution characteristics, namely, spatial, spectral, radiometric, and temporal resolution (Jensen, 2005).

Spatial resolution is a measurement of the minimum distance between two objects that will allow them to be differentiated from one another in an image and is a function of sensor altitude, detector size, focal size, and system configuration (Jensen, 2005). For aerial photography, spatial resolution is measured in resolvable line pairs per millimeter, whereas for other sensors, it refers to the dimensions (in meters) of the ground area that falls within the instantaneous field of view (IFOV) of a single detector within an array or pixel size (Jensen, 2005). Spatial resolution determines the level of spatial details that can be observed on the earth's surface. Coarse spatial resolution data may include a large number of mixed pixels, where more than one land-cover type can be found within a pixel. Whereas fine spatial resolution data considerably reduce the mixed-pixel problem, they may increase internal variation within the land-cover types. Higher resolution also means the need for greater data storage and higher cost and may introduce difficulties in image processing for a large study area. The relationship between the geographic scale of a study area and the spatial resolution of the remote-sensing image has been explored (Quattrochi and Goodchild, 1997). Generally speaking, on the local scale, high spatial-resolution imagery, such as that employing IKONOS and QuickBird data, is more effective. On the regional scale, medium-spatial-resolution imagery, such as that employing Landsat Thematic Mapper/Enhanced Thematic Mapping Plus (TM/ETM+) and Terra Advanced Spaceborne Thermal Emission and Reflection Radiometer (ASTER) data, is used most frequently. On the continental or global scale, coarse-spatial-resolution imagery, such as that employing Advanced Very High Resolution Radiometer (AHVRR) and Moderate Resolution Imaging Spectrometer (MODIS) data, is most suitable.

Each remote sensor is unique with regard to what portion(s) of the electromagnetic spectrum it detects. Different remote sensing instruments record different segments, or bands, of the electromagnetic spectrum. *Spectral resolution* of a sensor refers to the number and size of the bands it is able to record (Jensen, 2005). For example, AVHRR, onboard National Oceanographic and Atmospheric Administration's (NOAAs) Polar Orbiting Environmental Satellite (POES) platform, collects four or five broad spectral bands (depending on the individual instrument) in the visible (0.58–0.68 μm, red), near-IR (0.725–1.1 μm), mid-IR (3.55–3.93 μm), and thermal IR portions (10.3–11.3 and 11.5–12.5 μm) of the electromagnetic spectrum. AVHRR, acquiring image data at the spatial resolution of 1.1 km at nadir, has been used extensively for meteorologic studies, vegetation pattern analysis, and global modeling. The Landsat TM sensor collects seven spectral bands, including (1) 0.45–0.52 μm (blue), (2) 0.52–0.60 μm (green), (3) 0.63–0.69 μm (red), (4) 0.76–0.90 μm (near-IR), (5) 1.55–1.75 μm (short IR), (6) 10.4–12.5 μm (thermal IR), and (7) 2.08–2.35 μm (short IR). Its spectral resolution is higher than early instruments onboard Landsat such as the Multispectral Scanner (MSS) and the Return Beam Vidicon (RBV). Hyperspectral sensors (imaging spectrometers) are instruments that acquire images in many very narrow contiguous spectral bands throughout the visible, near-IR, mid-IR, and thermal IR portions of the spectrum. Whereas Landsat TM obtains only one data point corresponding to the integrated response over a spectral band 0.27 μm wide, a hyperspectral sensor, for example, is capable of obtaining many data points over this range using bands on the order of 0.01 μm wide. The National Aeronautics and Space Administration (NASA) Airborne Visible/Infrared Imaging Spectrometer (AVIRIS) collects 224 contiguous bands with wavelengths from 400–2500 nm. A broadband system can only discriminate general differences among material types, whereas a hyperspectral sensor affords the potential for detailed identification of materials and better estimates of their abundance. Another example of a hyperspectral sensor is MODIS, on both NASA's Terra and Aqua missions, and its follow-on to provide comprehensive data about land, ocean, and atmospheric processes simultaneously. MODIS has a 2-day repeat global coverage with spatial resolution (250, 500, or 1000 m depending on wavelength) in 36 spectral bands.

Radiometric resolution refers to the sensitivity of a sensor to incoming radiance, that is, how much change in radiance there must be on the sensor before a change in recorded brightness value takes place (Jensen, 2005). Coarse radiometric resolution would record a scene using only a few brightness levels, that is, at very high contrast, whereas fine radiometric resolution would record the same scene using many brightness levels. For example, the Landsat-1 Multispectral Scanner (MSS) initially recorded radiant energy in 6 bits (values ranging from 0 to 63) and later was expanded to 7 bits (values ranging from 0 to 127). In contrast, Landsat TM data are recorded in 8 bits; that is, the brightness levels range from 0 to 255.

Temporal resolution refers to the amount of time it takes for a sensor to return to a previously imaged location. Therefore, temporal resolution has an important implication in change detection and environmental monitoring. Many environmental phenomena constantly change over time, such as vegetation, weather, forest fires, volcanoes, and so on. Temporal resolution is an important consideration in remote sensing of vegetation because vegetation grows according to daily, seasonal, and annual phenologic cycles. It is crucial to obtain anniversary or near-anniversary images in change detection of vegetation. Anniversary images greatly minimize the effect of seasonal differences (Jensen, 2005). Many weather sensors have a high temporal resolution: the Geostationary Operational Environmental Satellite (GOES), 0.5/h; NOAA-9 AVHRR local-area coverage. 14.5/day; and Meteosat first generation, every 30 minutes.

In many situations, clear tradeoffs exist between different forms of resolution. For example, in traditional photographic emulsions, increases in spatial resolution are based on decreased size of film grain, which produces accompanying decreases in radiometric resolution; that is, the decreased sizes of grains in the emulsion portray a lower range of brightness values (Campbell, 2007). In multispectral scanning systems, an increase in spatial resolution requires a smaller IFOV, thus with less energy reaching the sensor. This effect may be compensated for by broadening the spectral window to pass more energy, that is, decreasing spectral resolution, or by dividing the energy into fewer brightness levels, that is, decreasing radiometric resolution (Campbell, 2007).

1.1.4 Remote Sensing Data Interpretation and Analysis

Remotely sensed data can be used to extract thematic and metric information, making it ready for input into GIS. Thematic information provides descriptive data about earth surface features. Themes can be as diversified as their areas of interest, such as soil, vegetation, water depth, and land cover. Metric information includes location, height, and their derivatives, such as area, volume, slope angle, and so on. Thematic information can be obtained through visual interpretation of remote sensing images (including photographs) or computer-based digital image analysis. Metric information is extracted by using the principles of photogrammetry.

Photographic/Image Interpretation and Photogrammetry

Photographic interpretation is defined as the act of examining aerial photographs/images for the purpose of identifying objects and judging their significance (Colwell, 1997). The activities of aerial photo/image interpreters may include (1) detection/identification, (2) measurement, and (3) problem solving. In the process of detection and identification, the interpreter identifies objects, features, phenomena, and processes in the photograph and conveys his or her response by labeling. These labels are often expressed in qualitative terms, for example, *likely, possible, probable,* or *certain.* The interpreter also may need to make quantitative measurements. Techniques used by the interpreter typically are not as precise as those employed by photogrammetrists. At the stage of problem solving, the interpreter identifies objects from a study of associated objects or complexes of objects from an analysis of their component objects, and this also may involve examining the effect of some process and suggesting a possible cause.

Seven elements are used commonly in photographic/image interpretation: (1) tone/color, (2) size, (3) shape, (4) texture, (5) pattern, (6) shadow, and (7) association. Tone/color is the most important element in photographic/image interpretation. *Tone* refers to each distinguishable variation from white to black and is a record of light reflection from the land surface onto the film. The more light received, the lighter is the image on the photograph. *Color* refers to each distinguishable variation on an image produced by a multitude of combinations of hue, value, and chroma. *Size* provides another important clue in discrimination of objects and features. Both the relative and absolute sizes of objects are important. An interpreter also should judge the significance of objects and features by relating to their background. The *shapes* of objects/features can provide diagnostic clues in identification. It is worthy to note that human-made features often have straight edges, whereas natural features tend not to. *Texture* refers to the frequency of change and arrangement in tones. The visual impression of smoothness or roughness of an area often can be a valuable clue in image interpretation. For example, water bodies typically are finely textured, whereas grass is medium and brush is rough, although there are always exceptions.

Pattern is defined as the spatial arrangement of objects. It is the regular arrangement of objects that can be diagnostic of features on the landscape. Human-made and natural patterns are often very different. Pattern also can be very important in geologic or geomorphologic analysis because it may reveal a great deal of information about the lithology and structural patterns in an area. *Shadow* relates to the size and shape of an object. Geologists like low-sun-angle photography because shadow patterns can help to identify objects. Steeples and smoke stacks can cast shadows that can facilitate interpretation. Tree identification can be aided by an examination of the shadows thrown. *Association* is one of the most helpful clues in identifying human-made installations. Some objects are commonly associated with one another. Identification of one tends to indicate or to confirm the existence of another. Smoke stacks, step buildings, cooling ponds, transformer yards, coal piles, and railroad tracks indicate the existence of a coal-fired power plant. Schools at different levels typically have characteristic playing fields, parking lots, and clusters of buildings in urban areas.

Photogrammetry traditionally is defined as the science or art of obtaining reliable measurements by means of photography (Colwell, 1997). Recent advances in computer and imaging technologies have transformed the traditional analog photogrammetry into digital (softcopy) photogrammetry, which uses modern technologies to produce accurate topographic maps, orthophotographs, and orthoimages employing the principles of photogrammetry. An *orthophotograph* is the reproduction of an aerial photograph with all tilts and relief displacements removed and a constant scale over the whole photograph. An *orthoimage* is the digital version of an orthophotograph, which can be produced from a stereoscopic pair of scanned aerial photographs or from a stereopair of satellite images (Lo and Yeung, 2002). The production of an orthophotograph or orthoimage requires the use of a digital elevation model (DEM) to register properly to the stereo model to provide the correct height data for differential rectification of the image (Jensen, 2005). Orthoimages are used increasingly to provide the base maps for GIS databases on which thematic data layers are overlaid (Lo and Yeung, 2002).

Photogrammetry for topographic mapping normally is applied to a stereopair of vertical aerial photograph (Wolf and Dewitt, 2000). An aerial photograph uses a central-perspective projection, causing an object on the earth's surface to be displaced away from the optical center (often overlaps with the geometric center) of the photograph depending on its height and location in the photograph. This relief displacement makes it possible to determine mathematically the height of the object by using a single photograph. To make geometrically corrected topographic maps out of aerial photographs, the relief displacement must be removed by using the theory of stereoscopic parallax with a stereopair of aerial photographs. Another type of error in a photograph is caused by tilts of the aircraft around the x, y, and z axes at the time of taking the photograph (Lo and Yeung, 2002). All the errors with photographs nowadays can be corrected by using a suite of computer programs.

Digital Image Preprocessing

In the context of digital analysis of remotely sensed data, the basic elements of image interpretation, although developed initially based on aerial photographs, also should be applicable to digital images. However, most digital image analysis methods are based on tone or color, which is represented as a digital number (i.e., brightness value) in each pixel of the digital image. As multisensor and high spatial-resolution data have become available, texture has been used in image classification, as well as contextual information, which describes the association of neighboring pixel values. Before main image analyses take place, preprocessing of digital images often is required. Image preprocessing may include detection and restoration of bad lines, geometric rectification or image registration, radiometric calibration and atmospheric correction, and topographic correction.

Geometric correction and atmospheric calibration are the most important steps in image preprocessing. *Geometric correction* corrects systemic and nonsystematic errors in the remote sensing system and during image acquisition (Lo and Yeung, 2002). It commonly involves (1) *digital rectification*, a process by which the geometry of an image is made planimetric, and (2) *resampling*, a process of extrapolating data values to a new grid by using such algorithms as nearest neighbor, bilinear, and cubic convolution. Accurate geometric rectification or image registration of remotely sensed data is a prerequisite, and many textbooks and articles have described them with details (e.g., Jensen, 2005).

If a single-date image is used for image classification, atmospheric correction may not be required (Song et al., 2001). When multitemporal or multisensor data are used, atmospheric calibration is mandatory.

This is especially true when multisensor or multiresolution data are integrated for image classification. A number of methods, ranging from simple relative calibration and dark-object subtraction to complicated model-based calibration approaches (e.g., 6S), have been developed for radiometric and atmospheric normalization or correction (Canty et al., 2004; Chavez, 1996; Gilabert et al., 1994; Du et al., 2002; Hadjimitsis et al., 2004; Heo and FitzHugh, 2000; Markham and Barker, 1987; McGovern et al., 2002; Song et al., 2001; Stefan and Itten, 1997; Tokola et al., 1999; Vermote et al., 1997).

In rugged or mountainous regions, shades caused by topography and canopy can seriously affect vegetation reflectance. Many approaches have been developed to reduce the shade effect, including (1) band ratio (Holben and Justice, 1981) and linear transformations such as principal component analysis and regression models (Conese et al., 1988, 1993; Naugle and Lashlee, 1992; Pouch and Campagna, 1990), (2) topographic correction methods (Civco, 1989; Colby, 1991), (3) integration of DEM and remote sensing data (Franklin et al., 1994; Walsh et al., 1990), and (4) slope/aspect stratification (Ricketts et al., 1993). Topographic correction is usually conducted before image classifications. More detailed information on topographic correction can be found in previous studies (Civco, 1989; Colby, 1991; Gu and Gillespie, 1998; Hale and Rock, 2003; Meyer et al., 1993; Richter, 1997; Teillet et al., 1982).

Image Enhancement and Feature Extraction

Various image-enhancement methods may be applied to enhance visual interpretability of remotely sensed data as well as to facilitate subsequent thematic information extraction. Image-enhancement methods can be roughly grouped into three categories: (1) contrast enhancement, (2) spatial enhancement, and (3) spectral transformation. *Contrast enhancement* involves changing the original values so that more of the available range of digital values is used, and the contrast between targets and their backgrounds is increased (Jensen, 2005). *Spatial enhancement* applies various algorithms, such as spatial filtering, edge enhancement, and Fourier analysis, to enhance low- or high-frequency components, edges, and textures. *Spectral transformation* refers to the manipulation of multiple bands of data to generate more useful information and involves such methods as band ratioing and differencing, principal components analysis, vegetation indices, and so on.

Feature extraction is often an essential step for subsequent thematic information extraction. Many potential variables may be used in image classification, including spectral signatures, vegetation indices, transformed images, textural or contextual information, multitemporal images, multisensor images, and ancillary data. Because of different capabilities in class separability, use of too many variables in a classification procedure may decrease classification accuracy (Price et al., 2002). It is important to select only the variables that are most effective for separating thematic classes. Selection of a suitable feature-extraction approach is especially necessary when hyperspectral data are used. This is so because the huge amount of data and the high correlations that exist among the bands of hyperspectral imagery and because a large number of training samples is required in image classification. Many feature-extraction approaches have been developed, including principal components analysis, minimum-noise fraction transform, discriminant analysis, decision-boundary feature extraction, nonparametric weighted-feature extraction, wavelet transform, and spectral mixture analysis (Asner and Heidebrecht, 2002; Landgrebe, 2003; Lobell et al., 2002; Myint, 2001; Neville et al., 2003; Okin et al., 2001; Rashed et al., 2001; Platt and Goetz, 2004).

Image Classification

Image classification uses spectral information represented by digital numbers in one or more spectral bands and attempts to classify each individual pixel based on the spectral information. The objective is to assign all pixels in the image to particular classes or themes (e.g., water, forest, residential, commercial, etc.) and to generate a thematic "map." It is important to differentiate between information classes and spectral classes. The former refers to the categories of interest that the analyst is actually trying to identify from the imagery, and the latter refers to the groups of pixels that are uniform (or near alike) with respect to their brightness values in the different spectral channels of the data. Generally, there are two approaches to image classification: supervised and unsupervised classification. In a *supervised* classification, the analyst identifies in the imagery homogeneous representative samples of different cover types (i.e., information classes) of interest to be used as training areas. Each pixel in the imagery then would be compared spectrally with the training samples to determine to which information class they should belong. Supervised classification employs such algorithms as minimum-distance-to-means, parallelepiped, and maximum likelihood classifiers (Lillesand et al., 2008). In an *unsupervised* classification, spectral classes are first grouped based solely on digital numbers in the imagery, which then are matched by the analyst to information classes.

In recent years, many advanced classification approaches, such as artificial neural network, fuzzy-set, and expert systems, have become widely applied for image classification. Table 1.1 lists the major advanced classification approaches that have appeared in the recent literature. A brief description of each category is provided in the following subsection. Readers who wish to have a detailed description of certain classification approaches should refer to cited references in the table.

Per-Pixel-Based Classification Most classification approaches are based on per-pixel information, in which each pixel is classified into one

Category	Advanced Classifiers	References
Per-pixel algorithms	Neural network	Atkinson and Tatnall, 1997; Chen et al., 1995; Erbek et al., 2004; Foody, 2002a, 2004; Foody and Arora, 1997; Foody et al., 1995; Kavzoglu and Mather, 2004; Ozkan and Erbeck, 2003; Paola and Schowengerdt, 1997; Verbeke et al., 2004
	Decision-tree classifier	DeFries and Chan, 2000; DeFries et al., 1998; Friedl et al., 1999; Friedl and Brodley, 1997; Hansen et al., 1996; Lawrence et al., 2004; Pal and Mather, 2003
	Spectral-angle classifier	Sohn and Rebello, 2002; Sohn et al., 1999
	Supervised iterative classification (multistage classification)	San Miguel-Ayanz and Biging, 1996, 1997
	Enhancement classification approach	Beaubien et al., 1999
	Multiple-forward-mode (MFM-5-scale) approach to running the 5-scale geometric optical reflectance model	Peddle et al., 2004
	Iterative partially supervised classification based on a combined use of a radial basis function network and a Markov random-field approach	Fernández-Prieto, 2002
	Classification by progressive generalization	Cihlar et al., 1998
	Support-vector machine	Brown et al., 1999; Foody and Mathur, 2004a, 2004b; Huang et al., 2002; Hsu and Lin, 2002; Keuchel et al., 2003; Kim et al., 2003; Mitra et al., 2004; Zhu and Blumberg, 2002

Table 1.1 Summary of Major Advanced Classification Methods

Category	Advanced Classifiers	References
	Unsupervised classification based on independent-components analysis mixture model	Lee et al., 2000; Shah et al., 2004
	Optimal iterative unsupervised classification	Jiang et al., 2004
	Model-based unsupervised classification	Koltunov and Ben-Dor, 2001, 2004
	Linear constrained discriminant analysis	Du and Chang, 2001; Du and Ren, 2003
	Multispectral classification based on probability-density functions	Erol and Akdeniz, 1996, 1998
	Layered classification	Jensen, 2005
	Nearest-neighbor classification	Collins et al., 2004; Haapanen et al., 2004; Hardin, 1994
	Selected-pixels classification	Emrahoglu et al., 2003
Subpixel algorithms	Imagine subpixel classifier	Huguenin et al., 1997
	Fuzzy classifier	Foody, 1996; Maselli et al., 1996; Shalan et al., 2003; Zhang and Foody, 2001
	Fuzzy expert system	Penaloza and Welch, 1996
	Fuzzy neural network	Foody, 1996, 1999; Kulkarni and Lulla, 1999; Mannan and Ray, 2003; Zhang and Foody, 2001
	Fuzzy-based multisensor data fusion classifier	Solaiman et al., 1999
	Rule-based machine-version approach	Foschi and Smith, 1997
	Linear regression or linear least squares inversion	Fernandes et al., 2004; Settle and Campbell, 1998

TABLE 1.1 Summary of Major Advanced Classification Methods (*Continued*)

Category	Advanced Classifiers	References
Per-field algorithms	Per-field or per-parcel classification	Aplin et al., 1999a; Dean and Smith, 2003; Lobo et al., 1996
	Per-field classification based on per-pixel or subpixel classified image	Aplin and Atkinson, 2001
	Parcel-based approach with two stages: per-parcel classification using conventional statistical classifier and then knowledge-based correction using contextual information	Smith and Fuller, 2001
	Map-guided classification	Chalifoux et al., 1998
	Object-oriented classification	Benz et al., 2004; Geneletti and Gorte, 2003; Gitas et al., 2004; Herold et al., 2003; Thomas et al., 2003; van der Sande et al., 2003; Walter, 2004
	Graph-based structural pattern recognition system	Barnsley and Barr, 1997
	Spectral shape classifier	Carlotto, 1998
Contextual-based approaches	Extraction and classification of homogeneous objects (ECHO)	Biehl and Landgrebe, 2002; Landgrebe, 2003; Lu et al., 2004
	Supervised relaxation classifier	Kontoes and Rokos, 1996
	Frequency-based contextual classifier	Gong and Howarth, 1992; Xu et al., 2003
	Contextual classification approaches for high- and low-resolution data, respectively, and a combination of both approaches	Kartikeyan et al., 1994; Sharma and Sarkar, 1998
	Contextual classifier based on region-growth algorithm	Lira and Maletti, 2002

Table 1.1 (*Continued*)

Category	Advanced Classifiers	References
	Fuzzy contextual classification	Binaghi et al., 1997
	Iterated conditional modes	Keuchel et al., 2003; Magnussen et al., 2004
	Sequential maximum a posteriori classification	Michelson et al., 2000
	Point-to-point contextual correction	Cortijo and de la Blanca, 1998
	Hierarchical maximum a posteriori classifier	Hubert-Moy et al., 2001
	Variogram texture classification	Carr, 1999
	Hybrid approach incorporating contextual information with per-pixel classification	Stuckens et al., 2000
	Two-stage segmentation procedure	Kartikeyan et al., 1998
Knowledge-based algorithms	Evidential reasoning classification	Franklin et al., 2002; Gong, 1996; Lein, 2003; Peddle, 1995; Peddle and Ferguson, 2002; Peddle et al., 1994; Wang and Civco, 1994
	Knowledge-based classification	Hung and Ridd, 2002; Kontoes and Rokos, 1996; Schmidt et al., 2004; Thomas et al., 2003
	Rule-based syntactical approach	Onsi, 2003
	Visual fuzzy classification based on use of exploratory and interactive visualization techniques	Lucleer and Kraak, 2004
	Decision fusion–based multitemporal classification	Jeon and Landgrebe, 1999
	Supervised classification with ongoing learning capability based on nearest-neighbor rule	Barandela and Juarez, 2002

Table 1.1 Summary of Major Advanced Classification Methods (*Continued*)

Category	Advanced Classifiers	References
Combinative approaches of multiple classifiers	Multiple classifier system (BAGFS: combines bootstrap aggregating with multiple feature subsets)	Debeir et al., 2002
	A consensus builder to adjust classification output (MLC, expert system, and neural network)	Liu et al., 2002b
	Integrated expert system and neural network classifier	Liu et al., 2002b
	Improved neuro-fuzzy image classification system	Qiu and Jensen, 2004
	Spectral and contextual classifiers	Cortijo and de la Blanca, 1998
	Mixed contextual and per-pixel classification	Conese and Maselli, 1994
	Combination of iterated contextual probability classifier and MLC	Tansey et al., 2004
	Combination of neural network and statistical consensus theoretical classifiers	Benediktsson and Kanellopoulos, 1999
	Combination of MLC and neural network using Bayesian techniques	Warrender and Augusteihn, 1999
	Combining multiple classifiers based on product rule, staked regression	Steele, 2000
	Combined spectral classifiers and GIS rule-based classification	Lunetta et al., 2003
	Combination of MLC and decision-tree classifier	Lu and Weng, 2004

TABLE 1.1 *(Continued)*

Category	Advanced Classifiers	References
	Combination of nonparametric classifiers (neural network, decision tree-classifier, and evidential reasoning)	Huang and Lees, 2004
	Combined supervised and unsupervised classification	Lo and Choi, 2004; Thomas et al., 2003

Adapted from Lu and Weng, 2007.

Table 1.1 Summary of Major Advanced Classification Methods (*Continued*)

category, and thematic classes are mutually exclusive. Traditional per-pixel classifiers typically develop a signature by combining the spectra of all training-set pixels for a given feature. The resulting signature contains the contributions of all materials present in the training pixels. Per-pixel-based classification algorithms may be parametric or nonparametric. The parametric classifiers assume that a normally distributed dataset exists and that statistical parameters (e.g., mean vector and covariance matrix) generated from the training samples are representative. However, the assumption of normal spectral distribution is often violated, especially with complex landscapes such as urban areas. In addition, insufficient, nonrepresentative, or multimode distributed training samples can introduce further uncertainty in the image-classification procedure. Another major drawback of the parametric classifiers lies in the difficulty in integrating spectral data with ancillary data.

With nonparametric classifiers, the assumption of a normal distribution of the dataset is not required. No statistical parameters are needed to generate thematic classes. Nonparametric classifiers thus are suitable for the incorporation of nonspectral data into a classification procedure. Much previous research has indicated that nonparametric classifiers may provide better classification results than parametric classifiers in complex landscapes (Foody, 2002b; Paola and Schowengerdt, 1995). Among commonly used nonparametric classification methods are neural-network, decision-tree, support-vector machine, and expert systems. Bagging, boosting, or a hybrid of both techniques may be used to improve classification performance in a nonparametric classification procedure. These techniques have been used in decision-tree (DeFries and Chan, 2000; Friedl et al., 1999; Lawrence et al., 2004) and support-vector machine (Kim et al., 2003) algorithms to enhance image classification.

Subpixel-Based Classification Owing to the heterogeneity of landscapes (particularly urban landscapes) and the limitation in spatial resolution of remote sensing imagery, mixed pixels are common in medium- and coarse-spatial-resolution data. The presence of mixed pixels has been recognized as a major problem that affects the effective use of remotely sensed data in per-pixel-based classifications (Cracknell, 1998; Fisher, 1997). Subpixel-based classification approaches have been developed to provide a more appropriate representation and accurate area estimation of land covers within the pixels, especially when coarse-spatial-resolution data are used (Binaghi et al., 1999; Foody and Cox, 1994; Ricotta and Avena, 1999; Woodcock and Gopal, 2000). A fuzzy representation, in which each location is decomposed of multiple and partial memberships of all candidate classes, is needed. Different approaches have been used to derive a soft classifier, including fuzzy-set theory, Dempster-Shafer theory, certainty factor (Bloch, 1996), softening the output of a hard classification from maximum likelihood (Schowengerdt, 1996), and neural networks (Foody, 1999; Kulkarni and Lulla, 1999; Mannan and Ray 2003). In addition to the fuzzy image classifier, other subpixel mapping approaches also have been applied. Among these approaches, the fuzzy-set technique (Foody 1996, 1998; Mannan et al., 1998; Maselli et al., 1996; Shalan et al., 2003; Zhang and Foody, 2001; Zhang and Kirby, 1999), ERDAS IMAGINE's subpixel classifier (Huguenin et al., 1997), and spectral mixture analysis (SMA)–based classification (Adams et al., 1995; Lu et al., 2003; Rashed et al., 2001; Roberts et al., 1998b) are the three most popular approaches used to overcome the mixed-pixel problem. An important issue for subpixel-based classifications lies in the difficulty in assessing classification accuracy.

Per-Field-Based Classification The heterogeneity in complex landscapes, especially in urban areas, results in high spectral variation within the same land cover class. With per-pixel classifiers, each pixel is individually grouped into a certain category, but the results may be noisy owing to high spatial frequency in the landscape. The per-field classifier is designed to deal with the problem of landscape heterogeneity and has been shown to be effective in improving classification accuracy (Aplin and Atkinson, 2001; Aplin et al., 1999a, 1999b; Dean and Smith, 2003; Lloyd et al., 2004). A per-field-based classifier averages out the noise by using land parcels (called *fields*) as individual units (Aplin et al., 1999a, 1999b; Dean and Smith 2003; Lobo et al., 1996; Pedley and Curran, 1991). GIS provides a means for implementing per-field classification through integration of vector and raster data (Dean and Smith 2003; Harris and Ventura, 1995; Janssen and Molenaar, 1995). The vector data are used to subdivide an image into parcels, and classification then is conducted based on the parcels, thus avoiding intraclass spectral variations. However, per-field classifications are often affected by such factors as the spectral and spatial properties of remotely sensed data, the size and shape of the fields, the definition of field boundaries, and land-cover classes chosen (Janssen and Molenaar, 1995). The difficulty in handling the dichotomy between vector and raster data models had an effect on the extensive use of the per-field classification approach. Remotely sensed data are acquired in the raster format, which represents regularly shaped patches of the earth's surface, whereas most GIS data are stored in vector format, representing geographic objects with points, lines, and polygons. With recent advances in GIS and image-processing software integration, the perceived difficulty is expected to lessen, and thus the per-field classification approach may become more popular.

Contextual Classification Contextual classifiers have been developed to cope with the problem of intraclass spectral variations (Flygare, 1997; Gong and Howarth, 1992; Kartikeyan et al., 1994; Keuchel et al., 2003; Magnussen et al., 2004; Sharma and Sarkar, 1998), in addition to object-oriented and per-field classifications. Contextual classification exploits spatial information among neighboring pixels to improve classification results (Flygare, 1997; Hubert-Moy et al., 2001; Magnussen et al., 2004; Stuckens et al., 2000). Contextual classifiers may be based on smoothing techniques, Markov random fields, spatial statistics, fuzzy logic, segmentation, or neural networks (Binaghi et al., 1997; Cortijo and de la Blanca, 1998; Kartikeyan et al., 1998; Keuchel et al., 2003; Magnussen et al. 2004). In general, presmoothing classifiers incorporate contextual information as additional bands, and a classification then is conducted using normal spectral classifiers, whereas postsmoothing classifiers use classified images that are developed previously using spectral-based classifiers. The Markov random-field-based contextual classifiers such as iterated conditional modes are the most frequently used approach in contextual classification (Cortijo and de la Blanca, 1998; Magnussen et al., 2004) and have proven to be effective in improving classification results.

Classification with Texture Information Many texture measures have been developed (Emerson et al., 1999; Haralick et al., 1973; He and Wang, 1990; Kashyap et al., 1982; Unser, 1995) and have been used for image classifications (Augusteijn et al., 1995; Franklin and Peddle, 1989; Gordon and Phillipson, 1986; Groom et al., 1996; Jakubauskas, 1997; Kartikeyan et al., 1994; Lloyd et al., 2004; Marceau et al., 1990; Narasimha Rao et al., 2002; Nyoungui et al., 2002; Podest and Saatchi, 2002). Franklin and Peddle (1990) found that gray-level co-occurrence matrix (GLCM)–based textures and spectral features of Le Systeme Pour l'Observation de la Terre (SPOT, or Earth Observation System) high resolution visible (HRV) images improved the overall classification accuracy. Gong and colleagues (1992) compared GLCM, simple statistical transformation (SST), and texture spectrum (TS) approaches with SPOT HRV data and found that some textures derived from

GLCM and SST improved urban classification accuracy. Shaban and Dikshit (2001) investigated GLCM, gray-level difference histogram (GLDH), and sum and difference histogram (SADH) textures from SPOT spectral data in an Indian urban environment and found that a combination of texture and spectral features improved the classification accuracy. Compared with the result obtained based solely on spectral features, about a 9 and 17 percent increases were achieved for an addition of one and two textures, respectively. Those authors further found that contrast, entropy, variance, and inverse difference moment provided higher accuracy and that the best size of moving window was 7 × 7 or 9 × 9. Use of multiple or multiscale texture images should be in conjunction with original image data to improve classification results (Butusov, 2003; Kurosu et al., 2001; Narasimha Rao et al., 2002; Podest and Saatchi, 2002; Shaban and Dikshit, 2001). Recently, geostatistical-based texture measures were found to provide better classification accuracy than using GLCM-based textures (Berberoglu et al., 2000; Lloyd et al., 2004). For a specific study, it is often difficult to identify a suitable texture because texture varies with the characteristics of the landscape under investigation and image data used. Identification of suitable textures involves determination of texture measure, image band, the size of the moving window, and other parameters (Chen et al., 2004; Franklin et al., 1996). The difficulty in identifying the best suitable textures and the computation cost for calculating textures limit extensive use of textures in image classification, especially in a large area.

1.2 Principles of GIS

1.2.1 Scope of Geographic Information System and Geographic Information Science

The appearance of geographic information systems (GIS) in the mid-1960s reflects the progress in computer technology and the influence of quantitative revolution in geography. GIS has evolved dramatically from a tool of automated mapping and data management in the early days into a capable spatial data-handling and analysis technology and, more recently, into geographic information science (GISc). The commercial success since the early 1980s has gained GIS an increasingly wider application. Therefore, to give GIS a generally accepted definition is difficult nowadays. An early definition by Calkins and Tomlinson (1977) states:

> A geographic information system is an integrated software package specifically designed for use with geographic data that performs a comprehensive range of data handling tasks. These tasks include data input, storage, retrieval and output, in addition to a wide variety of descriptive and analytical processes.

From the definition, it becomes clear that GIS handles geographic data, which include both spatial and attribute data that describe geographic features. Second, the basic functions of GIS include data input, storage, processing, and output. The basic concept of GIS is one of location and spatial distribution and relationship (Fedra, 1996). The backbone analytical function of GIS is overlay of spatially referenced data layers, which allows delineating their spatial relationships. However, it is also true that data used in GIS are predominantly static in nature, and most of them are collected on a single occasion and then archived.

GIS today is far broader and harder to define. Many people prefer to define its domain as geographic information science and technology (GIS&T), and it has become imbedded in many academic and practical fields. The GIS&T field is a loose coalescence of groups of users, managers, academics, and professionals all working with geospatial information. Each group has a distinct educational and "cultural" background. Each identifies itself with particular ways of approaching particular sets of problems.

Over the course of development, many disciplines have contributed to GIS. Therefore, GIS has many close and far "relatives." Disciplines that traditionally have researched geographic information technologies include cartography, remote sensing, geodesy, surveying, photogrammetry, etc. Disciplines that traditionally have researched digital technology and information include computer science in general and databases, computational geometry, image processing, pattern recognition, and information science in particular. Disciplines that traditionally have studied the nature of human understanding and its interactions with machines include particularly cognitive psychology, environmental psychology, cognitive science, and artificial intelligence.

In developing a core curriculum for GIS, the University Consortium for Geographic Information Science (UCGIS, 2003) suggests that "GIS&T should not be defined by merely linking segments of the traditional domains (e.g., cartography, remote sensing, statistical analysis, locational analysis, etc.), but rather by placing the emphasis directly upon concepts and methods for geographic problem solving in a computational environment." GIS should not be viewed merely as a collection of tools and techniques. Rather, the scope of GIS&T represents "a body of knowledge that focuses in an analytic fashion upon various aspects of spatial and spatiotemporal information and therefore constitutes, in some of its aspects, a science" (UCGIS, 2003).

GIS has been called or defined as an enabling technology because of the breadth of uses in the following disciplines as a tool. Disciplines that traditionally have studied the earth, particularly its surface and near surface in either physical or human aspect, include geology, geophysics, oceanography, agriculture, ecology, biogeography, environmental science, geography, global science, sociology, political science, epidemiology, anthropology, demography, and many more.

When the focus is largely on the utilization of GIS&T to attain solutions to real-world problems, GIS has more of an engineering flavor, with attention being given to both the creation and the use of complex tools and techniques that embody the concepts of GIS&T. Among the management and decision-making groups, for instance, GIS finds its intensive and extensive applications in resource inventory and management, urban planning, land records for taxation and ownership control, facilities management, marketing and retail planning, vehicle routing and scheduling, etc. Each application area of GIS requires a special treatment and must examine data sources, data models, analytical methods, problem-solving approaches, and planning and management issues.

1.2.2 Raster GIS and Capabilities

In GIS, data models provide rules to convert real geographic variation into discrete objects. There are generally two major types of data models: raster and vector. Figure 1.3 illustrates these two GIS data models. A *raster model* divides the entire study area into a regular grid of cells in specific sequence (similar to pixels in digital remote-sensing imagery), with each cell containing a single value. The raster model is a space-filling model because every location in the study area corresponds to a cell in the raster. A set of data describing a single characteristic for each location (cell) within a bounded geographic area forms a *raster data layer.* Within a raster layer, there may be numerous zones (also named *patches, regions,* and *polygons*), with each zone being a set of contiguous locations that exhibit the same value. All individual zones that have the same characteristics form a *class* of a raster layer. Each cell is identified by an ordered pair of coordinates (row and column numbers) and does not have an explicit topological relationship with its neighboring cells.

Important raster datasets used in GIS include digital land use and land cover data, DEMs of various resolutions, digital orthoimages, and digital raster graphics (DRGs) produced by the U.S. Geological Survey. Because remote sensing generates digital images in raster format, it is easier to interface with a raster GIS than any other type. It is believed that remote sensing images and information extracted from such images, along with GPS data, have become major data sources for raster GIS (Lillesand et al., 2008).

The analytical capabilities of a raster GIS result directly from the application of traditional statistics and algebra to process spatial data—raster data layers. Raster processing operations are commonly grouped into four classes based on the set of cells in the input layer(s) participating in the computation of a value for a cell in the output layer (Gao et al., 1996). They are (1) per-cell or local operation, (2) per-neighborhood or focal operation, (3) per-zone or zonal operation, and (4) per-layer or global operation. During a per-cell operation, the

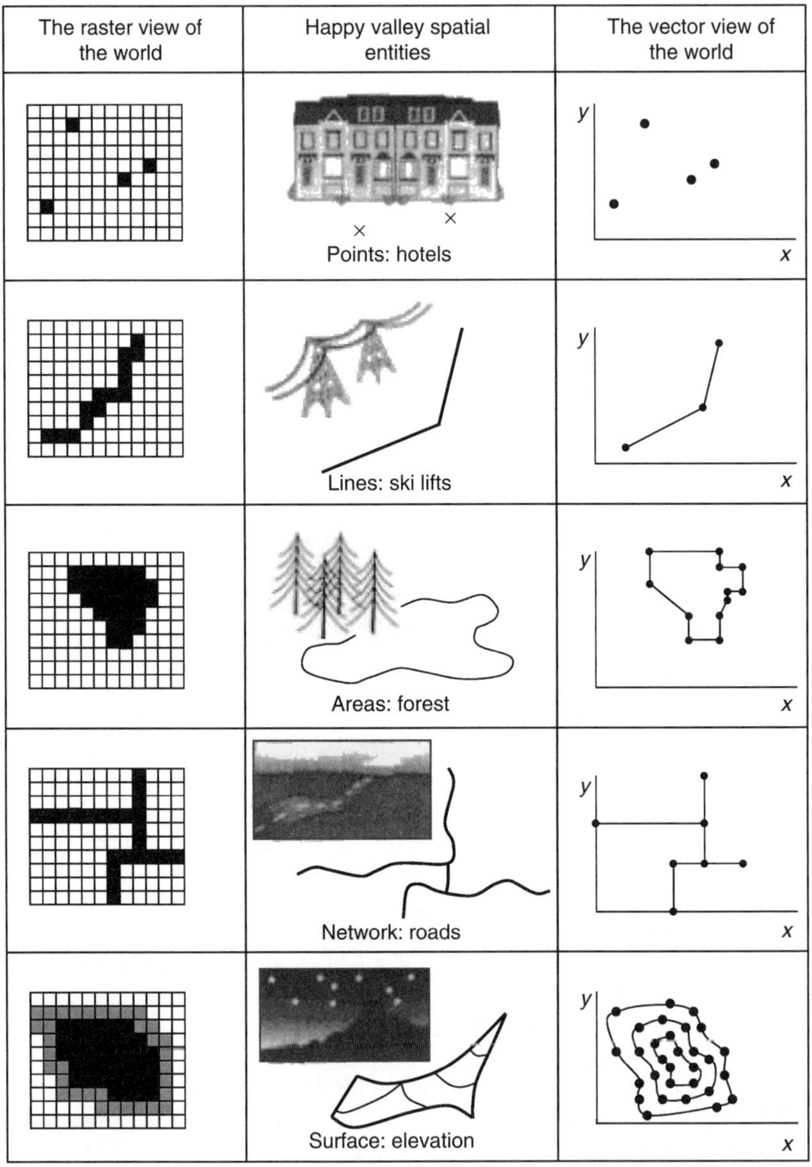

The raster view of the world	Happy valley spatial entities	The vector view of the world
	Points: hotels	y x
	Lines: ski lifts	y x
	Areas: forest	y x
	Network: roads	y x
	Surface: elevation	y x

Figure 1.3 Raster and vector data models. (*Adapted from Heywood et al. 1998; used with permission from Prentice Hall.*)

value of each new pixel is defined by the values of the same pixel on the input layer(s), and neighboring or distant pixels have no effect. In a per-neighborhood operation, the value of a pixel on the new layer is determined by the local neighborhood of the pixel on the old data layer. Per-zone operations treat each zone as an entity and apply algebra to the zones. Per-layer or global operations apply algebra to all cells in the old layer to the value of a pixel on the new layer.

A suite of raster processing operations may be logically organized in order to implement a particular data-analysis application. Cartographic modeling is such a technique, which builds models within a raster GIS by using *map algebra* (Tomlin, 1991). The method of map algebra, first introduced by Tomlin, involves the application of a sequence of processing commands to a set of input map layers to generate a desired output layer. Each processing operation is applied to all the cells in the layers, and the result is saved in a new layer. All the layers are referenced to a uniform geometry, usually that of a regular square grid. Map algebra is quite similar to traditional algebra. Most of the traditional mathematical capabilities plus an extensive set of advanced map-processing functions are used (Berry, 1993). Many raster GIS of the current generation have made a rich set of functions available to the user for cartographic modeling. Among these functions are overlay, distance and connectivity mapping, neighborhood analysis, topographic shading analysis, watershed analysis, surface interpolation, regression analysis, clustering, classification, and visibility analysis (Berry, 1993; Gao et al., 1996). The overlay functions, for example, may be categorized into two approaches: location-specific overlay and category-specific overlay. The first approach involves the vertical spearing of a set of layers; that is, the values assigned are a function of the point-by-point coincidence of the existing layers. Environmental modeling typically involves the manipulation of quantitative data using basic arithmetic operations, as well as simple descriptive statistical parameters. The type of function used for a overlaying may vary in the light of the nature of the data (e.g., nominal, ratio, etc.) being processed and use of the resulting data. An extreme case of this approach is dealing with Boolean images that only have values 1 and 0. Simple arithmetic operations in this case are used to perform logic or Boolean algebra operations. The second overlaying approach uses one layer to identify boundaries by which information is extracted from other layers; that is, the value assigned to a thematic region is considered as a function of the values on other layers that occur within its boundary. Summary statistics may be applied to this approach.

1.2.3 Vector GIS and Capabilities

The traditional vector data model is based on vectors. Its fundamental primitive is a point. Lines are created by connecting points with straight lines, and areas are defined by sets of lines. Encoding these features (i.e., points, lines, and polygons) is based on Cartesian coordinates using the principles of Euclidean geometry. The more advanced topological vector model, based on graph theory, encodes geographic features using nodes, arcs, and label points. The stretch of common

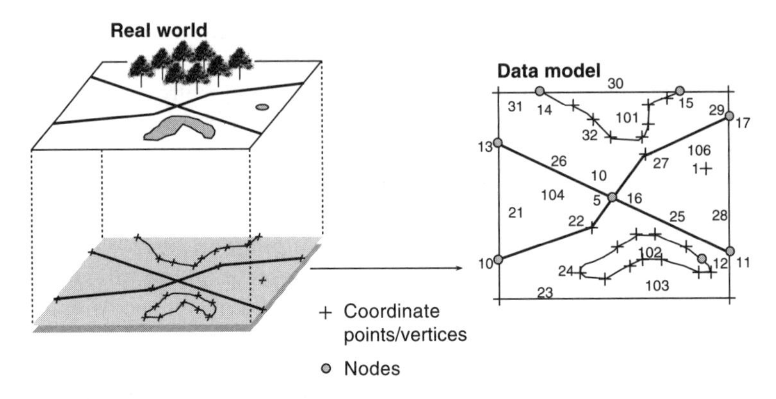

Point topology

Pt ID	F. code
1	10001

Pt ID	Coord.
1	*xy*

Arc topology

Arc ID	F. code
21	20001
22	20300
23	20001
24	20400
...	

Arc ID	F node	T node
21	13	10
22	10	16
23	10	11
24	12	12
...		

Arc ID	Start	Intermediate	End
21	*xy*	*xy,xy,xy...*	*xy*
22	*xy*	*xy,xy,xy...*	*xy*
23	*xy*	*xy,xy,xy...*	*xy*
...			
...			

Polygon topology

Poly ID	F. code
101	30200
102	40500
...	
...	

Poly ID	Arcs list
101	30,32
102	24

Arc ID	F node	T node	L-poly	R-poly
24	12	12	103	102
30	14	15	World	101
32	14	15	101	105

Arc ID	Start	Intermediate	End
24	*xy*	*xy,xy,xy...*	*xy*
30	*xy*	*xy,xy,xy...*	*xy*
32	*xy*	*xy,xy,xy...*	*xy*

FIGURE 1.4 The arc-node topological data model. (*Adapted from Lo and Yeung, 2002; used with permission from Prentice Hall.*)

boundary between two nodes forms an arc. Polygons are built by linking arcs or are stored as a sequence of coordinates. Figure 1.4 illustrates the arc-node topological data model.

The creation of a vector GIS database mainly involves three stages: (1) input of spatial data, (2) input of attribute data, and (3) linking of the spatial and attribute data. The spatial data are entered via digitized points and lines, scanned, and via vectorized lines or directly from other digital sources. In the process of spatial data generation, once points are entered and geometric lines are created, topology must be built, which is followed by editing, edge matching, and other graphic spatial correction procedures. *Topology* is the mathematical relationship among the points, lines, and polygons in a vector data layer. Building topology involves calculating and encoding relationships between the points, lines, and polygons. Attribute data are data that describe the spatial features, which can be keyed in or imported from other digital databases. Attribute data often are stored and manipulated in entirely separate ways from the spatial data, with a linkage between the two types of data by a corresponding object ID. Various database models are used for spatial and attribute data in vector GIS. The most widely used one is the georelational database model (Fig. 1.5), which uses a common-feature identifier to link the topological data model, representing spatial data and their relationships, to a relational database management system (DBMS), representing attribute data.

The analytical capabilities of vector GIS are not quite the same as those of raster GIS. There are more operations dealing with objects, and measures such as area have to be calculated from coordinates of objects instead of counting cells. The analytical functions can be broadly grouped into topological and nontopological ones. The former includes buffering, overlay, network analysis, and reclassification, whereas the latter includes attribute database query, address geocoding, area calculation, and statistical computation (Lo and Yeung, 2002).

Database query may be conducted according to the characteristics of the spatial or attribute data. Spatial queries may use points and arcs to display the locations of all objects stored or a subset of the data and may search within the radius of a point, within a bounding rectangle, or within an irregular polygon. Attributes and entity types can be displayed by varying colors, line patterns, or point symbols. Different vector GIS systems use different ways of formulating queries, but the Standard Query Language (SQL), including relational, arithmetic, and Boolean operators, is used by many systems.

Topological overlay is an important vector GIS analytical function. When two GIS layers are overlaid, the topology is updated for the new, combined map. The result may be information about new relationships for the old (input) maps rather than the creation of new objects. Figure 1.6 shows various types of topological overlays. The point-in-polygon overlay superimposes point objects on areas, thus computing "is contained in" relationship and resulting in a new attribute for each point. The line-on-polygon overlay superimposes line objects on area objects, thus computing "is contained in" relationships, and "containing area" is a new attribute of each output line. The polygon-on-polygon (i.e., polygon) overlay superimposes two

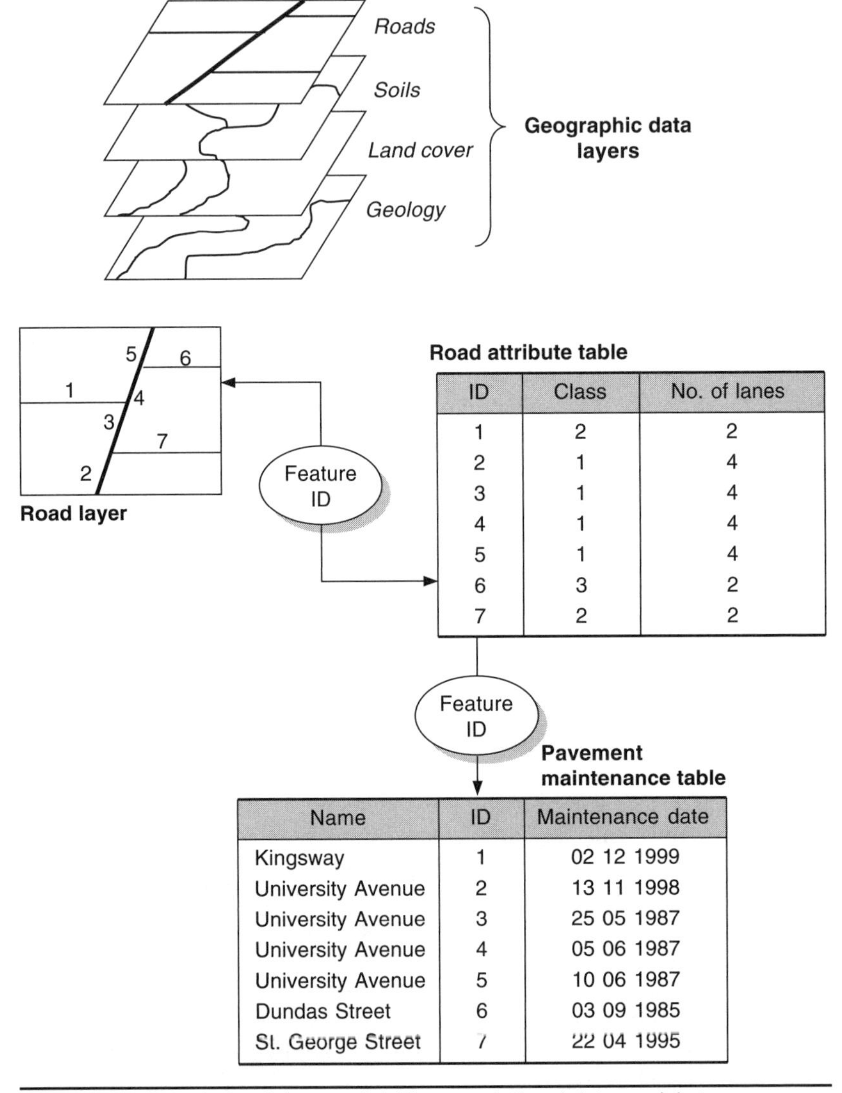

ID	Class	No. of lanes
1	2	2
2	1	4
3	1	4
4	1	4
5	1	4
6	3	2
7	2	2

Name	ID	Maintenance date
Kingsway	1	02 12 1999
University Avenue	2	13 11 1998
University Avenue	3	25 05 1987
University Avenue	4	05 06 1987
University Avenue	5	10 06 1987
Dundas Street	6	03 09 1985
St. George Street	7	22 04 1995

FIGURE 1.5 Georelational data model. The georelational data model stores graphical and attribute data separately. It makes use of common feature identifiers to link the data in different tables in the database. (*Adapted from Lo and Yeung, 2002; used with permission from Prentice Hall.*)

layers of area objects. After overlaying, either of the input layers can be recreated by dissolving and merging based on the attributes contributed by the input layer.

A buffer can be constructed around a point, line, or area, and buffering thus creates a new area enclosing the buffered object. Sometimes width of the buffer can be determined by an attribute of the

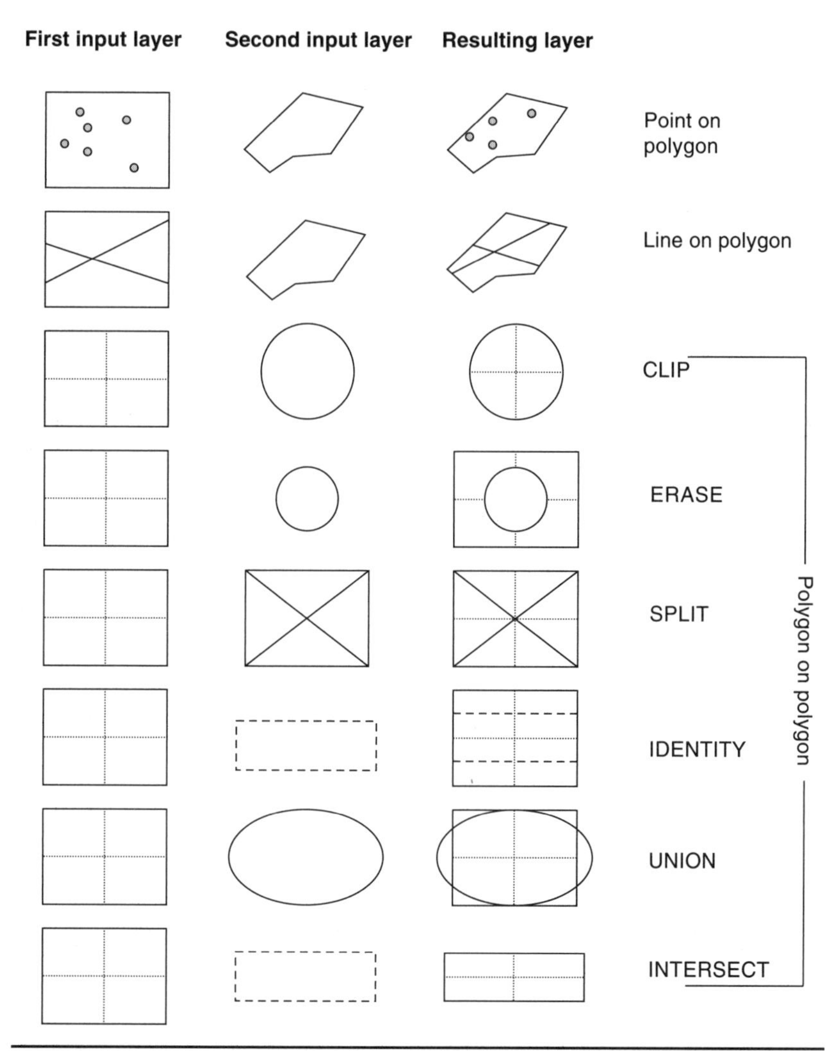

FIGURE 1.6 Types of topological overlay operations. (*Adapted from Lo and Yeung, 2002; used with permission from Prentice Hall.*)

object. Buffering may be applied to both vector and raster GIS and has found wide applications in transportation, forestry, resource management, etc.

1.2.4 Network Data Model

A *network* consists of a set of interconnected linear features. Examples of networks are roads, railways, air routes, rivers and streams, and utilities networks. In GIS, networks are modeled essentially by the arc-node topological data model, with attributes of links, nodes,

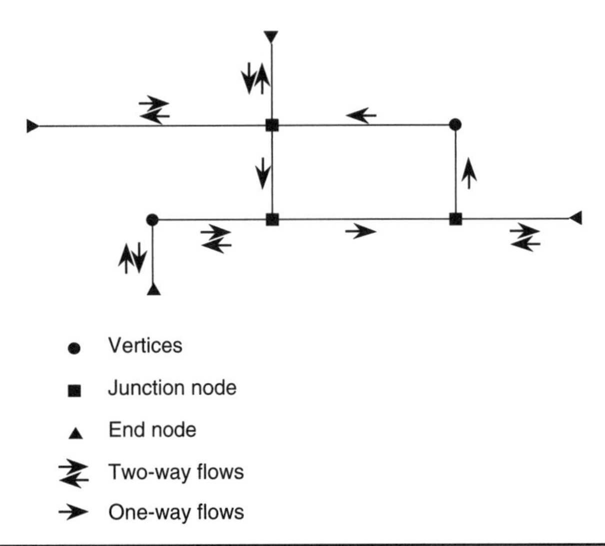

Figure 1.7 Network data model. (*Adapted from Heywood et al. 1998; used with permission from Prentice Hall.*)

stops, centers, and turns. To apply a network data model to real-world applications, additional attributes such as travel time, impedance (e.g., cost associated with passing a link, stop, center, turn, etc), the nature of flow (i.e., one-way flow, two-way flow, close, etc.), and supply and demand must be included. Figure 1.7 provides an example of a network data model. Correct topology is of foremost importance for network analysis, but correct geographic representation is not so long as key attributes are preserved (Heywood et al., 1998). Applications that use a network data model include shortest-path analysis, location-allocation analysis, transportation planning, spatial optimization, and so on.

1.2.5 Object-Oriented Data Model

The concept of object orientation embraces four components: (1) object-oriented user interface (e.g., icons, dialog boxes, glyphs, etc.), (2) object-oriented programming languages (e.g., Visual Basic and Visual C), (3) object-oriented analysis, and (4) object-oriented database management (Lo and Yeung, 2002). The object-oriented data model discussed in most GIS textbooks focuses on the last two components. Unlike the georelational data model, which separates spatial and attribute data and links them by using a common identifier, the object-oriented data model views the real world as a set of individual objects. An object has a set of properties and can perform operations on requests (Chang, 2009). Figure 1.8 illustrates the object-oriented representation approach. Features are no longer divided into separate layers; instead, they are

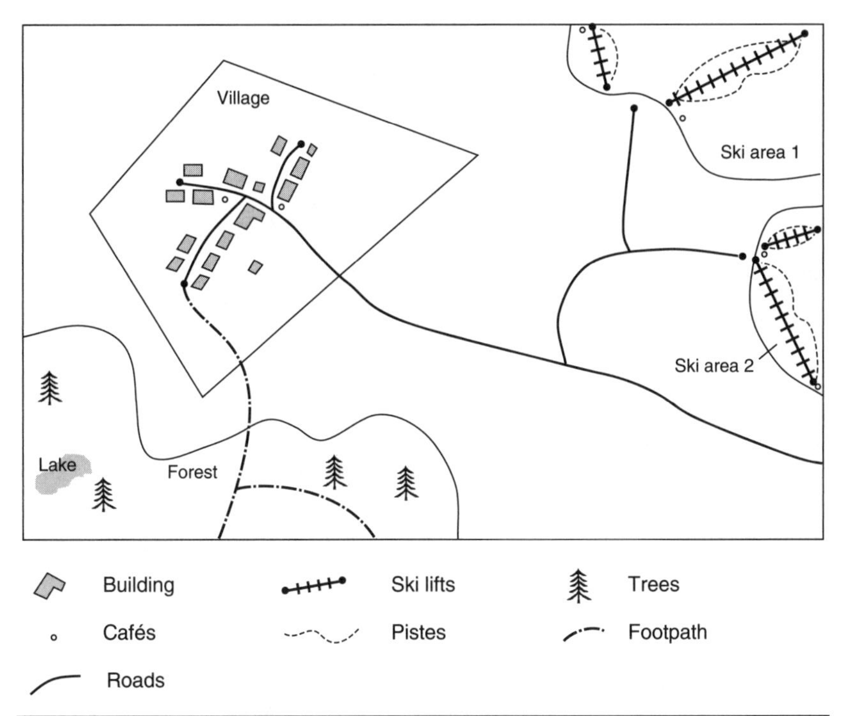

FIGURE 1.8 The object-oriented approach. (*Adapted from Heywood et al. 1998; used with permission from Prentice Hall.*)

grouped into classes and hierarchies of objects. The properties of each object include geographic information (e.g., size, shape, and location), topological information (i.e., one object relating to others), behaviors of the object, and behaviors of the objects in relation to one another. To apply the object-oriented data model to GIS, one must consider both the structural aspects of objects (Worboys, 1995) and behavior aspects of the objects (Egenhofer and Frank, 1992). Basic principles for explaining the structures of objects include association, aggregation, generalization, instantiation, and specialization, whereas those for explaining the behaviors of objects include inheritance, encapsulation, and polymorphism (Chang, 2009; Lo and Yeung, 2002).

References

Adams, J. B., Sabol, D. E., Kapos, V., et al. 1995. Classification of multispectral images based on fractions of endmembers: application to land cover change in the Brazilian Amazon. *Remote Sensing of Environment* **52,** 137–154.

Aplin, P., and Atkinson, P. M. 2001. Sub-pixel land cover mapping for per-field classification. *International Journal of Remote Sensing* **22,** 2853–2858.

Aplin, P., Atkinson, P. M., and Curran, P. J. 1999a. Per-field classification of land use using the forthcoming very fine spatial resolution satellite sensors: Problems and potential solutions. In *Advances in Remote Sensing and GIS Analysis*, edited by P. M. Atkinson and N. J. Tate, pp. 219–239. New York: Wiley.

Aplin, P., Atkinson, P. M., and Curran, P. J. 1999b. Fine spatial resolution simulated satellite sensor imagery for land cover mapping in the United Kingdom. *Remote Sensing of Environment* **68,** 206–216.

Asner, G. P., and Heidebrecht, K. B. 2002. Spectral unmixing of vegetation, soil and dry carbon cover in arid regions: Comparing multispectral and hyperspectral observations. *International Journal of Remote Sensing* **23,** 3939–3958.

Atkinson, P. M., and Tatnall, A. R. L. 1997. Neural networks in remote sensing. *International Journal of Remote Sensing* **18,** 699–709.

Augusteijn, M. F., Clemens, L.E., and Shaw, K. A. 1995. Performance evaluation of texture measures for ground cover identification in satellite images by means of a neural network classifier. *IEEE Transactions on Geoscience and Remote Sensing* **33,** 616–625.

Barandela, R., and Juarez, M. 2002. Supervised classification of remotely sensed data with ongoing learning capability. *International Journal of Remote Sensing* **23,** 4965–4970.

Barnsley, M. J., and Barr, S. L. 1997. Distinguishing urban land-use categories in fine spatial resolution land-cover data using a graph-based, structural pattern recognition system. *Computers, Environments and Urban Systems* **21,** 209–225.

Beaubien, J., Cihlar, J., Simard, G., and Latifovic, R. 1999. Land cover from multiple Thematic Mapper scenes using a new enhancement-classification methodology. *Journal of Geophysical Research* **104 D22,** 27909–27920.

Benediktsson, J. A., and Kanellopoulos, I. 1999. Classification of multisource and hyperspectral data based on decision fusion. *IEEE Transactions on Geoscience and Remote Sensing* **37,** 1367–1377.

Benz, U. C., Hofmann, P., Willhauck, G., et al. 2004. Multi-resolution, object-oriented fuzzy analysis of remote sensing data for GIS-ready information. *ISPRS Journal of Photogrammetry & Remote Sensing* **58,** 239–258.

Berberoglu, S., Lloyd, C. D., Atkinson, P. M., and Curran, P. J. 2000. The integration of spectral and textural information using neural networks for land cover mapping in the Mediterranean. *Computers and Geosciences* **26,** 385–396.

Berry, J. K. 1993. Cartographic modeling: The analytical capabilities of GIS. In *Environment Modeling with GIS*, edited by M. F. Goodchild, B. O. Parks, and L. T. Steyaert, pp. 58–74. Oxford, England: Oxford University Press.

Biehl, L., and Landgrebe, D. 2002. MultiSpec: A tool for multispectral-hyperspectral image data analysis. *Computers and Geosciences* **28,** 1153–1159.

Binaghi, E., Brivio, P. A., Ghezzi, P., and Rampini, A. 1999. A fuzzy set–based accuracy assessment of soft classification. *Pattern Recognition Letters* **20,** 935–948.

Binaghi, E., Madella, P., Montesano, M. G., and Rampini, A. 1997. Fuzzy contextual classification of multisource remote sensing images. *IEEE Transactions on Geoscience and Remote Sensing* **35,** 326–339.

Bloch, I. 1996. Information combination operators for data fusion: A comparative review with classification. *IEEE Transactions on Systems, Man, and Cybernetics* **26,** 52–67.

Brown, M., Gunn, S. R., and Lewis, H. G. 1999. Support vector machines for optimal classification and spectral unmixing. *Ecological Modeling* **120,** 167–179.

Butusov, O. B. 2003. Textural classification of forest types from Landsat 7 imagery. *Mapping Sciences and Remote Sensing* **40,** 91–104.

Calkins, H. W., and Tomlinson, R. F. 1977. *Geographic Information Systems: Methods and Equipment for Land Use Planning.* Ottawa, Canada: International Geographical Union, Commission of Geographical Data Sensing and Processing and U.S. Geological Survey.

Campbell, J. B. 2007. *Introduction to Remote Sensing,* 4th ed. New York: Guilford Press.

Canty, M. J., Nielsen, A. A., and Schmidt, M. 2004. Automatic radiometric normalization of multitemporal satellite imagery. *Remote Sensing of Environment* **91,** 441–451.

Carlotto, M. J. 1998. Spectral shape classification of Landsat Thematic Mapper imagery. *Photogrammetric Engineering and Remote Sensing* **64,** 905–913.

Carr, J. R. 1999. Classification of digital image texture using variograms. In *Advances in Remote Sensing and GIS Analysis,* edited by P. M. Atkinson and N. J. Tate, pp. 135–146. New York: Wiley.

Chalifoux, S., Cavayas, F., and Gray, J. T. 1998. Mapping-guided approach for the automatic detection on Landsat TM images of forest stands damaged by the spruce budworm. *Photogrammetric Engineering and Remote Sensing* **64,** 629–635.

Chang, K.-T. 2009. *Introduction to Geographic Information Systems,* 5th ed. New York: McGraw-Hill.

Chavez, P. S., Jr. 1996. Image-based atmospheric corrections: Revisited and improved. *Photogrammetric Engineering and Remote Sensing* **62,** 1025–1036.

Chen, D., Stow, D. A., and Gong, P. 2004. Examining the effect of spatial resolution and texture window size on classification accuracy: An urban environment case. *International Journal of Remote Sensing* **25,** 2177–2192.

Chen, K. S., Tzeng, Y. C., Chen, C. F., and Kao, W. L. 1995. Land-cover classification of multispectral imagery using a dynamic learning neural network. *Photogrammetric Engineering and Remote Sensing* **61,** 403–408.

Cihlar, J., Xiao, Q., Chen, J., et al. 1998. Classification by progressive generalization: A new automated methodology for remote sensing multispectral data. *International Journal of Remote Sensing* **19,** 2685–2704.

Civco, D. L. 1989. Topographic normalization of Landsat Thematic Mapper digital imagery. *Photogrammetric Engineering and Remote Sensing* **55,** 1303–1309.

Colby, J. D. 1991. Topographic normalization in rugged terrain. *Photogrammetric Engineering and Remote Sensing* **57,** 531–537.

Collins, M. J., Dymond, C., and Johnson, E. A. 2004. Mapping subalpine forest types using networks of nearest neighbor classifiers. *International Journal of Remote Sensing* **25,** 1701–1721.

Colwell, R. N. 1997. History and place of photographic interpretation. In *Manual of Photographic Interpretation,* 2nd ed., edited by W. R. Philipson. Bethesda, MD: American Association of Photogrammetry and Remote Sensing.

Conese, C., Maracchi, G., and Maselli, F. 1993. Improvement in maximum likelihood classification performance on highly rugged terrain using principal component analysis. *International Journal of Remote Sensing* **14,** 1371–1382.

Conese, C., Maracchi, G., Miglietta, F., et al. 1988. Forest classification by principal component analyses of TM data. *International Journal of Remote Sensing* **9,** 1597–1612.

Conese, C., and Maselli, F. 1994. Evaluation of contextual, per-pixel and mixed classification procedures applied to a subtropical landscape. *Remote Sensing Reviews* **9,** 175–186.

Cortijo, F. J., and de la Blanca, N. P. 1998. Improving classical contextual classification. *International Journal of Remote Sensing* **19,** 1591–1613.

Cracknell, A. P. 1998. Synergy in remote sensing: What's in a pixel? *International Journal of Remote Sensing* **19,** 2025–2047.

Dean, A. M., and Smith, G. M. 2003. An evaluation of per-parcel land cover mapping using maximum likelihood class probabilities. *International Journal of Remote Sensing* **24,** 2905–2920.

Debeir, O., van den Steen, I., Latinne, P., et al. 2002. Textural and contextual land-cover classification using single and multiple classifier systems. *Photogrammetric Engineering and Remote Sensing* **68,** 597–605.

DeFries, R. S., and Chan, J. C. 2000. Multiple criteria for evaluating machine learning algorithms for land cover classification from satellite data. *Remote Sensing of Environment* **74,** 503–515.

DeFries, R. S., Hansen, M., Townshend, J. R. G., and Sohlberg, R. 1998. Global land cover classification at 8-km spatial resolution: The use of training data derived from Landsat imagery in decision tree classifiers. *International Journal of Remote Sensing* **19,** 3141–3168.

Du, Q., and Chang, C. 2001. A linear constrained distance-based discriminant analysis for hyperspectral image classification. *Pattern Recognition* **34,** 361–373.

Du, Q., and Ren, H. 2003. Real-time constrained linear discriminant analysis to target detection and classification in hyperspectral imagery. *Pattern Recognition* **36,** 1–12.

Du, Y., Teillet, P. M., and Cihlar, J. 2002. Radiometric normalization of multitemporal high-resolution satellite images with quality control for land cover change detection. *Remote Sensing of Environment* **82,** 123–134.

Egenhofer, M. F., and Frank, A. 1992. Object-oriented modeling for GIS. *Journal of the Urban and Regional Information System Association* **4,** 3–19.

Emerson, C. W., Lam, N. S., and Quattrochi, D. A. 1999. Multi-scale fractal analysis of image texture and pattern. *Photogrammetric Engineering and Remote Sensing* **65,** 51–61.

Emrahoglu, N., Yegingil, I., Pestemalci, V., et al. 2003. Comparison of a new algorithm with the supervised classifications. *International Journal of Remote Sensing* **24,** 649–655.

Erbek, F. S., Ozkan, C., and Taberner, M. 2004. Comparison of maximum likelihood classification method with supervised artificial neural network algorithms for land use activities. *International Journal of Remote Sensing* **25,** 1733–1748.

Erol, H., and Akdeniz, F. 1996. A multi-spectral classification algorithm for classifying parcels in an agricultural region. *International Journal of Remote Sensing* **17,** 3357–3371.

Erol, H., and Akdeniz, F. 1998. A new supervised classification method for quantitative analysis of remotely sensed multi-spectral data. *International Journal of Remote Sensing* **19,** 775–782.

Fedra, K. 1996. Distributed models and embedded GIS. In *GIS and Environment Modeling: Progress and Research Issues,* edited by M. F. Goodchild, L. T. Steyaert, B. O. Parks, et al., pp. 413–417. Fort Collins, CO: GIS World, Inc.

Fernandes, R., Fraser, R., Latifovic, R., et al. 2004. Approaches to fractional land cover and continuous field mapping: A comparative assessment over the BOREAS study region. *Remote Sensing of Environment* **89,** 234–251.

Fernández-Prieto, D. 2002. An iterative approach to partially supervised classification problems. *International Journal of Remote Sensing* **23,** 3887–3892.

Fisher, P. 1997. The pixel: a snare and a delusion. *International Journal of Remote Sensing* **18,** 679–685.

Flygare, A.-M. 1997. A comparison of contextual classification methods using Landsat TM. *International Journal of Remote Sensing* **18,** 3835–3842.

Foody, G. M. 1996. Approaches for the production and evaluation of fuzzy land cover classification from remotely sensed data. *International Journal of Remote Sensing* **17,** 1317–1340.

Foody, G. M. 1998. Sharpening fuzzy classification output to refine the representation of sub-pixel land cover distribution. *International Journal of Remote Sensing* **19,** 2593–2599.

Foody, G. M., 1999. Image classification with a neural network: from completely-crisp to fully-fuzzy situation. In *Advances in Remote Sensing and GIS Analysis,* edited by P. M. Atkinson and N. J. Tate, pp. 17–37. New York: Wiley.

Foody, G. M. 2002a. Hard and soft classifications by a neural network with a non-exhaustively defined set of classes. *International Journal of Remote Sensing* **23,** 3853–3864.

Foody, G. M. 2002b. Status of land cover classification accuracy assessment. *Remote Sensing of Environment* **80,** 185–201.

Foody, G. M. 2004. Supervised image classification by MLP and RBF neural networks with and without an exhaustively defined set of classes. *International Journal of Remote Sensing* **25,** 3091–3104.

Foody, G. M., and Arora, M. K. 1997. An evaluation of some factors affecting the accuracy of classification by an artificial neural network. *International Journal of Remote Sensing* **18,** 799–810.

Foody, G. M., and Cox, D. P. 1994. Sub-pixel land cover composition estimation using a linear mixture model and fuzzy membership functions. *International Journal of Remote Sensing* **15,** 619–631.

Foody, G. M., and Mathur, A. 2004a. A relative evaluation of multiclass image classification by support vector machines. *IEEE Transactions on Geoscience and Remote Sensing* **42,** 1336–1343.

Foody, G. M., and Mathur, A. 2004b. Toward intelligent training of supervised image classifications: directing training data acquisition for SVM classification. *Remote Sensing of Environment* **93,** 107–117.

Foody, G. M., McCulloch, M. B., and Yates, W. B. 1995. Classification of remotely sensed data by an artificial neural network: issues related to training data characteristics. *Photogrammetric Engineering and Remote Sensing* **61,** 391–401.

Foschi, P. G., and Smith, D. K. 1997. Detecting subpixel woody vegetation in digital imagery using two artificial intelligence approaches. *Photogrammetric Engineering and Remote Sensing* **63,** 493–500.

Franklin, S. E., and Peddle, D. R. 1989. Spectral texture for improved class discrimination in complex terrain. *International Journal of Remote Sensing* **10,** 1437–1443.

Franklin, S. E., and Peddle, D. R. 1990. Classification of SPOT HRV imagery and texture features. *International Journal of Remote Sensing* **11,** 551–556.

Franklin, S. E., Connery, D. R., and Williams, J. A. 1994. Classification of alpine vegetation using Landsat Thematic Mapper, SPOT HRV and DEM data. *Canadian Journal of Remote Sensing* **20,** 49–56.

Franklin, S. E., Wulder, M. A., and Lavigne, M. B. 1996. Automated derivation of geographic window sizes for remote sensing digital image texture analysis. *Computers and Geosciences* **22,** 665–673.

Franklin, S. E., Peddle, D. R., Dechka, J. A., and Stenhouse, G. B. 2002. Evidential reasoning with Landsat TM, DEM and GIS data for land cover classification in support of grizzly bear habitat mapping. *International Journal of Remote Sensing* **23,** 4633–4652.

Friedl, M. A., and Brodley, C. E. 1997. Decision tree classification of land cover from remotely sensed data. *Remote Sensing of Environment* **61,** 399–409.

Friedl, M. A., Brodley, C. E., and Strahler, A. H. 1999. Maximizing land cover classification accuracies produced by decision trees at continental to global scales. *IEEE Transactions on Geoscience and Remote Sensing* **37,** 969–977.

Gao, P., Zhan, C., and Menon, S. 1996. An overview of cell-based modeling with GIS. In *GIS and Environment Modeling: Progress and Research Issues*, edited by M. F. Goodchild, L. T. Steyaert, B. O. Parks, et al., pp. 325–331. Fort Collins, CO: GIS World, Inc.

Geneletti, D., and Gorte, B. G. H. 2003. A method for object-oriented land cover classification combining Landsat TM data and aerial photographs. *International Journal of Remote Sensing* **24,** 1273–1286.

Gilabert, M. A., Conese, C., and Maselli, F. 1994. An atmospheric correction method for the automatic retrieval of surface reflectance from TM images. *International Journal of Remote Sensing* **15,** 2065–2086.

Gitas, I. Z., Mitri, G. H., and Ventura, G. 2004. Object-based image classification for burned area mapping of Creus Cape, Spain, using NOAA-AVHRR imagery. *Remote Sensing of Environment* **92,** 409–413.

Gong, P. 1996. Integrated analysis of spatial data from multiple sources: Using evidential reasoning and artificial neural network techniques for geological mapping. *Photogrammetric Engineering and Remote Sensing* **62,** 513–523.

Gong, P., and Howarth, P. J. 1992. Frequency-based contextual classification and gray-level vector reduction for land-use identification. *Photogrammetric Engineering and Remote Sensing* **58,** 423–437.

Gong, P., Marceau, D. J., and Howarth, P. J. 1992. A comparison of spatial feature extraction algorithms for land-use classification with SPOT HRV data. *Remote Sensing of Environment* **40,** 137–151.

Gordon, D. K., and Phillipson, W. R. 1986. A texture enhancement procedure for separating orchard from forest in Thematic Mapper imagery. *International Journal of Remote Sensing* **8,** 301–304.

Groom, G. B., Fuller, R. M., and Jones, A. R. 1996. Contextual correction: techniques for improving land cover mapping from remotely sensed images. *International Journal of Remote Sensing* **17,** 69–89.

Gu, D., and Gillespie, A. 1998. Topographic normalization of Landsat TM images of forest based on subpixel sun-canopy-sensor geometry. *Remote Sensing of Environment* **64,** 166–175.

Haapanen, R., Ek, A. R., Bauer, M. E., and Finley, A. O. 2004. Delineation of forest/nonforest land use classes using nearest neighbor methods. *Remote Sensing of Environment* **89,** 265–271.

Hadjimitsis, D. G., Clayton, C. R. I., and Hope, V. S. 2004. An assessment of the effectiveness of atmospheric correction algorithms through the remote sensing of some reservoirs. *International Journal of Remote Sensing* **25,** 3651–3674.

Hale, S. R., and Rock, B. N. 2003. Impacts of topographic normalization on land-cover classification accuracy. *Photogrammetric Engineering and Remote Sensing* **69,** 785–792.

Hansen, M., Dubayah, R., and DeFries, R. 1996. Classification trees: an alternative to traditional land cover classifiers. *International Journal of Remote Sensing* **17,** 1075–1081.

Haralick, R. M., Shanmugam, K., and Dinstein, I. 1973. Textural features for image classification. *IEEE Transactions on Systems, Man and Cybernetics* **SMC-3,** 610–620.

Hardin, P. J. 1994. Parametric and nearest-neighbor methods for hybrid classification: A comparison of pixel assignment accuracy. *Photogrammetric Engineering and Remote Sensing* **60,** 1439–1448.

Harris, P. M., and Ventura, S. J. 1995. The integration of geographic data with remotely sensed imagery to improve classification in an urban area. *Photogrammetric Engineering and Remote Sensing* **61,** 993–998.

He, D. C., and Wang, L. 1990. Texture unit, textural spectrum and texture analysis. *IEEE Transaction on Geoscience and Remote Sensing* **28,** 509–512.

Heo, J., and FitzHugh, T. W. 2000. A standardized radiometric normalization method for change detection using remotely sensed imagery. *Photogrammetric Engineering and Remote Sensing* **66,** 173–182.

Herold, M., Liu, X., and Clarke, K. C. 2003. Spatial metrics and image texture for mapping urban land use. *Photogrammetric Engineering and Remote Sensing* **69,** 991–1001.

Heywood, I., Cornelius, S., and Carver, S. 1998. *An Introduction to Geographical Information Systems.* Upper Saddle River, NJ: Prentice-Hall.

Holben, B. N., and Justice, C. O. 1981. An examination of spectral band ratioing to reduce the topographic effect on remotely sensed data. *International Journal of Remote Sensing* **2,** 115–123.

Hsu, C., and Lin, C. 2002. A comparison of methods for multi-class support vector machines. *IEEE Transactions on Neural Networks* **13,** 415–425.

Huang, C., Davis, L. S., and Townshend, J. R. G. 2002. An assessment of support vector machines for land cover classification. *International Journal of Remote Sensing* **23,** 725–749.

Huang, Z., and Lees, B. G. 2004. Combining non-parametric models for multisource predictive forest mapping. *Photogrammetric Engineering and Remote Sensing* **70,** 415–425.

Hubert-Moy, L., Cotonnec, A., Le Du, L., et al. 2001. A comparison of parametric classification procedures of remotely sensed data applied on different landscape units. *Remote Sensing of Environment* **75,** 174–187.

Huguenin, R. L., Karaska, M. A., Blaricom, D. V., and Jensen, J. R. 1997. Subpixel classification of bald cypress and tupelo gum trees in Thematic Mapper imagery. *Photogrammetric Engineering and Remote Sensing* **63,** 717–725.

Hung, M, and Ridd, M. K. 2002. A subpixel classifier for urban land-cover mapping based on a maximum-likelihood approach and expert system rules. *Photogrammetric Engineering and Remote Sensing* **68,** 1173–1180.

Jakubaukas, M. E. 1997. Effects of forest succession on texture in Landsat Thematic Mapper imagery. *Canadian Journal of Remote Sensing* **23,** 257–263.

Janssen, L., and Molenaar, M. 1995. Terrain objects, their dynamics and their monitoring by integration of GIS and remote sensing. *IEEE Transactions on Geoscience and Remote Sensing* **33,** 749–758.

Jensen, J. R. 2005. *Introductory Digital Image Processing: A Remote Sensing Perspective,* 3rd ed. Upper Saddle River, NJ: Prentice-Hall.
Jeon, B., and Landgrebe, D. A. 1999. Decision fusion approaches for multitemporal classification. *IEEE Transactions on Geoscience and Remote Sensing* **37,** 1227–1233.
Jiang, H., Strittholt, J. R., Frost, P. A., and Slosser, N. C. 2004. The classification of late seral forests in the Pacific Northwest USA using Landsat ETM+ imagery. *Remote Sensing of Environment* **91,** 320–331.
Kartikeyan, B., Gopalakrishna, B., Kalubarme, M. H., and Majumder, K. L. 1994. Contextual techniques for classification of high and low resolution remote sensing data. *International Journal of Remote Sensing* **15,** 1037–1051.
Kartikeyan, B., Sarkar, A., and Majumder, K. L. 1998. A segmentation approach to classification of remote sensing imagery. *International Journal of Remote Sensing* **19,** 1695–1709.
Kashyap, R. L., Chellappa, R., and Khotanzad, A. 1982. Texture classification using features derived from random field models. *Pattern Recognition Letters* **1,** 43–50.
Kavzoglu, T., and Mather, P. M. 2004. The use of backpropagating artificial neural networks in land cover classification. *International Journal of Remote Sensing* **24,** 4907–4938.
Keuchel, J., Naumann, S., Heiler, M., and Siegmund, A. 2003. Automatic land cover analysis for Tenerife by supervised classification using remotely sensed data. *Remote Sensing of Environment* **86,** 530–541.
Kim, H., Pang, S., Je, H., et al. 2003. Constructing support vector machine ensemble. *Pattern Recognition* **36,** 2757–2767.
Koltunov, A., and Ben-dor, E. 2001. A new approach for spectral feature extraction and for unsupervised classification of hyperspectral data based on the Gaussian mixture model. *Remote Sensing Reviews* **20,** 123–167.
Koltunov, A., and Ben-dor, E. 2004. Mixture density separation as a tool for high-quality interpretation of multi-source remote sensing data and related issues. *International Journal of Remote Sensing* **25,** 3275–3299.
Kontoes, C. C., and Rokos, D. 1996. The integration of spatial context information in an experimental knowledge based system and the supervised relaxation algorithm: Two successful approaches to improving SPOT-XS classification. *International Journal of Remote Sensing* **17,** 3093–3106.
Kulkarni, A. D., and Lulla, K. 1999. Fuzzy neural network models for supervised classification: Multispectral image analysis. *Geocarto International* **14,** 42–50.
Kurosu, T., Yokoyama, S., and Chiba, K. 2001. Land use classification with textural analysis and the aggregation technique using multi-temporal JERS-1 L-band SAR images. *International Journal of Remote Sensing* **22,** 595–613.
Landgrebe, D. A. 2003. *Signal Theory Methods in Multispectral Remote Sensing.* Hoboken, NJ: Wiley.
Lawrence, R., Bunn, A., Powell, S., and Zmabon, M. 2004. Classification of remotely sensed imagery using stochastic gradient boosting as a refinement of classification tree analysis. *Remote Sensing of Environment* **90,** 331–336.
Lee, T. W., Lewicki, M. S., and Sejnowski, T. J. 2000. ICA mixture models for unsupervised classification of non-Gaussian classes and automatic context switching in blind signal separation. *IEEE Transactions on Pattern Analysis and Machine Intelligence* **22,** 1078–1089.
Lein, J. K. 2003. Applying evidential reasoning methods to agricultural land cover classification. *International Journal of Remote Sensing* **24,** 4161–4180.
Lillesand, T. M., Kiefer, R. W., and Chipman, J. W. 2008. *Remote Sensing and Image Interpretation,* 6th ed. New York: Wiley.
Lira, J., and Maletti, G. 2002. A supervised contextual classifier based on a region-growth algorithm. *Computers and Geosciences* **28,** 951–959.
Liu, X., Skidmore, A. K., and Oosten, H. V. 2002. Integration of classification methods for improvement of land-cover map accuracy. *ISPRS Journal of Photogrammetry and Remote Sensing* **56,** 257–268.
Lloyd, C. D., Berberoglu, S., Curran, P. J., and Atkinson, P. M. 2004. A comparison of texture measures for the per-field classification of Mediterranean land cover. *International Journal of Remote Sensing* **25,** 3943–3965.

Lo, C. P., and Choi, J. 2004. A hybrid approach to urban land use/cover mapping using Landsat 7 Enhanced Thematic Mapper Plus (ETM+) images. *International Journal of Remote Sensing* **25,** 2687–2700.

Lo, C. P., and Yeung, A. K. W. 2002. *Concepts and Techniques of Geographic Information Systems.* Upper Saddle River, NJ: Prentice-Hall.

Lobell, D. B., Asner, G. P., Law, B. E., and Treuhaft, R. N. 2002. View angle effects on canopy reflectance and spectral mixture analysis of coniferous forests using AVIRIS. *International Journal of Remote Sensing* **23,** 2247–2262.

Lobo, A., Chic, O., and Casterad, A. 1996. Classification of Mediterranean crops with multisensor data: Per-pixel versus per-object statistics and image segmentation. *International Journal of Remote Sensing* **17,** 2385–2400.

Lu, D., and Weng, Q. 2004. Spectral mixture analysis of the urban landscapes in Indianapolis with Landsat ETM+ imagery. *Photogrammetric Engineering and Remote Sensing* **70,** 1053–1062.

Lu, D. and Weng, Q. 2007. A survey of image classification methods and techniques for improving classification performance. *International Journal of Remote Sensing* **28,** 823–870.

Lu, D., Mausel, P., Batistella, M., and Moran, E. 2004. Comparison of land-cover classification methods in the Brazilian Amazon Basin. *Photogrammetric Engineering and Remote Sensing* **70,** 723–731.

Lu, D., Moran, E., and Batistella, M. 2003, Linear mixture model applied to Amazonian vegetation classification. *Remote Sensing of Environment* **87,** 456–469.

Lucieer, A., and Kraak, M. 2004. Interactive and visual fuzzy classification of remotely sensed imagery for exploration of uncertainty. *International Journal of Geographic Information Science* **18,** 491–512.

Lunetta, R. S., Ediriwckrema, J., Iiames, J., et al. 2003. A quantitative assessment of a combined spectral and GIS rule-based land-cover classification in the Neuse River basin of North Carolina. *Photogrammetric Engineering and Remote Sensing* **69,** 299–310.

Magnussen, S., Boudewyn, P., and Wulder, M. 2004. Contextual classification of Landsat TM images to forest inventory cover types. *International Journal of Remote Sensing* **25,** 2421–2440.

Mannan, B., and Ray, A. K. 2003. Crisp and fuzzy competitive learning networks for supervised classification of multispectral IRS scenes. *International Journal of Remote Sensing* **24,** 3491–3502.

Mannan, B., Roy, J., and Ray, A. K. 1998. Fuzzy ARTMAP supervised classification of multi-spectral remotely-sensed images. *International Journal of Remote Sensing* **19,** 767–774.

Marceau, D. J., Howarth, P. J., Dubois, J. M., and Gratton, D. J. 1990. Evaluation of the grey-level co-occurrence matrix method for land-cover classification using SPOT imagery. *IEEE Transactions on Geoscience and Remote Sensing* **28,** 513–519.

Markham, B. L., and Barker, J. L. 1987. Thematic Mapper band pass solar exoatmospheric irradiances. *International Journal of Remote Sensing* **8,** 517–523.

Maselli, F., Rodolfi, A., and Conese, C. 1996. Fuzzy classification of spatially degraded Thematic Mapper data for the estimation of sub-pixel components. *International Journal of Remote Sensing* **17,** 537–551.

McGovern, E. A., Holden, N. M., Ward, S. M., and Collins, J. F. 2002. The radiometric normalization of multitemporal Thematic Mapper imagery of the midlands of Ireland: A case study. *International Journal of Remote Sensing* **23,** 751–766.

Meyer, P., Itten, K. I., Kellenberger, T., et al. 1993. Radiometric corrections of topographically induced effects on Landsat TM data in alpine environment. *ISPRS Journal of Photogrammetry and Remote Sensing* **48,** 17–28.

Michelson, D. B., Liljeberg, B. M., and Pilesjo, P. 2000. Comparison of algorithms for classifying Swedish land cover using Landsat TM and ERS-1 SAR data. *Remote Sensing of Environment* **71,** 1–15.

Mitra, P., Shankar, B. U., and Pal., S. K. 2004. Segmentation of multispectral remote sensing images using active support vector machines. *Pattern Recognition Letters* **25,** 1067–1074.

Myint, S. W. 2001. A robust texture analysis and classification approach for urban land-use and land-cover feature discrimination. *Geocarto International* **16,** 27–38.
Narasimha Rao, P. V., Sesha Sai, M. V. R., Sreenivas, K., et al. 2002. Textural analysis of IRS-1D panchromatic data for land cover classification. *International Journal of Remote Sensing* **23,** 3327–3345.
Naugle, B. I., and Lashlee, J. D. 1992. Alleviating topographic influences on land-cover classifications for mobility and combat modeling. *Photogrammetric Engineering and Remote Sensing* **48,** 1217–1221.
Neville, R. A., Levesque, J., Staene, K., et al. 2003. Spectral unmixing of hyperspectral imagery for mineral exploration: Comparison of results from SFSI and AVIRIS. *Canadian Journal of Remote Sensing* **29,** 99–110.
Nyoungui, A., Tonye, E., and Akono, A. 2002. Evaluation of speckle filtering and texture analysis methods for land cover classification from SAR images. *International Journal of Remote Sensing* **23,** 1895–1925.
Okin, G. S., Roberts, D. A., Murray, B., and Okin, W. J. 2001. Practical limits on hyperspectral vegetation discrimination in arid and semiarid environments. *Remote Sensing of Environment* **77,** 212–225.
Onsi, H. M. 2003. Designing a rule-based classifier usin syntactical approach. *International Journal of Remote Sensing* **24,** 637–647.
Ozkan, C., and Erbek, F. S. 2003. The comparison of activation functions for multispectral Landsat TM image classification. *Photogrammetric Engineering and Remote Sensing* **69,** 1225–1234.
Pal, M., and Mather, P. M. 2003. An assessment of the effectiveness of decision tree methods for land cover classification. *Remote Sensing of Environment* **86,** 554–565.
Paola, J. D., and Schowengerdt, R. A. 1995. A review and analysis of back propagation neural networks for classification of remotely sensed multispectral imagery. *International Journal of Remote Sensing* **16,** 3033–3058.
Paola, J. D., and Schowengerdt, R. A. 1997. The effect of neural-network structure on a multispectral land-use/land-cover classification. *Photogrammetric Engineering and Remote Sensing* **63,** 535–544.
Peddle, D. R. 1995. Knowledge formulation for supervised evidential classification. *Photogrammetric Engineering and Remote Sensing* **61,** 409–417.
Peddle, D. R., and Ferguson, D. T. 2002. Optimization of multisource data analysis: An example using evidential reasoning for GIS data classification. *Computers & Geosciences* **28,** 45–52.
Peddle, D. R., Foody, G. M., Zhang, A., et al. 1994. Multi-source image classification: II. An empirical comparison of evidential reasoning and neural network approaches. *Canadian Journal of Remote Sensing* **20,** 396–407.
Peddle, D. R., Johnson, R. L., Cihlar, J., and Latifovic, R. 2004. Large area forest classification and biophysical parameter estimation using the 5-scale canopy reflectance model in multiple-forward-mode. *Remote Sensing of Environment* **89,** 252–263.
Pedley, M. I., and Curran, P. J. 1991. Per-field classification: An example using SPOT HRV imagery. *International Journal of Remote Sensing* **12,** 2181–2192.
Penaloza, M. A., and Welch, R. M. 1996. Feature selection for classification of polar regions using a fuzzy expert system. *Remote Sensing of Environment* **58,** 81–100.
Platt, R. V., and Goetz, A. F. H. 2004. A comparison of AVIRIS and Landsat for land use classification at the urban fringe. *Photogrammetric Engineering and Remote Sensing* **70,** 813–819.
Podest, E., and Saatchi, S. 2002. Application of multiscale texture in classifying JERS-1 radar data over tropical vegetation. *International Journal of Remote Sensing* **23,** 1487–1506.
Pouch, G. W., and Campagna, D. J. 1990. Hyperspherical direction cosine transformation for separation of spectral and illumination information in digital scanner data. *Photogrammetric Engineering and Remote Sensing* **56,** 475–479.
Price, K. P., Guo, X., and Stiles, J. M. 2002. Optimal Landsat TM band combinations and vegetation indices for discrimination of six grassland types in eastern Kansas. *International Journal of Remote Sensing* **23,** 5031–5042.

Qiu, F., and Jensen, J. R. 2004. Opening the black box of neural networks for remote sensing image classification. *International Journal of Remote Sensing* **25,** 1749–1768.
Quattrochi, D. A., and Goodchild, M. F. 1997. *Scale in Remote Sensing and GIS.* New York: Lewis Publishers.
Rashed, T., Weeks, J. R., Gadalla, M. S., and Hill, A. G. 2001. Revealing the anatomy of cities through spectral mixture analysis of multispectral satellite imagery: A case study of the greater Cairo region, Egypt. *Geocarto International* **16,** 5–15.
Richter, R. 1997. Correction of atmospheric and topographic effects for high spatial resolution satellite imagery. *International Journal of Remote Sensing* **18,** 1099–1111.
Ricketts, T. H., Birnioe, R. W., Bryant, E. S., and Kimball, K. D. 1993. Landsat TM mapping of alpine vegetation to monitor global climate change. *Proceedings, ASPRS/ACSM/Resource Technology 92 Technical Papers,* Washington, DC, pp. 86–97.
Ricotta, C., and Avena, G. C. 1999. The influence of fuzzy set theory on the areal extent of thematic map classes. *International Journal of Remote Sensing* **20,** 201–205.
Roberts, D. A., Gardner, M., Church, R., et al. 1998. Mapping chaparral in the Santa Monica mountains using multiple endmember spectral mixture models. *Remote Sensing of Environment* **65,** 267–279.
San Miguel-Ayanz, J., and Biging, G. S., 1996. An iterative classification approach for mapping natural resources from satellite imagery. *International Journal of Remote Sensing* **17,** 957–982.
San Miguel-Ayanz, J., and Biging, G. S. 1997. Comparison of single-stage and multi-stage classification approaches for cover type mapping with TM and SPOT data. *Remote Sensing of Environment* **59,** 92–104.
Schmidt, K. S., Skidmore, A. K., Kloosterman, E. H., et al. 2004. Mapping coastal vegetation using an expert system and hyperspectral imagery. *Photogrammetric Engineering and Remote Sensing* **70,** 703–715.
Schowengerdt, R. A. 1996. On the estimation of spatial-spectral mixing with classifier likelihood functions. *Pattern Recognition Letters* **17,** 1379–1387.
Settle, J., and Campbell, N. 1998. On the errors of two estimators of subpixel fractional cover when mixing is linear. *IEEE Transactions on Geosciences and Remote Sensing* **36,** 163–169.
Shaban, M. A., and Dikshit, O. 2001. Improvement of classification in urban areas by the use of textural features: The case study of Lucknow City, Uttar Pradesh. *International Journal of Remote Sensing* **22,** 565–593.
Shah, C. A., Arora, M. K., and Varshney, P. K. 2004. Unsupervised classification of hyperspectral data: An ICA mixture model based approach. *International Journal of Remote Sensing* **25,** 481–487.
Shalan, M. A., Arora, M. K., and Ghosh, S. K. 2003. An evaluation of fuzzy classifications from IRS 1C LISS III imagery: A case study. *International Journal of Remote Sensing* **24,** 3179–3186.
Sharma, K. M. S., and Sarkar, A., 1998. A modified contextual classification technique for remote sensing data. *Photogrammetric Engineering and Remote Sensing* **64,** 273–280.
Smith, G. M., and Fuller, R. M. 2001. An integrated approach to land cover classification: An example in the Island of Jersey. *International Journal of Remote Sensing* **22,** 3123–3142.
Sohn, Y., and Rebello, N. S. 2002. Supervised and unsupervised spectral angle classifiers. *Photogrammetric Engineering and Remote Sensing* **68,** 1271–1281.
Sohn, Y., Moran, E., and Gurrl, F. 1999. Deforestation in north-central Yucatan (1985–1995): Mapping secondary succession of forest and agricultural land use in Sotuta using the cosine of the angle concept. *Photogrammetric Engineering and Remote Sensing* **65,** 947–958.
Solaiman, B., Pierce, L. E., and Ulaby, F. T. 1999. Multisensor data fusion using fuzzy concepts: Application to land-cover classification using ERS-1/JERS-1 SAR composites. *IEEE Transactions on Geoscience and Remote Sensing* **37,** 1316–1326.
Song, C., Woodcock, C. E., Seto, K. C., et al. 2001. Classification and change detection using Landsat TM data: When and how to correct atmospheric effect. *Remote Sensing of Environment* **75,** 230–244.

Steele, B. M. 2000. Combining multiple classifiers: An application using spatial and remotely sensed information for land cover type mapping. *Remote Sensing of Environment* **74,** 545–556.

Stefan, S., and Itten, K. I. 1997. A physically-based model to correct atmospheric and illumination effects in optical satellite data of rugged terrain. *IEEE Transactions on Geoscience and Remote Sensing* **35,** 708–717.

Stuckens, J., Coppin, P. R., and Bauer, M. E. 2000, Integrating contextual information with per-pixel classification for improved land cover classification. *Remote Sensing of Environment* **71,** 282–296.

Tansey, K. J., Luckman, A. J., Skinner, L., et al. 2004. Classification of forest volume resources using ERS tandem coherence and JERS backscatter data. *International Journal of Remote Sensing* **25,** 751–768.

Teillet, P. M., Guindon, B., and Goodenough, D. G. 1982. On the slope-aspect correction of multispectral scanner data. *Canadian Journal of Remote Sensing* **8,** 84–106.

Thomas, N., Hendrix, C., and Congalton, R. G. 2003. A comparison of urban mapping methods using high-resolution digital imagery. *Photogrammetric Engineering and Remote Sensing* **69,** 963–972.

Tokola, T., Löfman, S., and Erkkilä, A. 1999. Relative calibration of multitemporal Landsat data for forest cover change detection. *Remote Sensing of Environment* **68,** 1–11.

Tomlin, C. D. 1991. Cartographic modeling. In *Geographical Information Systems: Principles and Applications*, edited by D. J. Maguire, M. F. Goodchild, and D. W. Rhind. London: Longman.

UCGIS (University Consortium for Geographic Information Science). 2003. The Strawman Report: Development of Model Undergraduate Curricula for Geographic Information Science & Technology. Task Force on the Development of Model Undergraduate Curricula, University Consortium for Geographic Information Science.

Unser, M. 1995. Texture classification and segmentation using wavelet frames. *IEEE Transactions on Image Processing* **4,** 1549–1560.

van der Sande, C. J., de Jong, S. M., and de Roo, A. P. J. 2003. A segmentation and classification approach of IKONOS-2 imagery for land cover mapping to assist flood risk and flood damage assessment. *International Journal of Applied Earth Observation and Geoinformation* **4,** 217–229.

Verbeke, L. P. C., Vabcoillie, F. M. B., and de Wulf, R. R. 2004. Reusing backpropagating artificial neural network for land cover classification in tropical savannahs. *International Journal of Remote Sensing* **25,** 2747–2771.

Vermote, E., Tanre, D., Deuze, J. L., et al. 1997. Second simulation of the satellite signal in the solar spectrum, 6S: An overview. *IEEE Transactions on Geoscience and Remote Sensing* **35,** 675–686.

Walsh, S. J., Cooper, J. W., von Essen, I. E., and Gallagher, K. R. 1990. Image enhancement of Landsat Thematic Mapper data and GIS data integration for evaluation of resource characteristics. *Photogrammetric Engineering and Remote Sensing* **56,** 1135–1141.

Walter, V. 2004. Object-based classification of remote sensing data for change detection. *ISPRS Journal of Photogrammetry & Remote Sensing* **58,** 225–238.

Wang, Y., and Civco, D. L. 1994. Evidential reasoning-based classification of multisource spatial data for improved land cover mapping. *Canadian Journal of Remote Sensing* **20,** 380–395.

Warrender, C. E., and Augusteihn, M. F. 1999. Fusion of image classification using Bayesian techniques with Markov random fields. *International Journal of Remote Sensing* **20,** 1987–2002.

Wolf, P. R., and Dewitt, B. A. 2000. *Elements of Photogrammetry with Applications in GIS*. New York: McGraw-Hill.

Woodcock, C. E., and Gopal, S. 2000. Fuzzy set theory and thematic maps: Accuracy assessment and area estimation. *International Journal of Geographic Information Science* **14,** 153–172.

Worboys, M. F. 1995. *GIS: A Computing Perspective*. London: Taylor & Francis.

Xu, B., Gong, P., Seto, E., and Spear, R. 2003. Comparison of gray-level reduction and different texture spectrum encoding methods for land-use classification using a panchromatic IKONOS image. *Photogrammetric Engineering and Remote Sensing* **69,** 529–536.

Zhang, J., and Foody, G. M. 2001. Fully-fuzzy supervised classification of sub-urban land cover from remotely sensed imagery: Statistical neural network approaches. *International Journal of Remote Sensing* **22,** 615–628.

Zhang, J., and Kirby, R. P. 1999, Alternative criteria for defining fuzzy boundaries based on fuzzy classification of aerial photographs and satellite images. *Photogrammetric Engineering and Remote Sensing* **65,** 1379–1387.

Zhu, G., and Blumberg, D. G. 2002. Classification using ASTER data and SVM algorithms: The case study of Beer Sheva, Israel. *Remote Sensing of Environment* **80,** 233–240.

CHAPTER 2

Integration of Remote Sensing and Geographic Information Systems (GIS)

This chapter focuses on discussing theoretical and methodologic issues associated with the integration of remote sensing and geographic information systems (GIS). The chapter starts with an examination of integration methods and then addresses the theoretical bases of various approaches to integration. Next, technical impediments to integration and possible solutions are discussed. The chapter ends with an examination of prospects for the future development of integration. A close look will be taken of the impacts of computer, network, and telecommunication technologies; the availability of high-resolution satellite imaging and light detection and ranging (LiDAR) data, and finally, new image-analysis algorithms on integration.

2.1 Methods for the Integration between Remote Sensing and GIS

Wilkinson (1996) summarized three main ways in which remote sensing and GIS technologies can be combined to enhance each other: (1) remote sensing is used as a tool for gathering data for use in GIS, (2) GIS data are used as ancillary information to improve the products derived from remote sensing, and (3) remote sensing and GIS are used together for modeling and analysis. This section discusses each of the three methods, followed by a more detailed discussion on the urban applications of integration.

2.1.1 Contributions of Remote Sensing to GIS

Extraction of Thematic Information

Remotely sensed data can be used to extract thematic information to create GIS layers. There are three ways to incorporate so derived thematic layers (Campbell, 2007). First, manual interpretation of aerial photographs or satellite images produces a map or a set of maps that depict boundaries between a set of thematic categories (e.g., soil or land-use classes). These boundaries then are digitized to provide digital files suitable for entry into the GIS. Second, digital remote sensing data are analyzed or classified using automated methods to produce paper maps and images that are then digitized for entry into the GIS. Finally, digital remote sensing data are analyzed or classified using automated methods and then retained in digital format for entry into the GIS. Alternatively, digital remote sensing data are entered directly in their raw form for subsequent analyses. In more than three decades, the remote sensing community has been continuously making efforts to extract thematic information more effectively and efficiently from digital remote sensing imagery. Lu and Weng (2007) provided a comprehensive review of these efforts. Readers are encouraged to take a closer look at that review.

Extraction of Cartographic Information

The automated extraction of cartographic information has been another major application of remote sensing imagery as data input to GIS. Extraction of lines, polygons, and other geographic features has been achieved from satellite images through the use of pattern recognition, edge extraction, and segmentation algorithms (Ehlers et al., 1989; Hinton, 1996). Further developments in integration will produce smoother lines and boundaries (i.e., not having a stepped appearance), hiding their raster origin. Therefore, satellite images show a great potential in producing and revising base maps (Welch and Ehlers, 1988). The production of base maps by means of remote sensing imagery will make it easier to track error propagation in GIS layers because of reliable map metadata. In addition, the extracted cartographic information can be used to improve image classification. Image segmentation polygons derived from optical imagery, for instance, could be very useful for stratifying radar data, which traditionally are difficult to segment digitally owing to their noise (Hinton, 1996).

Aerial photography is a popular data source for feature extraction owing to its high resolution. Recently available high resolution satellite imagery, such as IKONOS and QuickBird imagery, has provided a potential to acquire effective and efficient feature information with automated extraction methods. These images have been applied to topographic mapping (Birk et al., 2003; Holland et al., 2006; Zanoni et al., 2003) and three-dimensional (3D) object reconstruction (Baltsavias et al.,

2001; Tao and Hu, 2002). Many GIS databases use these high-accuracy geopositioning images as base maps, providing both metric and thematic information. A new data source, light detection and ranging (LiDAR) data, is also being used increasingly in the extraction of cartographic information, and it provides land surface elevation information with vertical and horizontal accuracy to less than 1 m. Recent developments in sensor and computer technologies have made object oriented image analysis a new focus in digital image processing. When combined with high spatial-resolution imagery and/or LiDAR data, an object oriented analysis shows great potential for extracting the earth surface features, including both thematic and metric information.

Remotely Sensed Data Used to Update GIS

It is crucial to update GIS databases and maps in a timely manner. Remotely sensed data provide the most cost-effective source for such updates. Another area of the application of remotely sensed data as input to GIS is change detection. Ehlers et al. (1990) used Le Systeme Pour l'Observation de la Terre (SPOT) data in a GIS environment for regional analysis and local planning at a scale of 1:24,000 and achieved an accuracy of 93 percent for growth detection. Brown and Fletcher (1994) demonstrated that satellite images could be used to interactively update a land-use database by comparing the database with image statistics within the areas defined in a vector database. Today, the integration between remote sensing and GIS allows for querying raster pixels within vector areas and for performing analyses without format conversions and overlays. Image statistics within vector polygons then can be used to examine changes that have occurred and update maps.

Remotely Sensed Data as a Backdrop for GIS and Cartographic Representation

Cartographic representation is the fourth area of application of remote sensing imagery as an input to GIS. Terrain visualization using satellite images in association with digital elevation models (DEMs) has long been explored as a promising tool in environmental studies (Gugan, 1988). Progress in cartographic animation in recent years changes terrain visualization from a static to a dynamic state. DEM generation from satellite imagery using image correlation also has demonstrated its feasibility for deriving further topographic information for GIS applications. A DEM developed from Landsat Thematic Mapper (TM) images of a rugged terrain in north Georgia yielded a root-mean-square (rms) error in Z of ±42 m (Ehlers and Welch, 1987). With a more favorable base-to-height ratio and a higher resolution of 10 m, SPOT data produced better DEMs with rms errors between ±6 and ±18 m (Ehlers et al., 1990). Tong et al. (2009) demonstrated that with an appropriate bias correction method (e.g., an affine model), a stereo QuickBird image can yield a geopositioning accuracy as high as 0.5 m in planimetry and 1 m in height in Shanghai, China. In addition, the production of an orthophotograph or orthoimage requires the use of a DEM to register properly to the stereomodel so as to provide correct height data for differential rectification of the image (Jensen, 2005). Orthoimages are employed increasingly to provide the base maps for GIS databases on which thematic data layers are overlaid (Lo and Yeung, 2002). All these developments have implications for extending traditional two-dimensional (2D) to 3D GIS, which is crucial for applications in the fields of marine science, geology, soil modeling, and climatology.

2.1.2 Contributions of GIS to Remote Sensing

Use of Ancillary Data in Image Classification

GIS data can be used to enhance the functions of remote sensing image processing at various stages: selection of the area of interest for processing, preprocessing, and image classification. Ancillary data are those collected independently of remotely sensed data. Ancillary data have long been useful in the identification and delineation of features on manual interpretation of aerial photos; for digital remote sensing, ancillary data must be incorporated into the analysis in a structured, formalized manner that connects directly to the analysis of remotely sensed data (Campbell, 2007). Harris and Ventura (1995) and Williams (2001) suggested that ancillary data may be used to enhance image classification in three stages, namely, pre-classification stratification, classifier modification, and post-classification sorting. In the pre-classification method, ancillary data are used to assist the selection of training samples (Mesev, 1998) or to divide the study scene into smaller areas or strata based on some selected criteria or rules. When ancillary data are used during an image classification, two methods can be used. The first is a logical channel method that was introduced by Strahler et al. (1978). The ancillary data were incorporated into remote sensing images as additional channels. The second method is a classifier modification that involves altering *a priori* probabilities of the classes in a maximum likelihood classifier according to estimated areal composition or known relationships between classes and ancillary data (Harris and Ventura, 1995; Mesev, 1998). In the post-classification sorting approach, ancillary data are used to modify misclassified pixels based on established expert rules.

Table 2.1 summarizes major approaches for combining various ancillary data and remote sensing imagery for improving image classification. For example, previous research has shown that topographic data are valuable in improving the accuracy of land-cover classification, especially in mountainous regions (Franklin et al., 1994; Janssen et al., 1990; Meyer et al., 1993). This is so because land-cover distribution is related to topography. In addition to elevation, slope and aspect,

Method	Features	References
Use of ancillary data	DEM	Maselli et al., 2000
	Topography, land use, and soil maps	Baban and Yusof, 2001
	Road density	Zhang et al., 2002
	Road coverage	Epstein et al., 2002
	Census data	Harris and Ventura, 1995; Mesev 1998
Stratification	Based on topography	Bronge, 1999; Baban and Yusof, 2001
	Based on illumination and ecologic zone	Helmer et al., 2000
	Based on census data	Oetter et al., 2000
	Based on shape index of the patches	Narumalani et al., 1998
Post-classification processing	Kernel-based spatial reclassification	Barnsley and Barr, 1996
	Using zoning and housing-density data to modify the initial classification result	Harris and Ventura, 1995
	Using contextual correction	Groom et al., 1996
	Using co-occurrence matrix-based filtering	Zhang, 1999
	Using polygon and rectangular mode filters	Stallings et al., 1999
	Using an expert system to perform post-classification sorting	Stefanov et al., 2001
	Using a knowledge-based system to correct misclassification	Murai and Omatu, 1997

Table 2.1 Summary of Major Approaches to Using Ancillary Data for Improving Classification Accuracy

Method	Features	References
Use of multisource data	Spectral, texture, and ancillary data (such as DEM, soil, existing GIS-based maps)	Amarsaikhan and Douglas, 2004; Benediktsson and Kanellopoulos, 1999; Bruzzone et al., 1997, 1999; Franklin et al., 2002; Gong, 1996; Solberg et al., 1996; Tso and Mather, 1999

Adapted from Lu and Weng, 2007.

TABLE 2.1 Summary of Major Approaches to Using Ancillary Data for Improving Classification Accuracy (*Continued*)

which can be derived from DEM data, also have been employed in image classification. Topography data are useful at all three stages of image classification, that is, as a stratification tool in pre-classification, as an additional channel during the classification, and as a smoothing means in post-classification (Maselli et al., 2000; Senoo et al., 1990). Integration of DEM-related data and remotely sensed data for forest classification in mountainous areas has been demonstrated to be effective for improving classification accuracy (Franklin, 2001; Senoo et al., 1990). In urban studies, DEM data are rarely used to aid image classification owing to the fact that urban regions often locate in relatively flat areas. Other ancillary data, such as soil type, temperature, and precipitation, also can be used in assisting image classification.

Use of Ancillary Data in Image Preprocessing

At the stage of geometric and radiometric correction, GIS data such as vector points, area data, and DEMs are used increasingly for image rectification (Hinton, 1996). High resolution topographic data play an important part in radar image interpretation (Kwok et al., 1987). The impacts of varying topography on the radiometric characteristics of digital imagery can be corrected with the aid of DEMs (Hinton, 1996). The DEM-derived variables may be used at the stage of image preprocessing for topographic correction or normalization so that the impact of terrain on land-cover reflectance can be removed (Dymond and Shepherd, 1999; Ekstrand, 1996; Gu and Gillespie, 1998; Leprieur et al., 1988; Richter, 1997; Teillet et al., 1982; Tokola et al., 2001). Perhaps, the most frequently used vector data sets are ground control points in image rectification, where the most identifiable points are selected from an existing map with a defined coordinate system and used to register an image. Hinton (1996) suggested that with advances in pattern recognition and line-following techniques, extracted lines from satellite images could be registered to roads, rivers, and railways in vector datasets for more accurate image registration.

Use of Ancillary Data for Selection of the Area of Interest

The use of vector polygons to restrict the area of an image to be processed is possible in today's image processing software (e.g., ERDAS IMAGINE). This permits masking operations without raster masks, making image processing much more efficient because of faster processing times, no need to store intermediate data, and a reduction in data integrity problems.

However, there are a few practical and conceptual problems in using ancillary data in remote sensing image analysis (Campbell, 2007). Ideally, ancillary data should possess compatibility with remotely sensed imagery with respect to scale, level of detail, accuracy, geographic reference system, and date of acquisition. Sometimes ancillary data are presented as discrete classes (i.e., nominal or ordinal data), whereas remote sensing data represent ratio or interval data. The compatibility between the two types of data must be addressed.

Use of GIS to Organize Field/Reference Data for Remote Sensing Applications

In addition to enhancing the functions of remote sensing image processing at various stages, GIS technology provides a flexible environment for entering, analyzing, managing, and displaying digital data from the various sources necessary for remote sensing applications. Many remote sensing projects need to develop a GIS database to store, organize, and display aerial and ground photographs, satellite images, and ancillary, reference, and field data. Global Positioning Systems (GPS) technology also is essential when remote sensing projects need to collect *in situ* samples and observations (Gao, 2002).

2.1.3 Integration of Remote Sensing and GIS for Urban Analysis

The integration of remote sensing and GIS technologies has been applied widely and is recognized as an effective tool in urban analysis and modeling (Ehlers et al., 1990; Harris and Ventura, 1995; Treitz et al., 1992; Weng, 2002). Remotely sensed derived variables, GIS thematic layers, and census data are three essential data sources for urban analyses. Remote sensing collects multispectral, multiresolution, and multitemporal data and turns them into information that is valuable for understanding and monitoring urban land processes and for extracting urban biophysical and socioeconomic variables. GIS technology provides a flexible environment for entering, analyzing, and displaying digital data from the various sources necessary for urban feature identification, change detection, and database development (Weng, 2001). GIS data for urban applications frequently contain such thematic layers as transportation network, hydrographic features, land use and zoning, basic infrastructure, buildings, and administrative boundaries with different detail. Census data offer a wide range of demographic and socioeconomic information and have been used in various urban geography, planning, and environment-management studies. The advances in GIS technology provide an effective environment for spatial analysis of remotely sensed data and other sources of spatial data (Burrough, 1986; Donnay et al., 2001). Integration of remote sensing imagery and GIS (including census) data has received widespread attention in recent years.

Wilkinson (1996) summarized three main ways in which remote sensing and GIS technologies can be combined to enhance each other. Since census data collected within spatial units can be stored as GIS attributes, the combination of census and remote sensing data through GIS can be envisioned in the aforementioned three ways. Each of these ways has been related to urban analysis. First, remote sensing images have been used in extracting and updating transportation networks (Doucette et al., 2004; Harvey et al., 2004; Kim et al., 2004; Song and Civco, 2004) and buildings (Lee et al., 2003, 2008; Mayer, 1999; Miliaresis and Kokkas, 2007), providing land-use/cover data and biophysical attributes (Ehlers et al., 1990; Haack et al., 1987; Lu and Weng, 2004; Treitz et al., 1992; Weng and Hu, 2008; Weng et al., 2006) and detecting urban expansion (Cheng and Masser, 2003; Weng, 2002; Yeh and Li, 1997). Second, census data have been used to improve image classification in urban areas (Harris and Ventura, 1995; Mesev, 1998). Finally, the integration of remote sensing and census data has been applied to estimate population and residential density (Harris and Longley, 2000; Harvey, 2000, 2002; Langford et al., 1991; Li and Weng, 2005; Lo, 1995; Martin et al., 2000; Qiu et al., 2003; Sutton, 1997; Yuan et al., 1997), to assess socioeconomic conditions (Hall et al., 2001; Thomson, 2000), and to evaluate the quality of life (Li and Weng, 2007; Lo and Faber, 1997; Weber and Hirsch, 1992).

So far, most of the work in the integration of remote sensing and GIS has been implemented by converting vector GIS data (including census data) into raster format because of the similarity of remote sensing and raster GIS data models. Only in recent years has the improvement of image-analysis systems allowed extraction of image statistics based on GIS polygons. When remote sensing data are aggregated to a census unit with raster-to-vector conversion, values are assumed to be uniform throughout the census unit. This would lead to loss of spatial information in the remote sensing data. In addition, census data have different scales (levels), and thus integration of remote sensing data with different scales of census data would produce the so-called modifiable-area-unit problem (Liang and Weng, 2008). Therefore, finding desirable aggregation units is important in order to reduce the loss of spatial information from remote sensing. Another method of data integration is through vector-to-raster conversion by rasterization or by surface interpolation to produce a raster

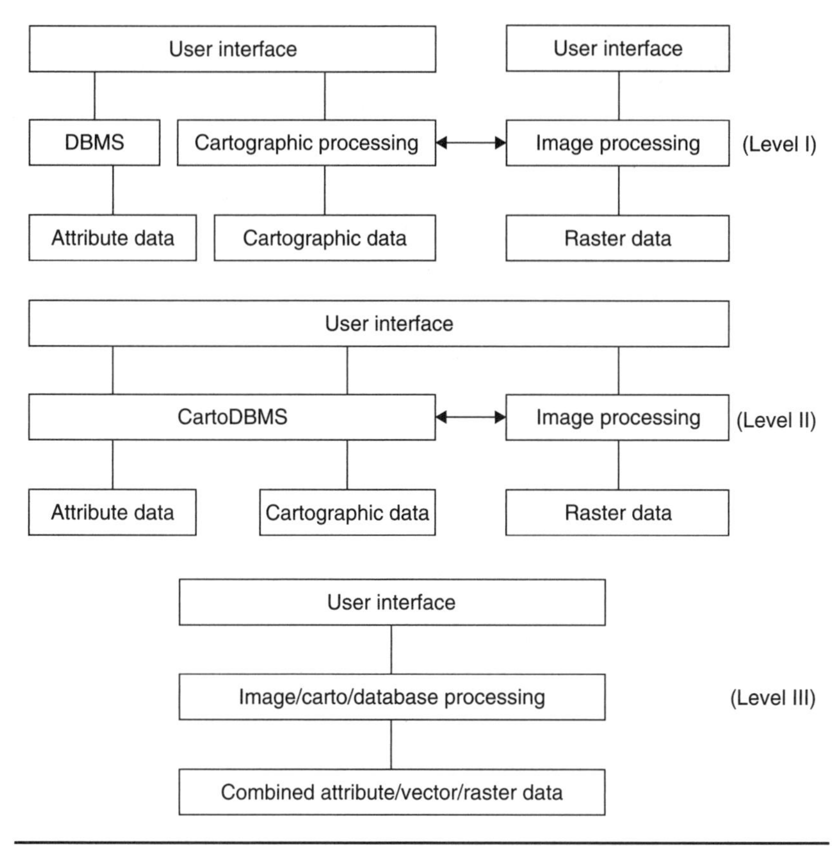

FIGURE 2.1 Integration levels. (*Adapted from Ehlers et al., 1989.*)

layer for each socioeconomic variable. More research is needed on disaggregating census data into individual pixels to match remote sensing data for the purpose of data integration.

2.2 Theories of the Integration

2.2.1 Evolutionary Integration

From an evolutionary perspective, Ehlers et al. (1989) proposed a three-stage process of integration between remote sensing and GIS described as "separate but equal," "seamless integration," and "total integration." Figure 2.1 illustrates these three levels of the integration.

- *Level I (separate but equal).* Here, there are separate databases. Two software modules, GIS and image processing, are linked only by data exchange. The integration at this level has the ability to move the results of low-level image processing (e.g., thematic maps, extracted lines, and so on) to the GIS and the results of GIS overlays and analysis to image processing software.

- *Level II (seamless integration).* Here, there are two software modules with a common user interface and simultaneous display. Such a system will allow for a "tandem raster-vector processing" (Ehlers et al., 1989), such as incorporating vector data directly into image processing, and entity-like control over remote sensing image components (e.g., themes). Moreover, the system will have the ability to accommodate hierarchical entities (e.g., "house" at one level, "block" at another, and "city" at another), and (spatially, radiometrically, spectrally, and temporally) inhomogeneous data in a coherent manner.
- *Level III (total integration).* Here, there is a single software unit with combined processing. In the fully integrated system, a single model will underlie all information in the GIS, which has the flexibility of handling both object- and field-based spacial representations. Remote sensing will become an integral part of the functionality of the GIS. The integrated GIS will be able to handle temporal and 3D information and thus play a more important role in earth system science.

2.2.2 Methodological Integration

Mesev (1997) proposed a hierarchical three-level integration scheme for urban analyses. Each successive level in the hierarchy deals with more detailed conceptual and operational factors and the issues of remote sensing–GIS linkage.

At the top level of the hierarchy (level I) are the issues of data unity, measurement conformity, potential integrity, statistical relationships, classification compatibility, and overall integration design. Each of these is more fully examined at the level II, tackling more complex linkages among the five components. Data unity, for example, is divided into factors such as information exchange, data availability, data accessibility, and data creation. Figure 2.2 illustrates the factors and links at level II. The level II factors are refined further to produce even more detailed factors at level III. For example, data availability is subdivided into awareness, publicity, search, data type, age, quality, and access (Table 2.2). Whereas other schemes have already been documented by others (e.g., Davies et al., 1991), Mesev (1997) attempted to itemize the common linkages and reexamined the relationships with the aim of providing a structured framework for direct data coupling between remote sensing and GIS.

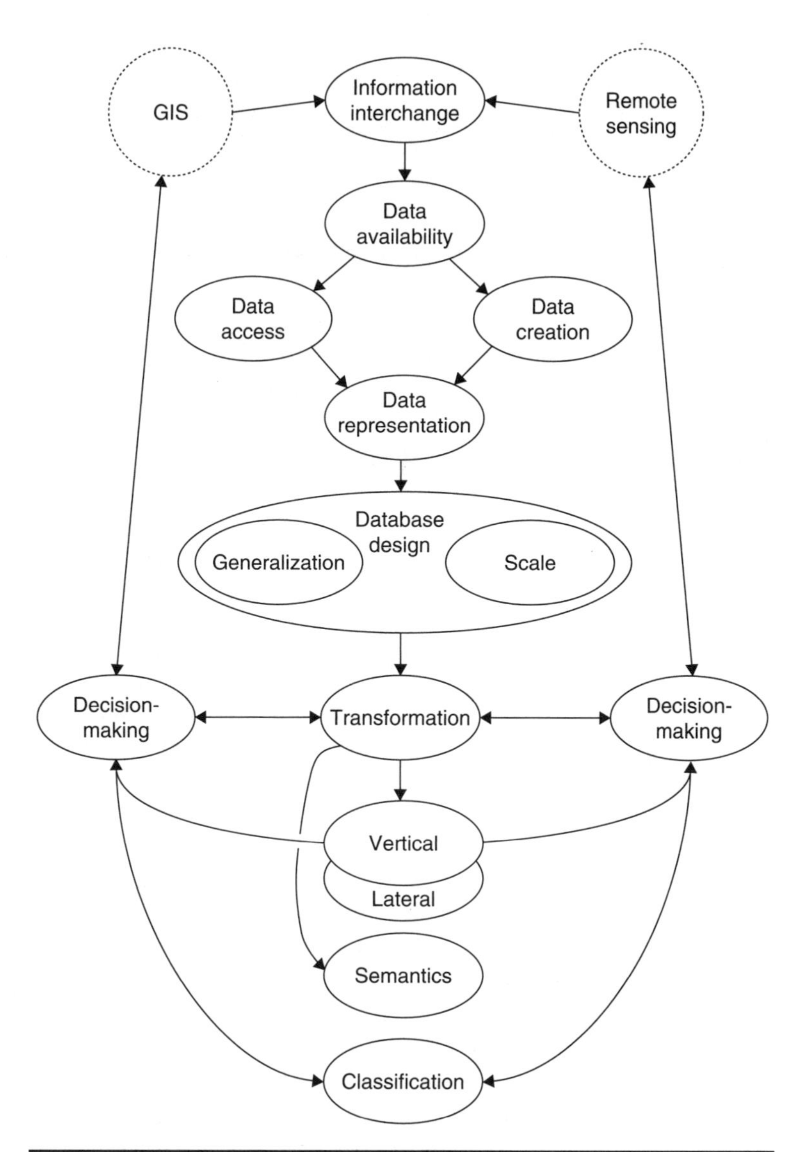

FIGURE 2.2 Diagrammatic representation of level II factors and their links. (*Adapted from Mesev, 1997.*)

2.2.3 The Integration Models

Gao (2002) reviewed the approaches to integration among remote sensing, GIS, and GPS and summarized them into four conceptual models: linear, interactive, hierarchical, and complex integration. These approaches are largely applied to resources management and environmental studies. Because of the focus of this book, the

Level II	Level III
Data unity (factors that bring remote sensing and GIS data together)	
Information interchange	Definition of integration, type of information needed, information harmony (spatial units and attributes)
Data availability	Awareness, publicity, search, data type, age, quality (access or create)
Data accessibility	Cost, agreements, exchanges, sharing, proprietary, resistance, confidentiality, liability
Data creation	Digitizing, scanning, survey information encoding, sampling, data transformation, GPS
Measurement conformity (factors that link data between remote sensing and GIS)	
Data representation	Data structures (vector, raster, quadtree, etc.), data type, level of measurement, field-based versus object-based modeling, interpolation
Database design	Type (relational, hybrid), scheme, data dictionary, implementation (query, testing)
Data transfer	Format, standards, precision, accuracy
Positional integrity (factors that spatially coordinate data between remote sensing and GIS)	
Generalization and scale	Spatial resolution, scale, data reduction and aggregation, fractals
Geometric transformation	Rectification, registration, resampling, coordinate system, projection, error evaluation
Statistical relationships (factors that measure links between remote sensing and GIS)	
Vertical	Boolean overlays, asymmetric mapping, areal interpolation, linear and nonlinear equations, time series, change detection
Lateral	Spatial searches, proximity analysis, textual properties
Classification compatibility (factors that harmonize information between remote sensing and GIS)	
Semantics	Classification scheme, levels, descriptions, class merging, standardization

TABLE 2.2 Level III Factors for Remote Sensing–GIS Integration

Level II	Level III
Classification	Stage (pre-, during, post-), level (pixel, subpixel) type (per-pixel, textual, contextual, neural nets, fuzzy sets), change detection, accuracy assessment
Integration design	
Objectives	Plan of integration, cost/benefit assessment, feasibility, alternatives to integration
Integration specifications	User requirements (intended use, level of training, education), system requirements (hardware, software, computing efficiency)
Decision making	Testing, visualization, ability to replicate integration, decision support, implementation or advocate alternatives, bidirectional updating and feedback into individual remote sensing and GIS projects

Adapted from Mesev, 1997.

TABLE 2.2 (*Continued*)

following discussions are limited to integration between remote sensing and GIS.

In the linear model, each of the three technologies is quite independent of the others. GPS is used to collect ground controls for aerial photographs and satellite images. After rectification, aerial photographs and satellite images are combined with other data into a GIS database and are used for subsequent analyses and modeling. As seen in Fig. 2.3, there is not a direct link between GPS and GIS. The integration between GPS and remote sensing comes in two temporal modes (Gao, 2008). While the independent mode suggests that GPS data and remotely sensed data are acquired independently, the simultaneous mode is applied when the two kinds of data are collected at the same time, in which the remote sensing platform is equipped with a GPS unit during the flights.

In the interactive model, data flow in two directions (Fig. 2.4) instead of one, as seen in the linear model. GPS can feed coordinate data to remote sensing but also can receive data from remote sensing,

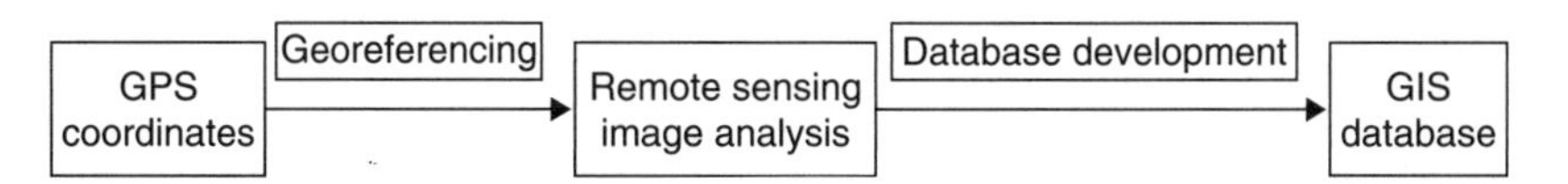

FIGURE 2.3 The linear model of integration with a unidirection of data form from GPS to remote sensing image analysis and ultimately to GIS. (*Adapted from Gao, 2002.*)

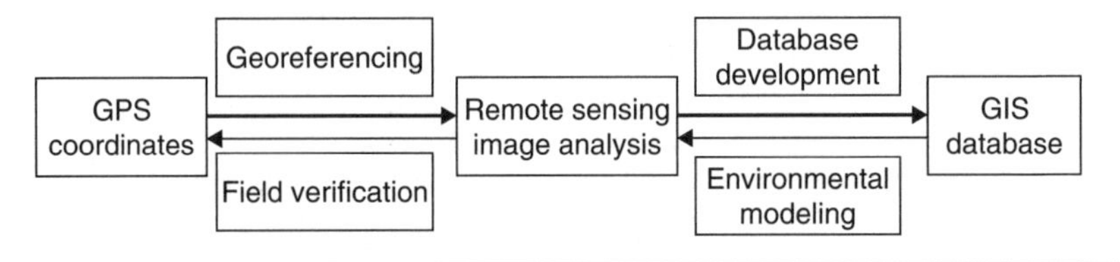

Figure 2.4 The interactive model of integration. (*Adapted from Gao, 2002.*)

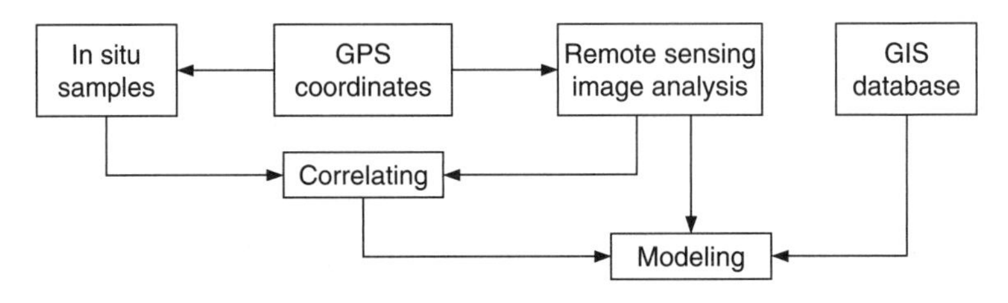

Figure 2.5 The hierarchical model of integration. (*Adapted from Gao, 2002.*)

mainly in the post-image analysis session when classified maps or multitemporal remotely sensed data need to be verified in the field. Similarly, there is two-way flow of data between remote sensing and GIS. The integration of remote sensing with GIS is best exemplified by land-use change detection through GIS overlay of historic and current maps, whereas integration of GIS with remote sensing facilitates image segmentation or classification by using the GIS database.

There are two tiers of integration in hierarchical integration (Fig. 2.5). The first tier of integration occurs between GPS and remote sensing via in situ samples. GPS provides ground control coordinates for digital photographs/satellite images, as well as positioning information for in situ samples, so that these samples are properly correlated spectrally with remotely sensed data. There is no a direct link between remote sensing and GIS; instead, both types of data are used in statistical or mathematical modeling, which should occur in an image analysis system or in a raster GIS environment.

The complex model of integration possesses all possible linkages between any two of the three technologies (Fig. 2.6). In addition to the links contained in the linear, interactive, and hierarchical models, there is an extra connection between GPS and GIS. Here, GPS data, in the form of point, linear, or areal data, may be exported into a GIS database for the purpose of updating. This type of integration is exemplified in precision farming, where GPS is used to determine coordinates associated with precision-farming variables, whereas GIS for data is used for integration, storage, and analysis. Because of the circular nature of the interrelationships among the three technologies, it is difficult to judge the relative importance of each technology in this model.

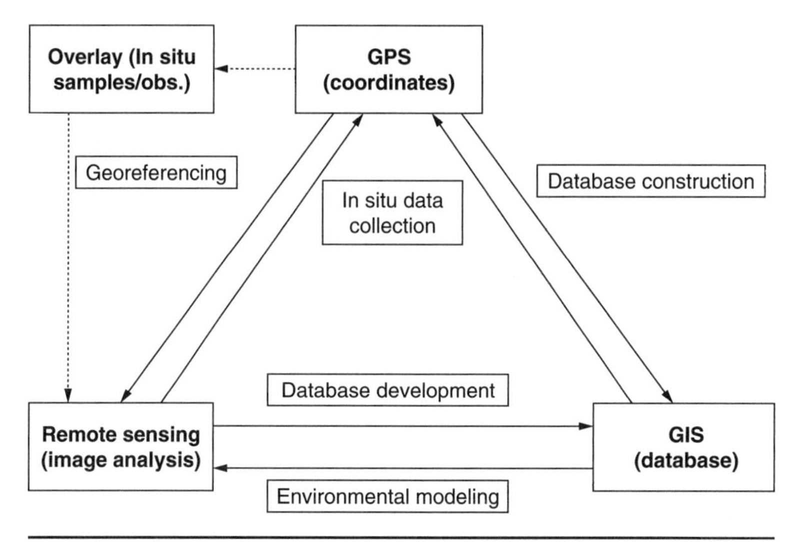

Figure 2.6 The complex model of integration. (*Adapted from Gao, 2002.*)

2.3 Impediments to Integration and Probable Solutions

Current impediments to the integration of remote sensing and GIS are related not only to technical development but also to conceptual issues, that is, our understanding of the phenomena under investigation and their representation in spatial databases. While computer technology in general and the technologies of remote sensing and GIS in particular have continued to improve over the past three decades, many needs for basic researches in integration remain untouched. This section begins with a review of conceptual impediments to *IGIS* (the term for a fully integrated system of remote sensing and GIS), followed by a more detailed discussion of technical impediments to integration and possible solutions.

2.3.1 Conceptual Impediments and Probable Solutions

Davies et al. (1991) identified two set of major impediments to integration that are conceptual in nature. One relates to defining appropriate strategies for data acquisition and spatial modeling, and the other relates to tracking and understanding the impact of data processing steps on output products. Examples of the major impediments include (Davies et al., 1991)

- "Use of multiple data layers varying in structure, level of preprocessing, and spatial consistency;
- Multiple (and often poorly known) measurement scales, ranging from 'points' to grids to irregular polygons;

- Unknown measurement errors for most variables;
- Unknown spatial dependencies in the data and their propagation through spatial models;
- Limited ability to verify or validate IGIS model outputs;
- Limited capability for model sensitivity analysis."

To overcome these obstacles, researchers are required to investigate the earth's surface features, patterns, and processes on multiple spatial and temporal scales to quantify the scale dependence of IGIS inputs and outputs, to develop robust procedures to account for scale dependence in IGIS modeling, and to establish theories and methods for quantifying and tracking data transformations and information flow in IGIS processing (Davies et al., 1991).

Having witnessed a slow progress in remote sensing–GIS integration, Mesev and Walrath (2007) suggested that the difference between GIS as a data-handling technology and remote sensing as a data-collection technology is nearly insurmountable and that the notion of a total integration should be defined as a high level of complementary information exchange and sharing of data processing functions. They further suggested that another major conceptual impediment lies in data modeling, that is, how geographic information is represented. Remote sensing instruments record continuous data representing the interaction of energy with the earth's surface features, whereas GIS characterizes geographic features with more defined and discrete boundaries. These two views of the real world are conflicting but also can be complementary (Mesev and Walrath, 2007). The two views are difficult to accommodate within one model or system, and thus we may leave the definition of integration as ambiguous. This approach would allow using the strengths of the two technologies and facilitating the data exchange (Mesev and Walrath, 2007).

In fact, early in 1989, Ehlers et al. (1989) pointed out that major technical impediments (such as raster-vector dichotomy and the data uniformity issue) were rooted in geographic information representation. Image data take a "field-based approach," through which geographic features are invisible without high-level processing. Cartographic data take an "object-based approach," through which data are obtained by abstracting some information about the earth's surface features and discarding the rest. Therefore, cartographic data have a higher level of abstraction. Integration of remote sensing and GIS requires us to understand the transition from one representation to another (Bruegger et al., 1989) (Fig. 2.7). Ehlers et al. (1989) further suggested that the relationship between the attribute data and cartographic data is also one of abstraction. That is to say, a GIS involves cartographic data at the lower level and attribute data at the higher level despite the fact that much of the attribute data comes from other data sources than cartography.

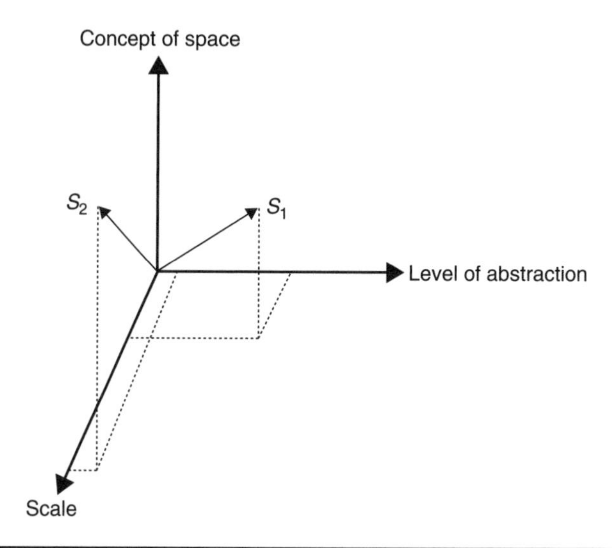

Figure 2.7 Different representations of spatial information. The representations may vary in concept of space, scale, level of abstraction, and other aspects. Some transformation from one representation to another may be reversible, e.g., from S_1 to S_2, whereas others may not be. (*Adapted from Ehlers et al., 1989.*)

According to Ehlers et al. (1989), research on human vision and perception has given rise to a model of spatial information extraction and object recognition from an image as a three-level integrated raster-vector processing system (Marr, 1982). Figure 2.8 illustrates this conceptual model of information extraction as a three-stage process. At the first level, gray values (i.e., raster data) are processed (low level), from which structures (i.e., features) can be extracted and manipulated as symbolic descriptors (midlevel). At the highest level, knowledge-based information, often coupled with spatiotemporal models, gives a predictive description (i.e., image understanding) of the "imaged" object (Pentland, 1985). This may lead to a hierarchical image-analysis system in which midlevel and high-level information is stored as vectors and/or objects, whereas low-level information is stored as raster data (i.e., gray values).

When data transformation (e.g., image capture, information extraction, and pattern recognition) is considered as different levels of abstraction, different models of geographic space can be conceptualized more easily. Figure 2.9 shows a typical path taken by data captured by satellite and then abstracted into a suitable form for GIS. The four stages involve four models of geographic space—field, image, theme, and object/feature—that are typical of models in the integration of GIS and remote sensing activities. The overall process of object extraction is sometimes referred to as *semantic abstraction* (Waterfeld

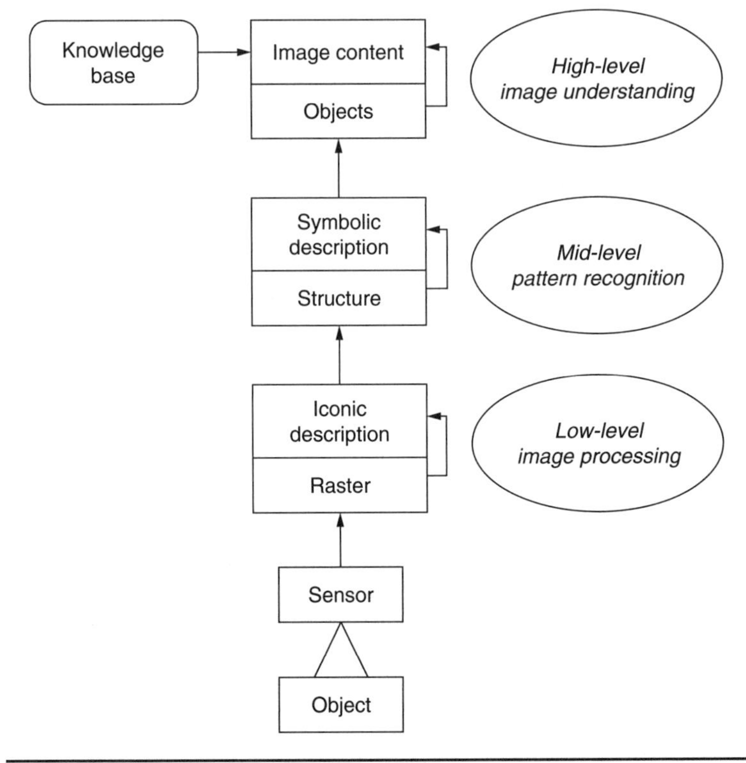

Figure 2.8 Vision concept of information extraction from an image as a three-stage process. (*Adapted from Ehlers et al. 1989.*)

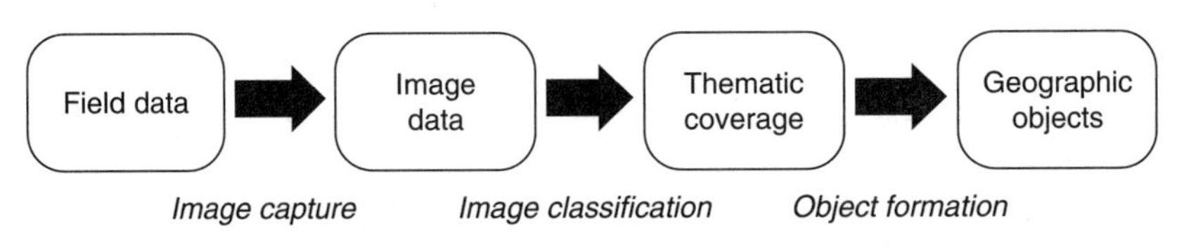

Figure 2.9 Continuum of abstraction from field model to object model. Continuously varying fields are quantized by the sensing device into image form, then classified and finally transformed into discrete geographic objects. (*Adapted from Gahegan and Ehlers, 2000.*)

and Schek, 1992) owing to the increasing semantic content of the data as they are manipulated into form (Gahegan and Ehlers, 2000). These models represent only the conceptual properties of the data and are considered to be independent of any particular data structure that might be used to encode and organize the data (Gahegan and Ehlers, 2000). Gahegan and Ehlers (2000) further suggested that there was an artificial barrier between GIS and remote sensing as a result of the traditional separation of the two activities into separate communities and separate software environments. Thus the integration of these two branches of science is to some extent an artificial problem (Gahegan and Ehlers, 2000). As a result, there is no easy flow of metadata between the systems, and interoperability is often restricted to the exchange of image files or object geometry, and the problem of managing uncertainty is compounded.

2.3.2 Technical Impediments and Probable Solutions

Data Structure

A fundamental impediment to integration is the vector-raster dichotomy. Remote sensing has predominantly used a raster approach to data acquisition and analysis. GIS, on the other hand, is predominantly vector-oriented, although there are also raster-based GIS. Ehlers et al. (1991) emphasized the need for a raster-vector intersection query to optimize the operation. This query would be capable of providing image statistics within vector polygons without carrying out any data format conversion or raster masking. The emergence of such an intersection query, therefore, would allow for a better selection of training areas for image classification by examining whether the areas display the spectral response characteristics of a class. Likewise, it would be possible to look at changes in image statistics within defined polygons directly without classifying the whole image and examining the result (Hinton, 1996). Moreover, the integration of remote sensing and GIS would allow vector data to be rasterized as "image planes" and incorporate into traditional classification and segmentation procedures (Ehlers et al., 1989). A closer integration is expected to be able to incorporate all GIS data layers in their native formats into an image classification. Then a two-way data flow between the raster image and the vector dataset comes true. This situation occurs only in *polygon classification* (Hinton, 1996), in which, image statistics are generated based on polygons and then return these statistics directly to the GIS database as attributes of the polygons. Figure 2.10 shows image classification based on polygon attributes rather than pixel values.

Today, many geospatial systems become capable of displaying and querying data in an increasing number of formats. Mature GIS systems can display and query raster images, whereas remote sensing digital image processing systems offer the similar capability of handling vector data. The raster-vector intersection query thus is possible, and so are some tandem data processing tasks. The polygon classification envisioned by Hinton (1996) now can be fulfilled in commercial software. However, even with all these technical improvements, no GIS or image analysis systems today can handle both data formats

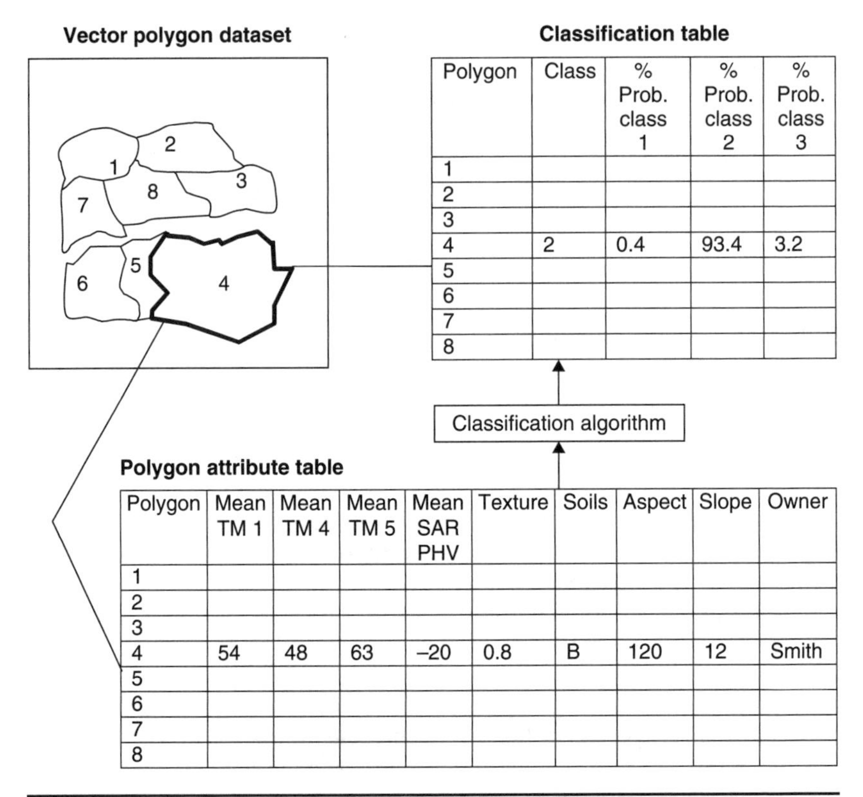

FIGURE 2.10 Polygon classification: image classification is based on polygon attributes rather than on pixel values. (*Adapted from Hinton, 1996.*)

equally well for all analyses (Gao, 2008). The hybrid processing of raster and vector data cannot be translated into a fully integrated GIS (Ehlers, 2007; Mesev and Walrath, 2007). The ideal goal would be that geographic objects can be extracted automatically from a remote sensing image to export into a GIS database, whereas GIS "intelligence" (e.g., object and analysis models) can be used to automate this object-extraction process (Ehlers, 2007). Figure 2.11 shows this flow of automatic extraction of GIS objects from remote sensing imagery. However, a complete extraction from remote sensing images nowadays can be performed only by manual interpretation (e.g., on-screen digitizing). Automatic extraction of geographic objects from remotely sensed data by using image segmentation algorithms has met with mixed success. With the advent of high spatial-resolution images, especially sub-meter resolution satellite images, we will see another push for object extraction and object-based analysis (more discussions on this issue in Sec. 2.4.3).

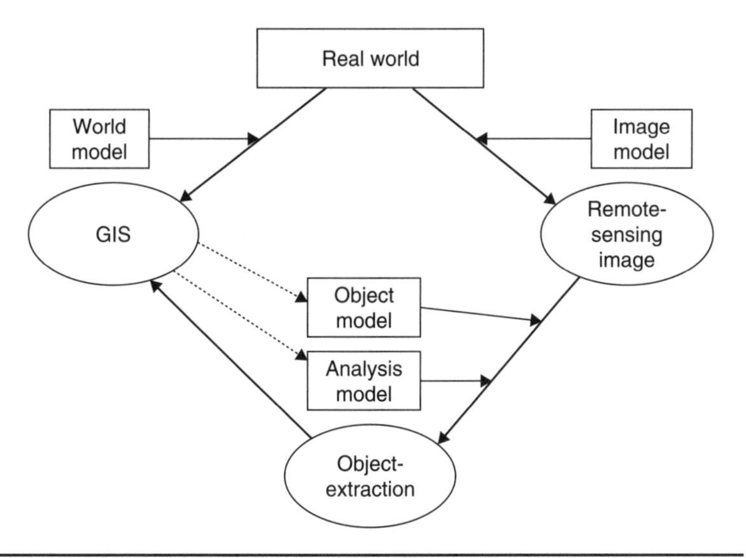

Figure 2.11 Concept for automatic extraction of GIS objects from remote sensing imagery. (*Adapted from Ehlers, 2007; used with permission from John Wiley and Sons.*)

One possible solution to the dichotomy is to establish a mixture of data structures (Ehlers et al., 1991). This involves the use of a high-level declarative language that supports queries and operations regardless of storage formats and a database management system (DBMS) that is capable of handling various types of data representations. Such a DBMS will spare the user repetitive data conversions and transference. Recent advancement in GIS technology has removed all technical obstacles to providing procedures that allow for an easy movement between various physical implementations of spatial data. The main issue now is providing users with an efficient and accessible means to do so that operates with known effects on the data's accuracy and precision (Ehlers et al., 1991). Research efforts on data structures in the past two decades included the introduction of alternative data structures such as quadtree and Voronoi polygons but have achieved limited success (Gao, 2008).

An example of a mixed data structure is a feature-based GIS (FBGIS) model proposed by Usery (1996). This conceptual model was proposed to solve the problem of the vector-raster dichotomy, as well as to permit 3D or higher-dimensional representation and temporal modeling. The model includes spatial, thematic, and temporal dimensions and constructs attributes and relationships for each dimension. Because each feature in the model is provided a direct access to spatial, thematic, and temporal attributes and their relationships, multiple representations and multiple geometries such as raster and vector are supported (Table 2.3).

	Space	Theme	Time
Attributes	Φ, λ, *Z* point, line, area, surface, volume, pixel, voxel, etc.	Color, size, shape, etc.	Date, duration, period, etc.
Relationships	Topology, direction, distance, etc.	Topology, is_a, kind_of, part_of, etc.	Topology, is_a, was_a, will_be, etc.

Adapted from Usery, 1996.

TABLE 2.3 Dimensions, Attributes, and Relationships in the Feature-Based GIS Conceptual Model

Data Uniformity and Uncertainty Analysis

The second technical impediment to the integration of remote sensing and GIS has to do with the issue of data uniformity (Ehlers et al., 1989). GIS rely on fairly uniform and predetermined data. Data collection in GIS tends to be separated from data processing, which makes tracking errors a difficult task. On the other hand, in remote sensing, data are collected first, and the user then has to decide how to use them. Data collection is uneven, and data coverage is far from being uniform. The integration of remotely sensed data requires that GIS be based on deeper, more complete models of the territory. A mixed GIS data structure, as described earlier, may be the solution for integration (Ehlers et al., 1989).

Uncertainty in geographic information has been a research focus since the early 1990s, influenced by the National Center for Geographic Information and Analysis (NCGIA)'s first research initiative. Most works have addressed the inherent errors present within specific types of data structures (e.g., raster or vector) or data models (e.g., field or object) (Gahegan and Ehlers, 2000). The effects of error propagation and analysis within various paradigms have been studied by Veregin (1989, 1995), Openshaw et al. (1991), Goodchild et al. (1992), Heuvelink and Burrough (1993), Ehlers and Shi (1997), Leung and Yan (1998), Shi (1998), Arbia et al. (1999), Zhang and Kirby (2000), and Shi et al. (2003). Uncertainty in the integration of remote sensing and GIS has generated fewer research works, even though it was identified as one of the major challenges in integration by NCGIA Initiative 12 (i.e., the integration of remote sensing and GIS initiative) (Lunetta et al., 1991). Previous remote sensing works are concerned mostly with the quantification of errors in the process of image geometric correction and thematic information extraction (Lu and Weng, 2007). Congalton and Green (1999) systematically reviewed the concept of basic accuracy assessment and some advanced topics involving fuzzy logic and

Categories	Major Approaches	References
Hard classification results	Error matrix or normalized error matrix, producer's and user's accuracies, kappa coefficient, modified Kappa coefficient, and Tau coefficient	Congalton, 1991; Janssen and van der Wel, 1994; Congalton and Green, 1999; Smits et al., 1999; Foody, 2002b, 2004a
Soft classification results	Fuzzy sets, conditional entropy and mutual information, symmetric index of information closeness, Renyi generalized entropy function, and parametric generalization of Morisita's index	Finn, 1993; Gopal and Woodcock, 1994; Foody, 1996; Binaghi et al., 1999; Woodcock and Gopal, 2000; Ricotta and Avena, 2002; Ricotta, 2004

Adapted from Lu and Weng, 2007.

Table 2.4 Major Approaches to Accuracy Assessment of Hard and Soft Classification Results

multilayer assessment and explained principles and practical considerations in designing and conducting accuracy assessment of remote sensing data. Table 2.4 summarizes major approaches to accuracy assessment of classified images using both hard and soft classifiers. The error matrix and associated derived parameters are the most common approach of accuracy assessment for categorical classes, but uncertainty analysis of classification results recently has gained more attention (McIver and Friedl, 2001; Liu et al., 2004).

Little attention has been given so far to the problem of modeling uncertainty as the data are transformed through different models of geographic space (Gahegan, 1996; Gahegan and Ehlers, 2000; Lunetta et al., 1991). Lunetta et al. (1991) provided a broad view of spatial data-error sources at each step of the data integration process that included data acquisition, processing, analysis, conversion, error assessment, and final product presentation. As a result, the areas listed below were recommended to receive top priorities for error quantification research:

- Development of standardized and more cost-effective remote sensing accuracy assessment procedures
- Development of field-verification data-collection guidelines
- Procedures for vector-to-raster and raster-to-vector conversions
- Assessment of scaling issues for the incorporation of elevation data in georeferencing
- Development of standardized geometric and thematic reliability legend diagrams

Gahegan and Ehlers (2000) noted that two fundamental issues need to be addressed when dealing with uncertainty as the data are transformed between different models of geographic space:

- How do the uncertainty characteristics of data change as data are transformed between models?
- How do the transformation methods used affect and combine the uncertainty present in the data?

Gahegan and Ehlers (2000) developed an integrated error-simulation model for the transition from field (raw remote sensing) data to geographic objects (see Fig. 2.9). The description of uncertainty follows the approach proposed by Sinton (1978), which covers the sources of error as they occur in remote sensing and GIS integration. Uncertainty is restricted to the following properties: (1) value (including measurement and label errors), (2) spatial, (3) temporal, (4) consistency, and (5) completeness. Of these, measurement and label errors, as well as uncertainty in space and time, can apply either individually to a single datum or to any set of data. The latter two properties of consistency and completeness can be applied only to a defined dataset because they are comparative (either internally among data or to some external framework). Table 2.5 summarizes the findings of the Gahegan and Ehlers (2000). In addition, other uncertainty issues in the integration of the two technologies can be related to scale and representation of the data or the provision of lineage information (Ehlers, 2007).

Analytical Interoperability

The slow advance toward an integrated GIS has to do with the data-integration approach over the past 20 years or so (Ehlers, 2007). As discussed earlier, some fundamental problems such as the raster-vector dichotomy and the issue of data uniformity remain to be solved in making more effective use of geospatial data and the combined use of the two technologies. Ehlers (2007) suggested that we should take an analysis integration approach rather than a data integration approach to develop a fully integrated system for remote sensing and GIS. The first step in the development of an analysis-integration approach is to examine all the analysis functions in GIS and those in remote sensing and then to develop system-independent functions for the IGIS. Current analysis functions in GIS or remote sensing software depend on the data structure, either for raster- or vector-based systems (Ehlers, 2007). Albrecht's universal GIS language is regarded as a step forward in moving toward the integration (Ehlers, 2007). Albrecht (1996) grouped 20 high-level data-structure- and system-independent GIS functions into six classes (Table 2.6) and designed a flowchart tool to communicate with the system.

	Field	Image	Thematic	Object
α	Measurement error and precision	Quantization of value in terms of spectral bands and dynamic range	Labeling uncertainty (classification error)	Identity error (incorrect assignment of object type), object definition uncertainty
β	Locational error and precision	Registration error, sampling precision	Combination effects when data represented by different spatial properties are combined	Object shape error, topological inconsistency, split-and-merge errors
χ	Temporal error and precision	(Temporal error and precision are usually negligible for image data)	Combination effects when data representing different times are combined	Combination effects when data representing different times are combined
δ	Samples or readings collected or measured in an identical manner	Image is captured identically for each pixel, but medium between satellite and ground is not consistent; inconsistent sensing, light falloff; shadows	Classifier strategies are usually consistent in their treatment of a dataset	Methods for object formation may be consistent but often are not; depends on extraction strategy
ε	Sampling strategy covers space, time, and attribute domains adequately	Image is complete, but parts of ground may be obscured	Completeness depends on the classification strategy (Is all the dataset classified or are only some classes extracted?)	Depends on extraction strategy; spatial and topological inconsistencies may arise as a result of object formation

Note: α = data or value; β = space; χ = time; δ = consistency; ε = completeness.
Adapted from Gahegan and Ehlers, 2000.

TABLE 2.5 Types of Uncertainty in Four Models of Geographic Space

Function Group	Subfunctions			
Search	Thematic search	Spatial search	Interpolation	(Re) classification
Location analysis	Buffer	Corridor	Overlay	Thiessen/ Voronoi
Terrain analysis	Slope/ aspect	Catchment basins	Drainage network	Viewshed analysis
Distribution, neighborhood	Cost, diffusion, spread	Proximity	Nearest neighbor	
Spatial analysis	Pattern dispersion	Centrality, connectivity	Shape	Multivariate analysis
Measurements	Distance	Area		

Adapted from Albrecht, 1996.

TABLE 2.6 Universal High-Level GIS Operators

In the field of remote sensing, many textbooks do not present a systematic, consistent taxonomy of image-analysis functions; instead, they mix hardware, sensors, systems, and operations (Ehlers, 2000). Having considered this inconsistency as an impediment to the development of an IGIS, Ehlers (2000) described 17 typical image processing functions and grouped them into four categories (Table 2.7) to be added to the Albrecht's table. Other functions, such as those for hybrid processing, polymorphic techniques, and 3D urban information systems, also need to be defined and described so as to develop a complete functionality for the IGIS.

2.4 Prospects for Future Developments

2.4.1 Impacts of Computer, Network, and Telecommunications Technologies

Faust et al. (1991) identified five major impediments that computer technology posed for the integration of remote sensing and GIS. First, for a fully integrated remote sensing–GIS system, the real-time processing must apply to relational access to multiple attributes for a geographic database class, as well as to complex overlay processing to answer "what if" questions. Second, a manager must be able to postulate questions to a system through a user interface that does not

Function Group	Subfunctions			
Preprocessing	Parametric radiometric sensor correction	Parametric geometric sensor correction		
Geometric registration	Deterministic techniques	Statistical techniques (interpolation)	Automated techniques (matching)	Error assessment
3D image analysis	Orthoimage generation	DEM extraction		
Atmospheric correction	Deterministic approaches	Histogram-based manipulations (point operations)	Filtering	Image enhancement
Feature/object extraction	Unsupervised techniques	Supervised techniques	Model-based techniques	Error assessment

Adapted from Ehlers, 2000.

TABLE 2.7 Universal Image Processing Functions for Integrated GIS

require sophisticated computer knowledge and to receive an answer instantaneously. Expert and learning systems must be incorporated. Third, there is only a handful of vendors of GIS and remote sensing systems whose development staff is well acquainted with the applications, algorithms, and interfaces necessary to provide a product to an increasingly diverse and less technically savvy user base. Standards for operating systems, programming languages, and user interfaces remain to be defined. Fourth, the most important limitation includes the cost of initially entering spatial and attribute data into a database and the cost of continuous updating of the database. Finally, a large impediment has been the sheer volume of data a single image can represent and the fact that processing often requires removing any radiometric or geometric distortions necessary to make the image useful in any given context. Moreover, image compression techniques need to be further developed.

Advances in computer technology have been so dramatic since 1991, when Faust et al. published the previously referenced article, that it is almost impossible to list all the major changes in computing power so as to evaluate the impacts of this technology on the integration of GIS and remote sensing. What we know is that those impediments are largely no longer considered major challenges, such as real-time processing, database updating, and handling of a single-image scene. For example, real-time integration of remotely sensed imagery and GIS data has been carried out with expert knowledge in hazard mapping, wildfire monitoring, and crop-disease surveillance (Kimes et al., 1991). It is routine, with the current technologies in computing (CPU), graphic user interface, visualization, and computer networking, to perform GIS and/or remote sensing image analysis on a desktop or laptop computer. With the advent of Internet technology in the mid-1990s (i.e., Web 1.0 and now Web 2.0), GIS can display, analyze, and manage data over the Web, making WebGIS a true reality. In conjunction with GPS and wireless communications technologies, mobile mapping is a fashion today. Two mobile-mapping techniques are gaining momentum in the commercial domain and in the everyday life of the general public. One relates to the development of a location-aware personal digital assistant (PDA) that consists of a GPS-equipped handheld computer or a palmtop GIS and may use such datasets as geographic features and attributes, aerial and ground photographs, satellite images, and digital pictures (Gao, 2002). Another prospect for mobile-mapping technology is distributed mobile GIS (DM-GIS) (Gao, 2002). In principle, DM-GIS is very similar to the PDA and typically consists of a palmtop computer with GPS and a camera (Karimi et al., 2000). DM-GIS communicate via wireless network with a cluster of backend servers where GIS and/or remote sensing image data are stored. Digital pictures taken in the field can be relayed to the servers to update the GIS database as frequently as needed (Gao, 2002). Xue et al. (2002) suggested that WebGIS, mobile mapping (they termed it *mobile geoprocessing*), and TeleGIS (an integration of GIS and other telecommunications techniques) are shaping a new field of study, namely, *telegeoprocessing*. It is based on a real-time spatial database updated regularly by means of telecommunications to support problem solving and decision making at any time and any place. Telegeoprocessing requires a seamless integration of the four components—remote sensing, GIS, GPS, and telecommunications—and real-time imaging and processing in remote sensing and real-time GIS (Xue et al., 2002). Real time in remote sensing imaging processing refers to the generation of images or other information to be made available for inspection simultaneously, or very nearly simultaneously, with their acquisitions (Sabins, 1987). To make a GIS real-timed, several key issues must be considered, such as real-time spatial data structures, real-time GIS indexing, interoperability, geographic data-interchange standards, parallel and distributed computing, networking, human-machine interface, client-server architect, multimedia and wireless communications, and real-time integration of remote sensing image and GIS data (Xue et al., 2002).

In response to Dangermond's comment (2007) on the emerging georeferenced information over the Web, Sui (2008) suggested that a series of basic GIS science questions may need to be reexamined in the age of Web 2.0, such as

- What kinds of new geographic concepts are needed to better understand and analyze the fully georeferenced world—both real and virtual?
- Do we need new models of geographic representation?
- Do we need new data mining techniques?
- Are existing GIS tools suitable for conducting analyses of the new geotagged world of both spatial and nonspatial information?

2.4.2 Impacts of the Availability of Very High Resolution Satellite Imagery and LiDAR Data

With recent advent of very high resolution satellite imagery (HRSI), such as IKONOS (launched 1999), QuickBird (2001), and OrbView (2003) images, great efforts have been made in the application of these remote sensing images in urban and environmental studies. HRSIs have been applied widely in urban land-cover mapping (Im et al., 2008; Lu and Weng, 2009; Thomas et al., 2003; Warner and Steinmaus, 2005; Wulder et al., 2008), 3D shoreline extraction and coastal mapping (Di et al., 2003; Ma et al., 2003), earthquake damage assessment (Al-Khudhairy et al., 2005; Miura and Midorikawa, 2006), digital terrain model (DTM) and DEM generation (Toutin 2004a, 2004b), 3D object reconstruction (Baltsavias et al., 2001; Tao and Hu, 2002), and topographic mapping and change detection (Birk et al., 2003; Holland et al., 2006; Zanoni et al., 2003). Many GIS databases also use high-accuracy geopositioning HRSI images as the base maps, providing both metric and thematic information.

These fine spatial-resolution images contain rich spatial information, providing a greater potential to extract much more detailed thematic information (e.g., land use and land cover, impervious surface, and vegetation), cartographic features (e.g., buildings and roads), and metric information with stereoimages (e.g., height and area) that are ready to be used in GIS. However, some new problems exist with HRSI data, notably shades caused by topography, tall buildings, or trees and the high spectral variation within the same land-cover class. The shade problem increases the difficulty of extracting both thematic and cartographic information. The shade problem and high spectral variation are common with the high degree of spectral heterogeneity in complex landscapes, such as urban areas (Lu and Weng, 2009). These disadvantages may lower image classification accuracy if the classifiers used cannot handle them effectively (Cushnie, 1987; Irons et al., 1985). In addition, the huge amount of data storage and computer display of HRSI images also can have an impact on image processing in general and the selection of classification algorithms in particular.

With respect to the extraction of cartographic features, HRSI data provide the great potential to achieve the effectiveness and efficiency of extraction through automated extraction methods. However, the issues of shade and image distortion can affect the resulting accuracy to a certain degree. A new data source, light detection and ranging (LiDAR) data, which provides land-surface-elevation information by emitting a laser pulse and providing high vertical and horizontal resolutions of less than 1 m, has been used increasingly in many geospatial applications owing to its high data resolution, short time consumption, and low cost compared with many traditional technologies. Unlike other remotely sensed data, LiDAR data focus solely on geometry rather than on radiometry. Some typical products derived include DEM, surface elevation model (SEM), triangulated irregular network (TIN), and intermediate return information. As a result, feature classification and extraction based on LiDAR data are conducted widely (Clode et al. 2007; Filin, 2004; Forlani et al., 2006; Lee et al., 2008). LiDAR data show a great potential in building and road extraction because elevation data can be derived quickly and at high resolution in comparison with photogrammetric techniques (Miliaresis and Kokkas, 2007). The DEM and associated products and LiDAR-derived cartographic information have become an important GIS data source in recent years. It also should be noted that many researchers have used LiDAR in conjunction with optical remote sensing and GIS data in urban environments and resource studies.

The management of such information requires that GIS technology be adapted, modified, and extended (Ehlers et al., 1989). These new data sources also require processing functionality that is not currently standard on many systems (Poulter, 1995). In order to use very high-dimensional image data in combination with preexisting GIS datasets, techniques such as virtual reality for visualization and projection pursuit for data reduction are needed (Wilkinson, 1996). Techniques for information/pattern extraction from such remote sensing and GIS datasets and data analysis need much development, for example, self-organizing neural networks, integrated spatial and temporal representation and analysis, and data mining (Wilkinson, 1996).

2.4.3 Impacts of New Image-Analysis Algorithms

Attribute Analysis and Classification of Remote Sensing and GIS Data

Knowledge Based Expert Systems This approach is now becoming increasingly attractive owing to its ability to accommodate multiple sources of data, such as satellite imagery and GIS data. GIS plays an important role in developing knowledge-based classification approaches because of its ability to manage different sources of data and spatial modeling. As different kinds of ancillary data, such as digital elevation models and soil mapping, housing and population density, road network, temperature, and precipitation data, become readily available, they may be incorporated into a classification procedure in different ways. One approach is to develop knowledge-based classifications based on expert knowledge of spatial distribution patterns of land-cover classes and selected ancillary data. For example, elevation, slope, and aspect are related to vegetation distribution in mountainous regions. Data on terrain features thus are useful for separation of vegetation classes. Population, housing, and road densities are related to urban land-use distribution and may be very helpful in distinguishing between commercial/industrial lands and high-intensity residential lands, between recreational grassland and pasture/crops, or between residential areas and forest land (Lu and Weng, 2006). Similarly, temperature, precipitation, and soil-type data are related to land-cover distribution on a large scale. Effective use of these relationships in a classification procedure has been shown to be very helpful in improving classification accuracy. Expert systems are considered to have a great potential for providing a general approach to the routine use of image ancillary data in image classification (Hinton, 1996). A critical step is to develop rules that can be used in an expert system or a knowledge-based classification approach. Three methods have been employed to build rules for image classification: (1) explicitly eliciting knowledge and rules from experts and then refining the rules (Hung and Ridd, 2002; Stefanov et al., 2001; Stow et al., 2003), (2) implicitly extracting variables and rules using cognitive methods (Hodgson, 1998; Lloyd et al., 2002), and (3) empirically generating rules from observed data with automatic induction methods (Hodgson et al., 2003; Huang and Jensen, 1997; Tullis and Jensen, 2003).

Artificial Neural Networks This approach has been applied increasingly in recent years. The neural network has several advantages, including its nonparametric nature, arbitrary decision-boundary capability, adaptation to different types of data and input structures, fuzzy output values, and generalization for use with multiple images (Paola and Schowengerdt, 1995). The fact that artificial neural networks (ANNs) behave as general pattern recognition systems and assume no prior statistical model for the input data makes them an excellent technique for integrating remote sensing and GIS data. Although many neural-network models have been developed, the multilayer perceptron (MLP) feed-forward neural network is used most frequently (Kavzoglu and Mather, 2004). The MLP has been applied in (1) land-use/land-cover classifications (Foody et al., 1997; Kavzoglu and Mather, 2004; Zhang and Foody, 2001), (2) impervious surface estimation (Weng and Hu, 2008), and (3) change detection (Li and Yeh, 2002). Other applications include water properties estimation (Corsini et al., 2003; Schiller and Doerffer, 1999; Zhang et al., 2002), forest structure mapping (Ingram et al., 2005), understory bamboo mapping (Linderman et al., 2004), cloud detection (Jae-Dong et al., 2006), and mean monthly ozone prediction (Chattopadhyay and Bandyopadhyay, 2007). Although the MLP has been applied widely, some drawbacks have been raised by previous researchers. For instance, how to design the number of hidden layers and the number of hidden-layer nodes in the model are still challenges. Another issue is that the MLP requires training sites to include both presence and absence data. The desired output must contain both true and false information so that the network can learn all kinds of patterns for a study area in order to classify accordingly (Li and Eastman, 2006). However, in some cases, absence data are not readily available. Finally, the MLP has the local minima problem in the training process, which may affect the accuracy of the result significantly.

Another neural-network approach, Kohonen's self-organizing map (SOM), has not been applied as widely as the MLP (Pal et al., 2005). The SOM can be used for both supervised and unsupervised classifications and has the properties for both vector quantization and projection (Li and Eastman, 2006). The SOM has been used for both "hard" classification and "soft" classification in previous studies (Ji, 2000; Lee and Lathrop, 2006). Ji (2000) compared the Kohonen self-organizing feature map (KSOFM) and the MLP for image classification at per-pixel level. Seven classes were identified, and the result showed that the SOM provided an excellent alternative to the MLP neural network in hard classification. Lee and Lathrop (2006) conducted an SOM-LVQ-GMM (Self-Organizing Map, Learning Vector Quantization, and Gaussian Mixture Model) to extract urban land-cover data from Landsat ETM+ imagery at subpixel level. It is found that the SOM can generate promising results in soft classification and that the SOM had several advantages over the MLP. Hu and Weng (2009) compared the two neural networks with three Advanced Spaceborne Thermal Emission and Reflection Radiometer (ASTER) images of Marion County, Indiana, and found that the SOM outperformed the MLP slightly for every season of image data, especially in residential areas.

Object-Based Image Analysis

The object-oriented concept was first introduced to the GIS community in the late 1980s (Egenhofer and Frank, 1992), and since then, especially after the 1990s, a great deal of research has been conducted involving the object-oriented approach (Bian, 2007). In remote sensing, image segmentation, which is usually applied before image classification, has a longer history and has its roots in industrial image processing but was not used extensively in the geospatial community in the 1980s and 1990s (Blaschke et al., 2004). Object oriented image analysis has been used increasingly in remote sensing applications owing to the advent of HRSI data and the emergence of commercial software such as eCognition (Benz et al., 2004; Wang et al., 2004). Image segmentation merges pixels into objects, and a classification then is implemented based on objects instead of individual pixels. In the process of creating objects, a scale determines the presence or absence of an object class, and the size of an object affects the classification result. This approach has proven to be able to provide better classification results than per-pixel-based approaches, especially for fine spatial-resolution data. As discussed earlier, object oriented image analysis holds great potential in the development of a fully integrated GIS (Ehlers, 2007). Two key difficulties lie in (1) algorithms for smoothing to produce acceptable vectors and (2) the automated assignment of attributes to vectors, points, and nodes (Faust et al., 1991).

Object Search in the Integrated Remote sensing and GIS Datasets

Data mining is a field that has developed by encompassing principles and techniques from statistics, machine learning, pattern recognition, numeric search, and scientific visualization to accommodate the new data types and data volumes being generated (Miller and Han, 2001). The tasks of data mining might vary, but the premise about discovering unknown information from large databases remains the same. In short, *data mining* can be defined as "analysis of (often large) observational datasets to find unsuspected relationships and to summarize the data in novel ways that are both understandable and useful to the data owners" (Hand et al., 2001). Over the last few years, the techniques of data mining have been pushed by three major technological factors that have advanced in parallel. First, the growth in the amount of data has led to the development of mass storage devices. Second, the problem of accessing this information has led to the development of advanced and improved processors. Third, the need for automating the tasks involved in data retrieval and processing led to advancement in statistic and machine-learning algorithms. Data-mining tasks can be classified broadly into five categories: segmentation, dependency analysis, outlier analysis, trend detection, and characterization. In order to do these tasks, various techniques, such as cluster analysis, neural networks, genetic algorithms, Bayesian networks, decision trees, etc., are applied. Some of these techniques are also good at executing more than one task and have their own advantages and disadvantages (Rajasekar et al., 2007). Data-mining techniques have been applied successfully to the combined datasets of GIS and remote sensing (Mennis, 2006; Mennis and Liu, 2005; Rajasekar and Weng, 2009a, 2009b) and may be a good approach to the processing and analysis of HRSI data or other image data of high volume.

Enhanced Environmental Mapping for Generation of New GIS Datasets

Data Fusion Images from different sensors contain distinctive features. Data fusion, or integration of multisensor or multiresolution data, takes advantage of the strengths of distinct image data for improvement of visual interpretation and quantitative analysis. In general, three levels of data fusion can be identified (Gong, 1994): pixel (Luo and Kay, 1989), feature (Jimenez et al., 1999), and decision (Benediktsson and Kanellopoulos, 1999). Data fusion involves two merging procedures: (1) geometric coregistration of two data sets and (2) mixture of spectral and spatial information contents to generate a new dataset that contains enhanced information from both datasets. Accurate registration between the two datasets is extremely important for precisely extracting information contents from both datasets, especially line features such as roads and rivers. Radiometric and atmospheric calibrations are also needed before multisensor data are merged.

Many methods have been developed to fuse spectral and spatial information in the previous literature (Chen and Stow, 2003; Gong, 1994; Lu and Weng, 2005; Pohl and van Genderen, 1998). Solberg et al. (1996) broadly divided data fusion methods into four categories: statistical, fuzzy logic, evidential reasoning, and neural network. Dai and Khorram (1998) presented a hierarchical data fusion system for vegetation classification. Pohl and van Genderen (1998) provided a literature review on the methods of multisensor data fusion. The methods, including color-related techniques [e.g., color composite, intensity-hue-saturation (IHS), and luminance-chrominance], statistical/numerical methods (e.g., arithmetic combination, principal components analysis, high-pass filtering, regression variable substitution, canonical variable substitution, component substitution, and wavelets), and various combinations of the preceding methods were examined. IHS transformation was identified to be the most frequently used method for improving visual display of multisensor data (Welch and Ehlers, 1987), but the IHS approach can employ only three image bands, and the resulting image may not be suitable for further quantitative analysis such as image classification. Principal component analysis (PCA) is often used for data fusion because it can produce an output that can better preserve the spectral integrity of the input dataset. In recent years, wavelet-merging techniques also have been shown to be an effective approach to enhance spectral and spatial information contents (Li et al., 2002; Simone et al., 2002; Ulfarsson et al., 2003). Previous research indicated that integration of Landsat TM and radar (Ban, 2003; Haack et al., 2002), SPOT high resolution visible (HRV) and Landsat TM (Munechika et al., 1993; Welch and Ehlers, 1987; Yocky, 1996), and SPOT multispectral and panchromatic bands (Garguet-Duport et al., 1996; Shaban and Dikshit, 2001) can improve classification results. An alternative way of integrating multiresolution images, such as Landsat TM (or SPOT) and Moderate Resolution Imaging Spectroradiometer (MODIS) [or Advanced Very High Resolution Radiometer (AVHRR)], is to refine the estimation of land use and land cover types from coarse spatial-resolution data (Moody, 1998; Price, 2003).

Hyperspectral Imaging The spectral characteristics of land surfaces are the fundamental principles for land-cover classification using remotely sensed data. The spectral features include the number of spectral bands, spectral coverage, and spectral resolution (or bandwidth). The number of spectral bands used for image classification can range from a limited number of multispectral bands (e.g., four bands in SPOT data and seven in for Landsat TM data), to a medium number of multispectral bands (e.g., ASTER with 14 bands and MODIS with 36 bands), to hyperspectral data (e.g., AVIRIS and EO-1 Hyperion images with 224 bands). The large number of spectral bands provides the potential to derive detailed information on the nature and properties of different surface materials on the ground, but it also means a difficulty in image processing and a large data redundancy owing to high correlation among the adjacent bands. High-dimension data also require a larger number of training samples for image classification. An increase in spectral bands may improve classification accuracy, but only when those bands are useful in discriminating the classes (Thenkabail et al., 2004b). In previous research, hyperspectral data have been used successfully for land-use/cover classification (Benediktsson et al., 1995; Hoffbeck and Landgrebe, 1996; Platt and Goetz, 2004, Thenkabail et al., 2004a, 2004b) and vegetation mapping (McGwire et al., 2000, Schmidt et al., 2004). As space-borne hyperspectral data such as EO-1 Hyperion data become available, research and applications with hyperspectral data will increase. Weng et al. (2008) found that an EO-1 Hyperion image was more powerful in discerning low-albedo surface materials, which has been a major obstacle for impervious surface estimation with medium-resolution multispectral images. A sensitivity analysis of the mapping of impervious surfaces using different scenarios of the EO-1 Hyperion band combinations suggested that the improvement in mapping accuracy in general and the better ability in discriminating low-albedo surfaces mainly came from additional bands in the mid-infrared region (Weng et al., 2008).

2.5 Conclusions

Having realized a great deal of benefits and wide applications, the remote sensing and GIS communities have to continue to aim their enthusiasm at the integration between remote sensing and GIS. However, in examining a large volume of literature dealing with other topics and issues in remote sensing and GIS, we have seen a dearth of publications since 1991 specifically concerned with the theories and methods of integration. Many applications of integration in the past decade can be described as having an ad hoc approach driven by project demands (Ehlers, 2007; Mesev and Walrath, 2007). The temptation to take advantage of the opportunity to combine ever-increasing computational power, modern telecommunications technologies, more plentiful and capable digital data, and more advanced algorithms has resulted in a new round of attention to the integration of remote sensing and GIS (as well as with GPS) for environmental, resources, and urban studies. Despite continuous push by researchers from various fields, the state of the art of the issue is still the data integration approach, not through an analysis integration approach. Current trends in the integrated analysis of remote sensing and GIS data include (1) attribute analysis/classification of remote sensing and GIS data by more powerful ANN models and knowledge-based expert systems, (2) object-based image analysis, (3) object search in the integrated databases by data-mining techniques, and (4) enhanced environmental mapping via data fusion of different sensors and hyperspectral imaging.

References

Albrecht, J. 1996. Universal Analytical GIS Operations. Ph.D. thesis, ISPA-Mitteilungen 23, University of Vechta, Germany.

Al-Khudhairy, D. H. A., Caravaggi, I., and Glada, S. 2005. Structural damage assessments from IKONOS data using change detection, object-oriented segmentation, and classification techniques. *Photogrammetric Engineering and Remote Sensing* **71,** 825–837.

Amarsaikhan, D., and Douglas, T. 2004. Data fusion and multisource image classification. *International Journal of Remote Sensing* **25,** 3529–3539.

Arbia, G., Griffith, D., and Haining, R. 1999. Error propagation modeling in raster GIS: Adding and ratioing operations. *Cartography and Geographic Information Science* **26,** 297–315.

Baban, S. M. J., and Yusof, K. W. 2001. Mapping land use/cover distribution on a mountainous tropical island using remote sensing and GIS. *International Journal of Remote Sensing* **22,** 1909–1918.

Baltsavias, E., Pateraki, M., and Zhang, L. 2001. Radiometric and geometric evaluation of IKONOS geo images and their use for 3D building modeling. *Proceedings of Joint ISPRS Workshop "High Resolution Mapping from Space 2001,"* Hannover, Germany, pp. 15–35.

Ban, Y. 2003. Synergy of multitemporal ERS-1 SAR and Landsat TM data for classification of agricultural crops. *Canadian Journal of Remote Sensing* **29,** 518–526.

Barnsley, M. J., and Barr, S. L. 1996. Inferring urban land use from satellite sensor images using kernel-based spatial reclassification. *Photogrammetric Engineering and Remote Sensing* **62,** 949–958.

Benediktsson, J. A., and Kanellopoulos, I. 1999. Classification of multisource and hyperspectral data based on decision fusion. *IEEE Transactions on Geoscience and Remote Sensing* **37,** 1367–1377.

Benediktsson, J. A., Sveinsson, J. R., and Arnason, K. 1995. Classification and feature extraction of AVIRIS data. *IEEE Transactions on Geoscience and Remote Sensing* **33,** 1194–1205.

Benz, U. C., Hofmann, P., Willhauck, G., et al. 2004. Multi-resolution, object-oriented fuzzy analysis of remote sensing data for GIS-ready information. *ISPRS Journal of Photogrammetry and Remote Sensing* **58,** 239–258.

Bian, L. 2007. Object-oriented representation of environmental phenomena: Is everything best represented as an object? *Annals of the Association of American Geographers* **97,** 266–280.

Binaghi, E., Brivio, P. A., and Rampini, A. 1999, A fuzzy set-based accuracy assessment of soft classification. *Pattern Recognition Letters* **20**, 935–948.

Birk, R. J., Stanley, T., Snyder, G. I., et al. 2003. Government programs for research and operational uses of commercial remote sensing data. *Remote Sensing of Environment* **88,** 3–16.

Blaschke, T., Burnett, C., and Pekkarinen, A. 2004. New contextual approaches using image segmentation for object-based classification. In *Remote Sensing Image Analysis: Including the Spatial Domain,* edited by F. De Meer and S. de Jong, pp. 211–236. Dordrecht, The Netherlands: Kluwer Academic Publishers.

Bronge, L. B. 1999. Mapping boreal vegetation using Landsat TM and topographic map data in a stratified approach. *Canadian Journal of Remote Sensing* **25,** 460–474.

Brown, R., and Fletcher, P. 1994. Satellite images and GIS: Making it work. *Mapping Awareness* **8,** 20–22.

Bruegger, B. P., Barrera, R., Frank, A. U., et al. 1989. Research topics on multiple representation. In *Proceedings, NCGIA Specialist Meeting on Multiple Representation,* Technical Paper 89-3, NCGIA, Buffalo, NY, pp. 53–67.

Bruzzone, L., Prieto, D. F., and Serpico, S. B. 1999. A neural-statistical approach to multitemporal and multisource remote sensing image classification. *IEEE Transactions on Geoscience and Remote Sensing* **37,** 1350–1359.

Bruzzone, L., Conese, C., Maselli, F., and Roli, F. 1997. Multisource classification of complex rural areas by statistical and neural-network approaches. *Photogrammetric Engineering and Remote Sensing* **63,** 523–533.

Burrough, P. A. 1986. *Principals of Geographical Information Systems for Land Resource Assessment.* New York: Clarendon Press.

Campbell, J. B. 2007. *Introduction to Remote Sensing,* 4th ed. New York: Guilford Press.

Chattopadhyay, S., and Bandyopadhyay, G. 2007. Artificial neural network with backpropagation learning to predict mean monthly total ozone in Arosa, Switzerland. *International Journal of Remote Sensing* **28,** 4471–4482.

Chen, D., and Stow, D. A. 2003. Strategies for integrating information from multiple spatial resolutions into land-use/land-cover classification routines. *Photogrammetric Engineering and Remote Sensing* **69,** 1279–1287.

Cheng, J., and Masser, I. 2003. Urban growth pattern modeling: A case study of Wuhan city, PR, China. *Landscape and Urban Planning* **62,** 199–217.

Clode, S., Rottensteinerb, F., Kootsookosc, P., and Zelniker, E. 2007. Detection and vectorization of roads from Lidar data. *Photogrammetric Engineering and Remote Sensing* **73,** 517–535.

Congalton, R.G. 1991. A review of assessing the accuracy of classification of remotely sensed data. *Remote Sensing of Environment* **37,** 35–46.

Congalton, R. G., and Green, K. 1999. *Assessing the Accuracy of Remotely Sensed Data: Principles and Practices.* Boca Raton, FL: Lewis Publishers.

Corsini, G., Diani, M., Grasso, R., et al. 2003. Radial basis function and multilayer perceptron neural networks for sea water optically active parameter estimation in case II waters: A comparison. *International Journal of Remote Sensing* **24,** 3917.

Cushnie, J. L. 1987. The interactive effect of spatial resolution and degree of internal variability within land-cover types on classification accuracies. *International Journal of Remote Sensing* **8,** 15–29.

Dai, X., and Khorram, S. 1998. A hierarchical methodology framework for multisource data fusion in vegetation classification. *International Journal of Remote Sensing* **19,** 3697–3701.

Dangemond, J. 2007. Comment made during the workshop on volunteered geographic information (VGI), Santa Barbara, CA, December 13–14, 2007.

Davies, F. W., Quattrochi, D. A., Ridd, M. K., et al. 1991. Environmental analysis using integrated GIS and remotely sensed data: Some research needs and priorities. *Photogrammetric Engineering and Remote Sensing* **57,** 689–697.

Di, K., Ma, R., and Li, R. 2003. Geometric processing of IKONOS geostereo imagery for coastal mapping applications. *Photogrammetric Engineering and Remote Sensing* **69,** 873–879.

Donnay, J. P., Barnsley, J. M., and Longley, A. P. 2001. Remote sensing and urban analysis. In *Remote Sensing and Urban Analysis,* edited by J. P. Donnay, M. J. Barnsley, and P. A. Longley. New York: Taylor & Francis.

Doucette, P., Agoouris, P., and Stefanidis, A. 2004. Automated road extraction from high resolution multispectral imagery. *Photogrammetric Engineering and Remote Sensing* **70,** 1405–1416.

Dymond, J. R., and Shepherd, J. D. 1999. Correction of the topographic effect in remote sensing. *IEEE Transactions on Geoscience and Remote Sensing* **37,** 2618–2620.

Egenhofer, M. F., and Frank, A. 1992. Object-oriented modeling for GIS. *Journal of the Urban and Regional Information System Association* **4,** 3–19.

Ehlers, M. 2000. Integrated GIS: From data integration to integrated analysis. *Surveying World* **9,** 30–33.

Ehlers, M. 2007. Integration taxonomy and uncertainty. In *Integration of GIS and Remote Sensing,* edited by V. Mesev, pp. 17–42. West Sussex, England: Wiley.

Ehlers, M., and Shi, W. Z. 1997. Error modeling for integrated GIS. *Cartographica* **33,** 11–21.

Ehlers, M., and Welch, R. 1987. Stereocorrelation of Landsat TM images. *Photogrammetric Engineering and Remote Sensing* **53,** 1231–1237.

Ehlers, M., Edwards, G., and Bedard, Y. 1989. Integration of remote sensing with geographic information systems: A necessary evolution. *Photogrammetric Engineering and Remote Sensing* **55,** 1619–1627.

Ehlers, M., Greenlee, D., Terrence, S., and Star, J. 1991. Integration of remote sensing and GIS: Data and data access. *Photogrammetric Engineering and Remote Sensing* **57,** 669–675.

Ehlers, M., Jadkowski, M. A., Howard, R. R., and Brostuen, D. E. 1990. Application of SPOT data for regional growth analysis and local planning. *Photogrammetric Engineering and Remote Sensing* **56,** 175–180.
Ekstrand, S. 1996. Landsat TM-based forest damage assessment: Correction for topographic effects. *Photogrammetric Engineering and Remote Sensing* **62,** 151–161.
Epstein, J., Payne, K., and Kramer, E. 2002. Techniques for mapping suburban sprawl. *Photogrammetric Engineering and Remote Sensing* **68,** 913–918.
Faust, N. L., Anderson, W. H., and Star, J. L. 1991. Geographic information systems and remote sensing future computing environment. *Photogrammetric Engineering and Remote Sensing* **57,** 655–668.
Filin, S. 2004. Surface classification from airborne laser scanning data. *Computers and Geosciences* **30,** 1033–1041.
Finn, J.T. 1993. Use of the average mutual information index in evaluating classification error and consistency. *International Journal of Geographical Information Systems* **7,** 349–366.
Foody, G. M., Lucas, R. M., Curran, P. J., and Honzak, M. 1997. Nonlinear mixture modelling without end-members using an artificial neural network. *International Journal of Remote Sensing* **18,** 937–953.
Foody, G.M. 1996. Approaches for the production and evaluation of fuzzy land cover classification from remotely-sensed data. *International Journal of Remote Sensing* **17,** 1317–1340.
Foody, G.M. 2002b. Status of land cover classification accuracy assessment. *Remote Sensing of Environment* **80,** 185–201.
Foody, G.M. 2004a. Thematic map comparison: evaluating the statistical significance of differences in classification accuracy. *Photogrammetric Engineering and Remote Sensing* **70,** 627–633.
Forlani, G., Nardinocchi, C., Scaioni, M., and Zingaretti, P. 2006. Complete classification of raw LiDAR data and 3D reconstruction of buildings. *Pattern Analysis and Applications* **8,** 357–374.
Franklin, S. E. 2001. *Remote Sensing for Sustainable Forest Management.* New York: Lewis Publishers.
Franklin, S. E., Connery, D. R., and Williams, J. A. 1994. Classification of alpine vegetation using Landsat Thematic Mapper, SPOT HRV and DEM data. *Canadian Journal of Remote Sensing* **20,** 49–56.
Franklin, S. E., Peddle, D. R., Dechka, J. A., and Stenhouse, G. B. 2002. Evidential reasoning with Landsat TM, DEM and GIS data for land cover classification in support of grizzly bear habitat mapping. *International Journal of Remote Sensing* **23,** 4633–4652.
Gahegan, M. N. 1996. Specifying the transformations within and between geographic data models. *Transactions in GIS* **1,** 137–152.
Gahegan, M., and Ehlers, M. 2000. A framework for the modeling of uncertainty between remote sensing and geographic information systems. *ISPRS Journal of Photogrammetry and Remote Sensing* **55,** 176–188.
Gao, J. 2002. Integration of GPS with remote sensing and GIS: Reality and prospect. *Photogrammetric Engineering and Remote Sensing* **68,** 447–453.
Gao, J. 2008. *Digital Analysis of Remotely Sensed Imagery*. New York: McGraw-Hill.
Garguet-Duport, B., Girel, J., Chassery, J., and Pautou, G. 1996. The use of multiresolution analysis and wavelet transform for merging SPOT panchromatic and multispectral image data. *Photogrammetric Engineering and Remote Sensing* **62,** 1057–1066.
Gong, P. 1994. Integrated analysis of spatial data from multiple sources: an overview. *Canadian Journal of Remote Sensing* **20,** 349–359.
Gong, P. 1996. Integrated analysis of spatial data from multiple sources: Using evidential reasoning and artificial neural network techniques for geological mapping. *Photogrammetric Engineering and Remote Sensing* **62,** 513–523.
Goodchild, M. F., Sun, G., and Yang, S. 1992. Development and test of an error model of categorical data. *International Journal of Geographical Information System* **6,** 87–104.

Gopal, S., and Woodcock, C. 1994. Theory and methods for accuracy assessment of thematic maps using fuzzy sets. *Photogrammetric Engineering and Remote Sensing* **60**, 181–188.
Groom, G. B., Fuller, R. M., and Jones, A. R. 1996. Contextual correction: Techniques for improving land cover mapping from remotely sensed images. *International Journal of Remote Sensing* **17**, 69–89.
Gu, D., and Gillespie, A. 1998. Topographic normalization of Landsat TM images of forest based on subpixel sun-canopy-sensor geometry. *Remote Sensing of Environment* **64**, 166–175.
Gugan, D. J. 1988. Satellite imagery as an integrated GIS component. *Proceedings of GIS/LIS 88*, San Antorio, TX, pp. 741–750.
Haack, B., Bryant, N., and Adams, S. 1987. Assessment of Landsat MSS and TM data for urban and near-urban land cover digital classification. *Remote Sensing of Environment* **21**, 201–213.
Haack, B. N., Solomon, E. K., Bechdol, M. A., and Herold, N. D. 2002. Radar and optical data comparison/integration for urban delineation: A case study. *Photogrammetric Engineering and Remote Sensing* **68**, 1289–1296.
Hall, G. B., Malcolm, N.W., and Piwowar, J. M. 2001. Integration of remote sensing and GIS to detect pockets of urban poverty: the case of Rosario, Argentina. *Transactions in GIS* **5**, 235–253.
Hand, D., Mannila, H., and Smyth, P. 2001. *Principles of Data Mining*, Vol. 1: *A Bradford Book*. Boston: MIT Press.
Harris, P. M., and Ventura, S. J. 1995. The integration of geographic data with remotely sensed imagery to improve classification in an urban area. *Photogrammetric Engineering and Remote Sensing* **61**, 993–998.
Harris, R. J., and Longley, P. A. 2000. New data and approaches for urban analysis: Modeling residential densities. *Transactions in GIS* **4**, 217–234.
Harvey, J. 2000. Small area population estimation using satellite imagery. *Transactions in GIS* **4**, 611–633.
Harvey, J. T. 2002. Estimation census district population from satellite imagery: Some approaches and limitations. *International Journal of Remote Sensing* **23**, 2071–2095.
Harvey, W., McGlone, J. C., McKeown, D. M., and Irvine, J. M. 2004. User-centric evaluation of semi-automated road network extraction. *Photogrammetric Engineering and Remote Sensing* **70**, 1353–1364.
Helmer, E. H., Brown, S., and Cohen, W. B. 2000. Mapping montane tropical forest successional stage and land use with multi-date Landsat imagery. *International Journal of Remote Sensing* **21**, 2163–2183.
Heuvelink, G. B. M., and Burrough, P. A. 1993. Error propagation in cartographic modeling using Boolean logic and continuous classification. *International Journal of Geographical Information System* **7**, 231–246.
Hinton, J. C. 1996. GIS and remote sensing integration for environmental applications. *International Journal of Geographic Information Systems* **10**, 877–890.
Hodgson, M. E. 1998. What size window for image classification? A cognitive perspective. *Photogrammetric Engineering and Remote Sensing* **64**, 797–808.
Hodgson, M. E., Jensen, J. R., Tullis, J. A., et al. 2003. Synergistic use lidar and color aerial photography for mapping urban parcel imperviousness. *Photogrammetric Engineering and Remote Sensing* **69**, 973–980.
Hoffbeck, J. P., and Landgrebe, D. A. 1996. Classification of remote sensing having high spectral resolution images. *Remote Sensing of Environment* **57**, 119–126.
Holland, D. A., Boyd, D.S., and Marshall, P. 2006. Updating topographic mapping in Great Britain using imagery from high-resolution satellite sensors. *ISPRS Journal of Photogrammetry and Remote Sensing* **60**, 212–223.
Hu, X., and Weng, Q. 2009. Estimating impervious surfaces from medium spatial resolution imagery using the self-organizing map and multi-layer perceptron neural networks, *Remote Sensing of Environment*. Accepted on May 21, 2009.
Huang, X., and Jensen, J. R. 1997. A machine-learning approach to automated knowledge-base building for remote sensing image analysis with GIS data. *Photogrammetric Engineering and Remote Sensing* **63**, 1185–1194.

Hung, M, and Ridd, M. K. 2002. A subpixel classifier for urban land-cover mapping based on a maximum-likelihood approach and expert system rules. *Photogrammetric Engineering and Remote Sensing* **68,** 1173–1180.

Im, J., Jensen, J. R., and Hodgson, M. E. 2008. Optimizing the binary discriminant function in change detection applications. *Remote Sensing of Environment* **112,** 2761–2776.

Ingram, J. C., Dawson, T. P., and Whittaker, R. J. 2005. Mapping tropical forest structure in southeastern Madagascar using remote sensing and artificial neural networks. *Remote Sensing of Environment* **94,** 491–507.

Irons, J. R., Markham, B. L., Nelson, R. F., et al. 1985. The effects of spatial resolution on the classification of Thematic Mapper data. *International Journal of Remote Sensing* **6,** 1385–1403.

Jae-Dong, J., Viau, A. A., Francois, A., and Bartholome, E. 2006. Neural network application for cloud detection in SPOT VEGETATION images. *International Journal of Remote Sensing* **27,** 719–736.

Janssen, L. F., Jaarsma, M. N., and van der Linden, E. T. M. 1990. Integrating topographic data with remote sensing for land-cover classification. *Photogrammetric Engineering and Remote Sensing* **56,** 1503–1506.

Janssen, L.F.J., and van der Wel, F.J.M.1994. Accuracy assessment of satellite derived land-cover data: a review. *Photogrammetric Engineering and Remote Sensing* **60,** 419–426.

Jensen, J. R. 2005. *Introductory Digital Image Processing: A Remote Sensing Perspective,* 3rd ed. Upper Saddle River, NJ: Prentice-Hall.

Ji, C. Y. 2000. Land-use classification of remotely sensed data using Kohonen self-organizing feature map neural networks. *Photogrammetric Engineering and Remote Sensing* **66,** 1451–1460.

Jimenez, L. O., Morales-Morell, A., and Creus, A. 1999. Classification of hyperdimensional data based on feature and decision fusion approaches using projection pursuit, majority voting, and neural networks. *IEEE Transactions on Geoscience and Remote Sensing* **37,** 1360–1366.

Karimi, H. A., Krishnamurthy, P., Banerjee, S., and Chrysanthis, P. K. 2000. Distributed mobile GIS: Challenges and architecture for integration of GIS, GPS, mobile computing and wireless communications. *Geomatics Info Magazine* **14,** 80–83.

Kavzoglu, T., and Mather, P. M. 2004. The use of backpropagating artificial neural networks in land cover classification. *International Journal of Remote Sensing* **24,** 4907–4938.

Kim, T., Park, S., Kim, M., et al. 2004. Tracking road centerlines from high resolution remote sensing images by least squares correlation matching. *Photogrammetric Engineering and Remote Sensing* **70,** 1417–1422.

Kimes, D. S., Harrison, P. R., and Ratcliffe, P. A. 1991. A knowledge-based expert system for inferring vegetation characteristics. *International Journal of Remote Sensing* **12,** 1987–2020.

Kwok, R., Curlander, C., and Pang, S. S. 1987. Rectification of terrain induced distortions in radar imagery. *Photogrammetric Engineering and Remote* Sensing **53,** 507–513.

Langford, M., Maguire, D. J., and Unwin, D. J. 1991. The areal interpolation problem: Estimating population using remote sensing in a GIS framework. In *Handling Geographical Information: Methodology and Potential Applications,* edited by L. Masser and M. Blakemore. New York: Longman Scientific and Technical, copublished in the United States with Wiley.

Lee, D. H., Lee, K. M., and Lee, S. U. 2008. Fusion of lidar and imagery for reliable building extraction. *Photogrammetric Engineering and Remote Sensing* **74,** 215–225.

Lee, D. S., Shan, J., and Bethel, J. S. 2003. Class-guided building extraction from IKONOS imagery. *Photogrammetric Engineering and Remote Sensing* **69,** 143–150.

Lee, S., and Lathrop, R. G. 2006. Subpixel analysis of Landsat ETM+ using self-organizing map (SOM) neural networks for urban land cover characterization. *IEEE Transactions on Geoscience and Remote Sensing* **44,** 1642–1654.

Leprieur, C. E., Durand, J. M., and Peyron, J. L. 1988. Influence of topography on forest reflectance using Landsat Thematic Mapper and digital terrain data. *Photogrammetric Engineering and Remote Sensing* **54,** 491–496.

Leung, Y., and Yan, J. 1998. A locational error model for spatial features. *International Journal of Geographical Information System* **12,** 607–620.

Li, G., and Weng, Q. 2005. Using Landsat ETM+ imagery to measure population density in Indianapolis, Indiana, USA. *Photogrammetric Engineering and Remote Sensing* **71,** 947–958.

Li, G., and Weng, Q. 2007. Measuring the quality of life in city of Indianapolis by integration of remote sensing and census data. *International Journal of Remote Sensing* **28,** 249–267.

Li, S., Kwok, J. T., and Wang, Y. 2002. Using the discrete wavelet frame transform to merge Landsat TM and SPOT panchromatic images. *Information Fusion* **3,** 17–23.

Li, X., and Yeh, A. G.-O. 2002. Neural-network-based cellular automata for simulating multiple land use changes using GIS. *International Journal of Geographical Information Science* **16,** 323–343.

Li, Z., and Eastman, J. R. 2006. Commitment and typicality measurements for the self-organizing map. In *Proceedings of SPIE: The International Society for Optical Engineering,* Bellingham, WA, pp. 64201I-1–64201I-4.

Liang, B., and Weng, Q. 2008. A multi-scale analysis of census-based land surface temperature variations and determinants in Indianapolis, United States. *Journal of Urban Planning and Development* **134,** 129–139.

Linderman, M., Liu, J., Qi, J., et al. 2004. Using artificial neural networks to map the spatial distribution of understorey bamboo from remote sensing data. *International Journal of Remote Sensing* **25,** 1685–1700.

Liu, W., Gopal, S., and Woodcock, C. E. 2004. Uncertainty and confidence in land cover classification using a hybrid classifier approach. *Photogrammetric Engineering and Remote Sensing* **70,** 963–972.

Lloyd, R. E., Hodgson, M. E., and Stokes, A. 2002. Visual categorization with aerial photography. *Annals of the Association American Geographers* **92,** 241–266.

Lo, C. P. 1995. Automated population and dwelling unit estimation from high resolution satellite images: A GIS approach. *International Journal of Remote Sensing* **16,** 17–34.

Lo, C. P., and Faber, B. J. 1997. Integration of Landsat Thematic Mapper and census data for quality of life assessment. *Remote Sensing of Environment* **62,** 143–157.

Lo, C. P., and Yeung, A. K. W. 2002. *Concepts and Techniques of Geographic Information Systems.* Upper Saddle River, NJ: Prentice-Hall.

Lu, D., and Weng, Q. 2004. Spectral mixture analysis of the urban landscapes in Indianapolis with Landsat ETM+ imagery. *Photogrammetric Engineering and Remote Sensing* **70,** 1053–1062.

Lu, D., and Weng, Q. 2005. Urban land-use and land-cover mapping using the full spectral information of Landsat ETM+ data in Indianapolis, Indiana. *Photogrammetric Engineering and Remote Sensing* **71,** 1275–1284.

Lu, D., and Weng, Q. 2006. Use of impervious surface in urban land use classification. *Remote Sensing of Environment* **102,** 146–160.

Lu, D., and Weng, Q. 2007. A survey of image classification methods and techniques for improving classification performance. *International Journal of Remote Sensing* **28,** 823–870.

Lu, D., and Weng, Q. 2009. Extraction of urban impervious surfaces from IKONOS imagery. *International Journal of Remote Sensing* **30**, 1297–1311.

Lunetta, R. S., Congalton, R. G., Fenstermaker, L. K., et al. 1991. Remote sensing and geographic information system data integration: Error sources and research issues. *Photogrammetric Engineering and Remote Sensing* **57,** 677–687.

Luo, R. C., and Kay, M. G. 1989. Multisensor integration and fusion for intelligent systems. *IEEE Transactions on Systems, Man, and Cybernetics* **19,** 901–931.

Ma, R., Di, K., and Li, R. 2003. 3D shoreline extraction from IKONOS satellite. *Journal of Marine Geodesy* **26,** 107–115.

Marr, D. 1982. *Vision.* San Francisco: W. H. Freeman.

Martin, D., Tate, N. J., and Langford, M. 2000. Refining population surface models: Experiments with Northern Ireland census data. *Transactions in GIS* **4,** 343–360.

Maselli, F., Rodolfi, A., Bottai, L., et al. 2000. Classification of Mediterranean vegetation by TM and ancillary data for the evaluation of fire risk. *International Journal of Remote Sensing* **21,** 3303–3313.

Mayer, H. 1999. Automatic object extraction from aerial imagery: A survey focusing on building. *Computer Vision and Image Understanding* **74,** 138–139.

McGwire, K., Minor, T., and Fenstermaker, L. 2000. Hyperspectral mixture modeling for quantifying sparse vegetation cover in arid environments. *Remote Sensing of Environment* **72,** 360–374.

McIver, D. K., and Friedl, M. A. 2001. Estimating pixel-scale land cover classification confidence using nonparametric machine learning methods. *IEEE Transactions on Geoscience and Remote Sensing* **39,** 1959–1968.

Mennis, J. 2006. Socioeconomic-vegetation relationships in urban, residential land: The case of Denver, Colorado. *Photogrammetric Engineering and Remote Sensing* **72,** 911–921.

Mennis, J. and Liu, J. W. 2005. Mining association rules in spatio-temporal data: An analysis of urban socioeconomic and land cover change. *Transactions in GIS* **9,** 5–17.

Mesev, V. 1997. Remote sensing of urban systems: hierarchical integration with GIS. *Computer, Environment, and Urban Systems* **21,** 175–187.

Mesev, V. 1998. The use of census data in urban image classification. *Photogrammetric Engineering and Remote Sensing* **64,** 431–438.

Mesev, V., and Walrath, A. 2007. GIS and remote sensing integration: In search of a definition. In *Integration of GIS and Remote Sensing,* edited by V. Mesev, pp. 1–16. West Sussex, England: Wiley.

Meyer, P., Itten, K. I., Kellenberger, T., et al. 1993. Radiometric corrections of topographically induced effects on Landsat TM data in alpine environment. *ISPRS Journal of Photogrammetry and Remote Sensing* **48,** 17–28.

Miliaresis, G., and Kokkas, N. 2007. Segmentation and object-based classification for the extraction of the building class from LiDAR DEMs. *Computers and Geosciences* **33,** 1076–1087.

Miller, H. J., and Han, J. 2001. *Geographic Data Mining and Knowledge Discovery.* London: Taylor & Francis.

Miura, H., and Midorikawa, S. 2006. Updating GIS building inventory data using high-resolution satellite images for earthquake damage assessment: Application to metro Manila, Philippines, *Earthquake Spectra* **22,** 151–168.

Moody, A. 1998. Using Landsat spatial relationships to improve estimates of land-cover area from coarse resolution remote sensing. *Remote Sensing of Environment* **64,** 202–220.

Munechika, C. K., Warnick, J. S., Salvaggio, C., and Schott, J. R. 1993. Resolution enhancement of multispectral image data to improve classification accuracy. *Photogrammetric Engineering and Remote Sensing* **59,** 67–72.

Murai, H., and Omatu, S. 1997. Remote sensing image analysis using a neural network and knowledge-based processing. *International Journal of Remote Sensing* **18,** 811–828.

Narumalani, S., Zhou, Y., and Jelinski, D. E. 1998. Utilizing geometric attributes of spatial information to improve digital image classification. *Remote Sensing Reviews* **16,** 233–253.

Oetter, D. R., Cohen, W. B., Berterretche, M., et al. 2000. Land cover mapping in an agricultural setting using multiseasonal Thematic Mapper data. *Remote Sensing of Environment* **76,** 139–155.

Openshaw, S., Charlton, M., and Carver, S. 1991. Error propagation: A Monte Carlo simulation. In *Handling Geographic Information,* edited by I. Masser and M. Blakemore, pp. 78–101. Harlow, UK: Longmans.

Pal, N. R., Laha, A., and Das, J. 2005. Designing fuzzy rule based classifier using self-organizing feature map for analysis of multispectral satellite images. *International Journal of Remote Sensing* **26,** 2219–2240.

Paola, J. D., and Schowengerdt, R. A. 1995. A review and analysis of back propagation neural networks for classification of remotely sensed multispectral imagery. *International Journal of Remote Sensing* **16,** 3033–3058.

Pentland, A. P. 1985. *From Pixels to Predicates*. Norwood, NJ: Alex Publishing.

Platt, R. V., and Goetz, A. F. H. 2004. A comparison of AVIRIS and Landsat for land use classification at the urban fringe. *Photogrammetric Engineering and Remote Sensing* **70,** 813–819.

Pohl, C., and van Genderen, J. L. 1998. Multisensor image fusion in remote sensing: Concepts, methods, and applications. *International Journal of Remote Sensing* **19,** 823–854.

Poulter, M. 1995. Integrating remote sensing and GIS: Designs for an operational future. In *Proceedings of a Seminar on Integrated GIS and High Resolution Satellite Data,* edited by M. Palmer. Farnborough: Defense Research Agency [DRA/CIS/(CSC2)/5/26/8/1/PRO/1].

Price, J. C. 2003. Comparing MODIS and ETM+ data for regional and global land classification. *Remote Sensing of Environment* **86,** 491– 499.

Qiu, F., Woller, K. L., and Briggs, R. 2003. Modeling urban population growth from remotely sensed imagery and TIGER GIS road data. *Photogrammetric Engineering and Remote Sensing* **69,** 1031–1042.

Rajasekar, U., and Weng, Q. 2009a. Urban heat island monitoring and analysis by data mining of MODIS imageries. *ISPRS Journal of Photogrammetry and Remote Sensing* **64,** 86–96.

Rajasekar, U., and Weng, Q. 2009b. Application of association rule mining for exploring the relationship between urban land surface temperature and biophysical/social parameters. *Photogrammetric Engineering and Remote Sensing* **75,** 385–396.

Rajasekar, U., Bijker, W., and Stein, A. 2007. Image mining for modeling of forest fires from meteosat images. *IEEE Transactions of Remote Sensing* **45,** 246–253.

Richter, R. 1997. Correction of atmospheric and topographic effects for high spatial resolution satellite imagery. *International Journal of Remote Sensing* **18,** 1099–1111.

Ricotta, C. and Avena, G.C. 2002. Evaluating the degree of fuzziness of thematic maps with a generalized entropy function: a methodological outlook. *International Journal of Remote Sensing,* **23,** 4519–4523.

Ricotta, C. 2004. Evaluating the classification accuracy of fuzzy thematic maps with a simple parametric measure. *International Journal of Remote Sensing,* **25,** 2169–2176.

Sabins, F. F. 1987. *Remote Sensing: Principles and Interpretation.* New York: W. H. Freeman.

Schiller, H., and Doerffer, R. 1999. Neural network for emulation of an inverse model operational derivation of case II water properties from MERIS data. *International Journal of Remote Sensing* **20,** 1735–1746.

Schmidt, K. S., Skidmore, A. K., Kloosterman, E. H., et al. 2004. Mapping coastal vegetation using an expert system and hyperspectral imagery. *Photogrammetric Engineering and Remote Sensing* **70,** 703–715.

Senoo, T., Kobayashi, F., Tanaka, S., and Sugimura, T. 1990. Improvement of forest type classification by SPOT HRV with 20 m mesh DTM. *International Journal of Remote Sensing* **11,** 1011–1022.

Shaban, M. A., and Dikshit, O. 2001. Improvement of classification in urban areas by the use of textural features: The case study of Lucknow city, Uttar Pradesh. *International Journal of Remote Sensing* **22,** 565–593.

Shi, W. Z. 1998. A generic statistical approach for modeling errors of geometric features in GIS. *International Journal of Geographical Information Science* **12,** 131–143.

Shi, W. Z., Cheung, C. K., and Zhu, C. Q. 2003. Modeling error propagation in vector-based GIS. *International Journal of Geographical Information Science* **17,** 251–271.

Simone, G., Farina, A., Morabito, F. C., et al. 2002. Image fusion techniques for remote sensing applications. *Information Fusion* **3,** 3–15.

Sinton, D. 1978. The inherent structure of information as a constraint to analysis: Mapped thematic data as a case study. In *Harvard Papers on Geographic Information Systems,* Vol. 6, edited by G. Dutton. Reading, MA: Addison-Wesley.

Smits, P.C., Dellepiane, S.G., and Schowengerdt, R.A. 1999. Quality assessment of image classification algorithms for land-cover mapping: a review and a proposal for a cost-based approach. *International Journal of Remote Sensing* **20,** 1461–1486.

Solberg, A. H. S., Taxt, T., and Jain, A. K. 1996. A Markov random field model for classification of multisource satellite imagery. *IEEE Transactions on Geoscience and Remote Sensing* **34,** 100–112.

Song, M., and Civco, D. 2004. Road extraction using SVM and image segmentation. *Photogrammetric Engineering and Remote Sensing* **70,** 1365–1372.

Stallings, C., Khorram, S., and Huffman, R. L. 1999. Incorporating ancillary data into a logical filter for classified satellite imagery. *Geocarto international* **14,** 42–51.

Stefanov, W. L., Ramsey, M. S., and Christensen, P. R. 2001. Monitoring urban land cover change: an expert system approach to land cover classification of semi-arid to arid urban centers. *Remote Sensing of Environment* **77,** 173–185.

Stow, D., Coulter, L., Kaiser, J., et al. 2003. Irrigated vegetation assessment for urban environments. *Photogrammetric Engineering and Remote Sensing* **69,** 381–390.

Strahler, A. H., Logan, T. L., and Bryant, N. A. 1978. Improving forest cover classification accuracy from Landsat by incorporating topographic information. In *Proceedings of the Twelfth International Symposium on Remote Sensing of Environment,* Environmental Research Institute of Michigan, Ann Arbor, MI, pp. 927–942.

Sui, D. Z. 2008. The wikification of GIS and its consequences: Or Angelina Jolie's new tattoo and the future of GIS. *Computers, Environment, and Urban Systems* **32,** 1–5.

Sutton, P. 1997. Modeling population density with nighttime satellite imagery and GIS. *Computers, Environment and Urban Systems* **21,** 227–244.

Tao, C. V., and Hu, Y. 2002. 3D reconstruction methods based on the rational function model. *Photogrammetric Engineering and Remote Sensing* **68,** 705–714.

Teillet, P. M., Guindon, B., and Goodenough, D. G. 1982. On the slope-aspect correction of multispectral scanner data. *Canadian Journal of Remote Sensing* **8,** 84–106.

Thenkabail, P. S., Enclona, E. A., Ashton, M. S., and van der Meer, B. 2004a. Accuracy assessments of hyperspectral waveband performance for vegetation analysis applications. *Remote Sensing of Environment* **91,** 354–376.

Thenkabail, P. S., Enclona, E. A., Ashton, M. S., et al. 2004b. Hyperion, IKONOS, ALI, and ETM+ sensors in the study of African rainforests. *Remote Sensing of Environment* **90,** 23–43.

Thomas, N., Hendrix, C., and Congalton, R. G. 2003. A comparison of urban mapping methods using high-resolution digital imagery. *Photogrammetric Engineering and Remote Sensing* **69,** 963–972.

Thomson, C. N. 2000. Remote sensing/GIS integration to identify potential low-income housing sites. *Cities* **17,** 97–109.

Tokola, T., Löfman, S., and Erkkilä, A. 1999. Relative calibration of multitemporal Landsat data for forest cover change detection. *Remote Sensing of Environment* **68,** 1–11.

Tokola, T., Sarkeala, J., and van der Linden, M. 2001. Use of topographic correction in Landsat TM-based forest interpretation in Nepal. *International Journal of Remote Sensing* **22,** 551–563.

Tong, X., Liu, S., and Weng, Q. 2009. Geometric processing of QuickBird stereo imagery for urban land use mapping: A case study in Shanghai, China. *IEEE Journal of Selected Topics in Applied Earth Observations and Remote Sensing* **2,** 61–66.

Toutin, T. 2004a. DTM generation from IKONOS in-track stereo images using a 3D physical model. *Photogrammetric Engineering and Remote Sensing* **70,** 695–702.

Toutin, T. 2004b. Comparison of stereo-extracted DTM from different high-resolution sensors: SPOT-5, EROS-A, IKONOS-II and Quickbird. *IEEE Transactions on Geosciences and Remote Sensing* **42,** 2121–2129.

Treitz, P. M., Howard, P. J., and Gong, P. 1992. Application of satellite and GIS technologies for land-cover and land-use mapping at the rural-urban fringe: A case study. *Photogrammetric Engineering and Remote Sensing* **58,** 439–448.

Tso, B. C. K., and Mather, P. M. 1999. Classification of multisource remote sensing imagery using a genetic algorithm and Markov random fields. *IEEE Transactions on Geoscience and Remote Sensing* **37,** 1255–1260.

Tullis, J. A., and Jensen, J. R. 2003. Export system house detection in high spatial resolution imagery using size, shape, and context. *Geocarto International* **18,** 5–15.

Ulfarsson, M. O., Benediktsson, J. A., and Sveinsson, J. R. 2003. Data fusion and feature extraction in the wavelet domain. *International Journal of Remote Sensing* **24,** 3933–3945.

Usery, E. L. 1996. A feature-based geographic information system model. *Photogrammetric Engineering and Remote Sensing* **62,** 833–838.

Veregin, H. 1989. Error modeling for the map overlay operation. In *Accuracy of Spatial Databases,* edited by M. F. Goodchild and S. Gopal, pp. 3–19. London: Taylor & Francis.

Veregin, H. 1995. Developing and testing of an error propagation model for GIS overlay operations. *International Journal of Geographical Information Systems* **9,** 595–619.

Wang, L., Sousa, W. P., Gong, P., and Biging, G. S. 2004. Comparison of IKONOS and QuickBird images for mapping mangrove species on the Caribbean coast of panama. *Remote Sensing of Environment* **91,** 432–440.

Warner, T. A., and Steinmaus, K. 2005. Spatial classification of orchards and vineyards with high spatial resolution panchromatic imagery, *Photogrammetric Engineering and Remote Sensing* **71,** 179–187.

Waterfeld, W., and Schek, H. J. 1992. The DASBDS Geokernel: An extensible database system for GIS. In *Three-Dimensional Modeling with Geoscientific Information System,* edited by A. K. Turner, pp. 45–55. Amsterdam: Kluwer Academic Publishers.

Weber, C., and Hirsch, J. 1992. Some urban measurements from SPOT data: Urban life quality indices. *International Journal of Remote Sensing* **13,** 3251–3261.

Welch, R., and Ehlers, M. 1987. Merging multi-resolution SPOT HRV and Landsat TM data. *Photogrammetric Engineering and Remote Sensing* **53,** 301–303.

Welch, R., and Ehlers, M. 1988. Cartographic feature extraction from integrated SIR-B and Landsat TM images. *International Journal of Remote Sensing* **9,** 873–889.

Weng, Q. 2001. A remote sensing-GIS evaluation of urban expansion and its impact on surface temperature in the Zhujiang Delta, China. *International Journal of Remote Sensing* **22,** 1999–2014.

Weng, Q. 2002. Land use change analysis in the Zhujiang Delta of China using satellite remote sensing, GIS, and stochastic modeling. *Journal of Environmental Management* **64,** 273–284.

Weng, Q., and Hu, X. 2008. Medium spatial resolution satellite imagery for estimating and mapping urban impervious surfaces using LSMA and ANN. *IEEE Transaction on Geosciences and Remote Sensing* **46,** 2397–2406.

Weng, Q., Hu, X., and Lu, D. 2008. Extracting impervious surface from medium spatial resolution multispectral and hyperspectral imagery: A comparison. *International Journal of Remote Sensing* **29,** 3209–3232.

Weng, Q., Lu, D., and B. Liang. 2006. Urban surface biophysical descriptors and land surface temperature variations. *Photogrammetric Engineering & Remote Sensing* **72,** 1275–1286.

Weng, Q., Lu, D., and Schubring J. 2004. Estimation of land surface temperature-vegetation abundance relationship for urban heat island studies. *Remote Sensing of Environment* **89,** 467–483.

Woodcock, C.E. and Gopal, S. 2000. Fuzzy set theory and thematic maps: accuracy assessment and area estimation. *International Journal of Geographic Information Science* **14,** 153–172.

Wilkinson, G. G. 1996. A review of current issues in the integration of GIS and remote sensing data. *International Journal of Geographical Information Systems* **10,** 85–101.

Williams, J. 2001. *GIS Processing of Geocoded Satellite Data,* p. 327. Chichester, UK: Springer and Praxis Publishing.

Wulder, M. A., White, J. C., Coops, N. C., and Butson, C. R. 2008. Multi-temporal analysis of high spatial resolution imagery for disturbance monitoring. *Remote Sensing of Environment* **112,** 2729–2740.

Xue, Y., Cracknell, A. P., and Guo, H. D. 2002. Telegeoprocessing: The integration of remote sensing, geographic information system (GIS), Global Positioning System (GPS) and telecommunication. *International Journal of Remote Sensing* **23,** 1851–1893.

Yeh, A. G. O., and Li, X. 1997. An integrated remote sensing and GIS approach in the monitoring and evaluation of rapid urban growth for sustainable development in the Pearl River Delta, China. *International Planning Studies* **2,** 193–210.

Yocky, D. A. 1996. Multiresolution wavelet decomposition image merger of Landsat Thematic Mapper and SPOT panchromatic data. *Photogrammetric Engineering and Remote Sensing* **62,** 1067–1074.

Yuan, Y., Smith, R. M., and Limp, W. F. 1997. Remodeling census population with spatial information from Landsat imagery. *Computers, Environment and Urban Systems* **21,** 245–258.

Zanoni, V., Stanley, T., Ryan, R., et al. 2003. The Joint Agency Commercial Imagery Evaluation team: Overview and IKONOS joint characterization approach. *Remote Sensing of Environment* **88,** 17–22.

Zhang, J. and Foody, G. M. 2001. Fully-fuzzy supervised classification of sub-urban land cover from remotely sensed imagery: Statistical neural network approaches. *International Journal of Remote Sensing* **22,** 615–628.

Zhang, J. and Kirby, R. P. 2000. A geostatistical approach to modeling positional errors in vector data. *Transactions in GIS* **4,** 145–159.

Zhang, Q., Wang, J., Peng, X., et al. 2002. Urban built-up land change detection with road density and spectral information from multitemporal Landsat TM data. *International Journal of Remote Sensing* **23,** 3057–3078.

Zhang, Y. 1999. Optimisation of building detection in satellite images by combining multispectral classification and texture filtering. *ISPRS Journal of Photogrammetry and Remote Sensing* **54,** 50–60.

Zhang, Y., Pulliainen, J., Koponen, S., and Hallikainen, M. 2002. Application of an empirical neural network to surface water quality estimation in the Gulf of Finland using combined optical data and microwave data. *Remote Sensing of Environment* **81,** 327–336.

CHAPTER 3

Urban Land Use and Land Cover Classification

Urban land use and land cover (LULC) datasets are very important sources for many applications, such as socioeconomic studies, urban management and planning, and urban environmental evaluation. The increasing population and economic growth have resulted in rapid urban expansion in the past decades. Therefore, timely and accurate mapping of urban LULC is often required. Although many approaches for remote sensing image classification have been developed (Lu and Weng, 2007), urban LULC classification is still a challenge because of the complex urban landscape and limitations in remote sensing data.

Conventional survey and mapping methods do not provide the necessary information in a timely and cost-effective manner. Remotely sensed data, with their advantages in spectral, spatial, and temporal resolution, have demonstrated their power in providing information about the physical characteristics of urban areas, including size, shape, and rates of change, and have been used widely for mapping and monitoring of urban biophysical features (Haack et al., 1997; Jensen and Cowen, 1999). Geographic information system (GIS) technology provides a flexible environment for entering, analyzing, and displaying digital data from various sources that are necessary for urban feature identification, change detection, and database development. The integration of remote sensing and GIS technologies has been applied widely and has been recognized as an effective tool in urban-related research (Ehlers et al., 1990; Harris and Ventura, 1995; Treitz et al., 1992; Weng, 2002). Because of the confusion of spectral signatures in some land cover types, such as between impervious surface and soil and between low-density residential area and forest, ancillary data have become an important source for improving urban LULC classification accuracy. However, in urban areas, the separation of different densities of residential areas, the distinction between residential areas and forest or grass, and the separation of commercial/industrial areas from residential areas are important. Census data, such as those on housing density distribution, are closely related to urban LULC patterns. This chapter explores, based on a case study of Indianapolis, Indiana, the use of housing information in different stages in the image classification procedure in order to identify a suitable method for improvement of urban LULC classification performance.

3.1 Incorporation of Ancillary Data for Improving Image Classification Accuracy

Generating a satisfactory classification image from remote sensing data is not a straightforward task. Many factors contribute to this difficulty, for example, the characteristics of a study area, the availability of suitable remote sensing data and ancillary and ground-reference data, proper use of variables and classification algorithms, and the analyst's experience (Lu and Weng, 2007). Many approaches have been explored to improve classification accuracy, including incorporation of geographic data (Harris and Ventura, 1995), census data (Mesev, 1998), texture features (Lu and Weng, 2005; Myint, 2001; Shaban and Dikshit, 2001), and structure or contextual information (Gong and Howarth, 1990; Stuckens et al., 2000) into remote sensing spectral data. Moreover, expert systems (Hung and Ridd, 2002; Stefanov et al., 2001), fuzzy classification (Zhang and Foody, 2001), and merged multisensor data such as those between radar and Thematic Mapper (TM) data (Haack et al., 2002) and between Le Systeme Pour l'Observation de la Terre (SPOT) and TM data (Gluch, 2002) have been applied. Lu and Weng (2007) have summarized major classification approaches and discussed some potential techniques to improve classification performance. Hence this section focuses only on the integration of ancillary data into remote sensing data for improvement of classification accuracy.

The advantages of ancillary data in improving image classification accuracy have long been recognized (Catlow et al., 1984; Janssen et al., 1990; Kenk et al., 1988). Many techniques using ancillary data have been developed in an attempt to improve classification accuracy. Hutchinson (1982) categorized them into three approaches according to the stages where the ancillary data are incorporated in a classification procedure: before or pre-classification stratification, during the classification, and post-classification sorting.

In the pre-classification method, ancillary data are used to assist the selection of class training samples (Mesev, 1998) or are used to divide the study scene into smaller areas or strata based on some selected criteria or rules. When ancillary data are used during a classification, two methods can be employed. The first approach is a logical-channel method that was introduced by Strahler et al. (1978). The ancillary data were incorporated into remote sensing images as additional channels. The second approach is a classifier modification that involves altering *a priori* probabilities of classes in a maximum likelihood classifier according to estimated areal composition or known relationships between classes and ancillary data (Harris and Ventura, 1995; Mesev, 1998). In the post-classification sorting approach, ancillary data are used to refine the misclassified pixels based on developed decision rules.

The integration of ancillary data into remote sensing image classification is often applied to uneven terrains, complicated landscapes, and urban areas. The commonly used ancillary data include GIS data such as soil, terrain [digital elevation model (DEM)], zoning, and census data. There are a few successful experiments in the previous studies using ancillary data to improve classification.

Fleming and Hoffer (1979) developed two techniques for mapping forest cover types in Colorado using topographic data. One technique was a logical-channel approach in which the topographically stratified random samplings were used to derive both spectral and topographic training statistics for a supervised classification. This method provided a more accurate classification result than spectral data alone. Another technique applied topographically stratified random sampling in a layered approach for classification. Based on sample data, a topographic distribution model was developed to determine the probability of each forest cover type occurring at any given elevation, slope, and aspect, whereas spectral training statistics were developed independently using the modified-clustering technique. Spectral training statistics were used to classify major forest types, and the topographic statistics were used to further subdivide forest types into the species level. It was found that the use of ancillary data in the classification process improved accuracy, but the time and labor for the data analysis were increased significantly.

Franklin (1987) incorporated five geomorphometric variables that were extracted from the DEM as additional channels into Landsat images in a mountainous area of Canada's Yukon. An improvement from 46 to 75 percent for nine land cover classes was achieved. In another study, Franklin (1989) used two examples of integrated datasets to illustrate the computation and interpretation of a covariance matrix for remote sensing and ancillary data that was applied to classification of complex terrain. One was geological analysis based on Landsat Multispectral Scanner (MSS) and geophysical data in which geophysical data were used to stratify images into different regions; another was land-system mapping based on Landsat MSS and the DEM. Six variables, including elevation, slope, aspect, relief, downslope convexity, and cross-slope convexity, that were derived from the DEM were used as discriminators in the classification. Average class accuracies approached 85 percent when all available discriminators were used compared with 40 percent when Landsat data were used alone.

Franklin and Wilson (1991) incorporated the DEM into SPOT multispectral images for a mountainous region in southwestern Yukon. They devised a three-stage classifier that consisted of a quadtree-based segmentation operator, a Gaussian minimum-distance-to-means test, and a final test involving topographic data and a spectral curve measure. The overall accuracy was improved significantly compared with the multispectral technique alone (accuracy of 87.64 percent compared with 71.86 percent). This method required less time than the per-pixel maximum likelihood classifier in producing an output map and used a minimum of field or training data that might be difficult and expensive to acquire in complex terrain.

Maselli and colleagues (1995) incorporated ancillary data (i.e., elevation, slope, aspect, and soil) with Landsat TM images in a classification procedure through modification of the prior probabilities in a rugged area of Tuscany, Italy. The kappa value of 0.744 based on only TM images was increased to 0.910 based on integration of the ancillary data and TM images. A similar study was conducted by Pedroni (2003) in Costa Rica. Richetti (2000) combined topographic data such as slope maps with remote sensing images in a classification process for geological purposes. Logical-channel and stratification methods were applied and compared with spectral classification. Results demonstrated an increase in accuracy for the logical-channel method, but not for the stratification method.

Besides using topographic data for mapping land cover or vegetation in mountainous regions, there have been other attempts to integrate GIS and census data to improve LULC classification accuracy in urban landscapes. Harris and Ventura (1995) incorporated zoning and housing data into a TM image classification for non-point-source pollution modeling in the city of Beaver Dam, Wisconsin. They developed a post-classification model based on zoning and housing data to help identify and correct the areas that appeared to be confused on TM classification. The results indicated that zoning and housing data improved the overall classification accuracy by about 10 percent, and the number of identifiable classes that were identified (e.g., shopping center, strip commercial, hospital, etc.) increased.

Mesev (1998) integrated census data into SPOT and TM images to improve urban land use classification for four settlements in the United Kingdom. In this study, census data were incorporated in three stages. First, census tract data were used to assist the selection of training samples; second, tract data were converted to surface data and integrated with SPOT and TM images by modifying *a priori* probability for each spectral class. Finally, the census data were used for post-classification sorting. The study showed that the use of census data significantly improved classification accuracy, especially for residential and non-residential classes.

Although many efforts have been made to integrate ancillary data into remote sensing image classification, most of them have shown varying degrees of success in producing accurate classification results depending on the methods applied and the nature of the study areas. Use of ancillary data to stratify images into different regions based on selected criteria reduces variation within land cover types and enables analysts to accurately label the land cover types. This is easily implemented and effective, but incorrect stratification or selection of training samples may corrupt the classification accuracy. Simply adding ancillary data to increase channels of observation is easy but not reliable in improving classification accuracy owing to the fact that statistics based on ancillary data usually do not satisfy the requirements of the maximum likelihood classifier about a Gaussian distribution. Classification through modification of *a priori* probability for the maximum likelihood classifier is efficient but requires extracting many samples to find the relationships between ancillary data and spectral data to obtain *a priori* probability estimates. Post-classification sorting has several advantages too. It is simple, quick, and easy to implement, and it only deals with "problem" classes. In addition, it is possible to incorporate several types of ancillary data in developing decision rules. However, the deterministic decision rules need be developed carefully.

3.2 Case Study: Landsat Image-Housing Data Integration for LULC Classification in Indianapolis

The objectives of this case study is to improve urban LULC classification accuracy by incorporating ancillary data (specifically census data) and Landsat imagery at different classification stages and to develop a suitable procedure for effective integration of these two types of data.

3.2.1 Study Area

The study area is located at 39°46'N and 86°09'W, Marion County (the city proper of Indianapolis), Indiana (Fig. 3.1). According to the U.S. Census Bureau, the county has a total area of 1044 km^2, including 1026 km^2 of land and 18 km^2 of water. The average annual temperature is 11.7°C, with the highest temperature of 24.6°C in July and the lowest temperature of –1.9°C in January. Annual precipitation is evenly distributed throughout the year, and average annual rainfall is 1021 mm, with the least amount of precipitation occurring in February. It is the core of the Indianapolis metropolitan area. The county seat—Indianapolis—was called "plain city" because of its flat topography (elevation ranges from 218 to 276 m above sea level), which provides the possibility of expansion in all directions. In recent decades, the city has been experiencing areal expansion through encroachment on agricultural land and other nonurban land owing

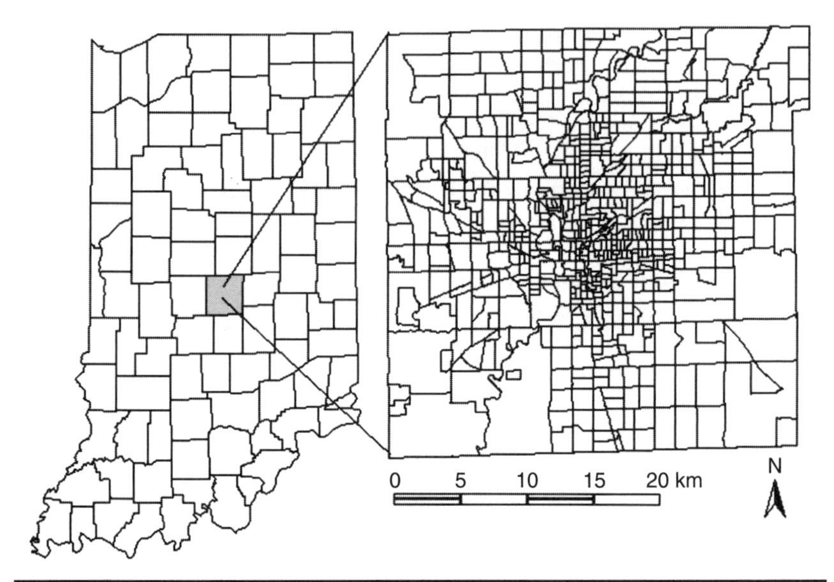

Figure 3.1 Study area—Marion County, Indiana.

to population increases and economic growth (the population of Marion County increased by approximately 7.9 percent from 1990 to 2000).

According to the 2000 Census, there were 860,454 people with a density of 838/km^2. There were 387,183 housing units at an average density of 377/km^2. With its large population, Indianapolis ranks as the twelfth largest city in the United States. In 1999, the median household income in the county was $40,421, and the per capita income was $21,789, with 11.40 percent of the population and 8.7 percent of families below the poverty line.

Indianapolis has the highest concentration of major employers and manufacturing, professional, technical, and educational services in the state. With its moderate climate, rich history, excellent education, and abundant social services, arts, leisure, and recreation, Indianapolis was named as one of America's "Best Places to Live and Work" (*Employment Reviews,* August 1996). In 1996, it ranked fifth on the list of the 30 best cities for small business by *Entrepreneur* magazine, and it was one of the top 10 metropolitan areas in the nation for business success in a 1996 study by Cognetics, a research firm in Cambridge, Massachusetts.

3.2.2 Datasets Used

The primary data sources used in this research are Census 2000 data and Landsat Enhanced Thematic Mapping Plus (ETM+) images (Fig. 3.2). The Landsat 7 ETM+ image (L1G product of path 21, row 32)

FIGURE 3.2 Landsat ETM+ image (bands 4, 5, and 3 as red, green, and blue, respectively) of Indianapolis acquired on June 22, 2000. See also color insert.

used in this study was acquired on June 22, 2000, under clear-sky conditions. The data were converted radiometrically to at-sensor reflectance using an image-based correction method (Markham and Barker, 1987). Although the L1G ETM+ data were corrected geometrically, the geometric accuracy was not high enough for combining them with other high-resolution datasets. Hence the image was further rectified to a common Universal Transverse Mercator (UTM) coordinate system based on 1:24,000 scale topographic maps. A root-mean-square error of less than 0.5 pixels was obtained in the rectification.

Census 2000 data from the U.S. Bureau of the Census include (1) tabular data stored in Summary File 1, which contains information about the population, families, households and housing unit, etc. and (2) spatial data called topologically integrated geographic encoding and referencing (TIGER) data that contains data representing the position and boundaries of legal and statistical entities. These two types of data are linked by Census geographic entity code. The U.S. Census has a hierarchical structure consisting of 10 basic levels: United States, region, division, state, county, county subdivision, place, Census tract, block group, and block. For population estimation, the block-group level was selected as the work level, whereas the block level was used as the work level for classification. Other ancillary data used in this research included (1) aerial photographs taken in the summer 2003 that were downloaded from the Indiana Spatial Data Portal and (2) GIS zoning data that were created by an Indianapolis and Marion County GIS team with the state plane coordinate system. All data, including housing, population, aerial photograph, zoning, and ETM+ images, were georegistered to UTM coordinates before data integration.

3.2.3 Methodology

Landsat ETM+ image, zoning, and Census housing data were combined for use in LULC classification in this research. The strategy of the LULC classification procedure is illustrated in Fig. 3.3. Zoning and housing data were used in different stages of image classification (e.g., pre-classification, during classification, and post classification) to identify a suitable procedure for improving LULC classification accuracy.

Housing information was extracted from the Census data at block level, and housing density (Fig. 3.4) was calculated by dividing housing units by block area. In order to incorporate housing information into the Landsat ETM+ image, housing data must be converted to raster format. Two methods are often used to this conversion. One method is to conduct rasterization, and the other is to generate a

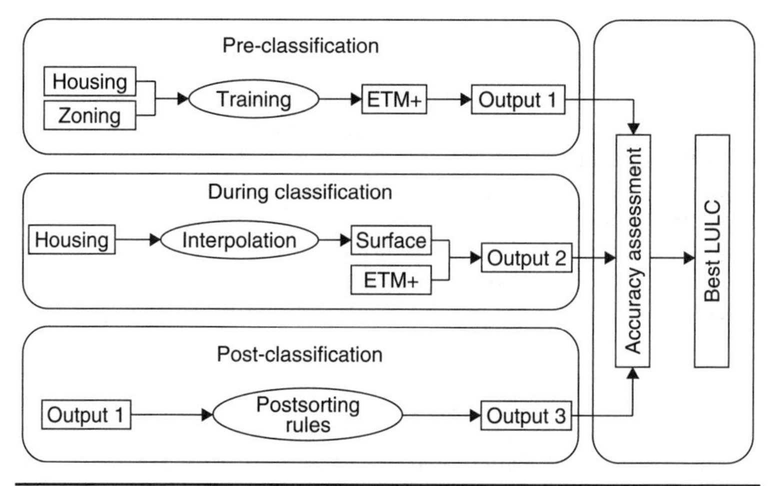

Figure 3.3 Strategies of LULC classification by incorporating zoning and housing datasets for improving classification performance.

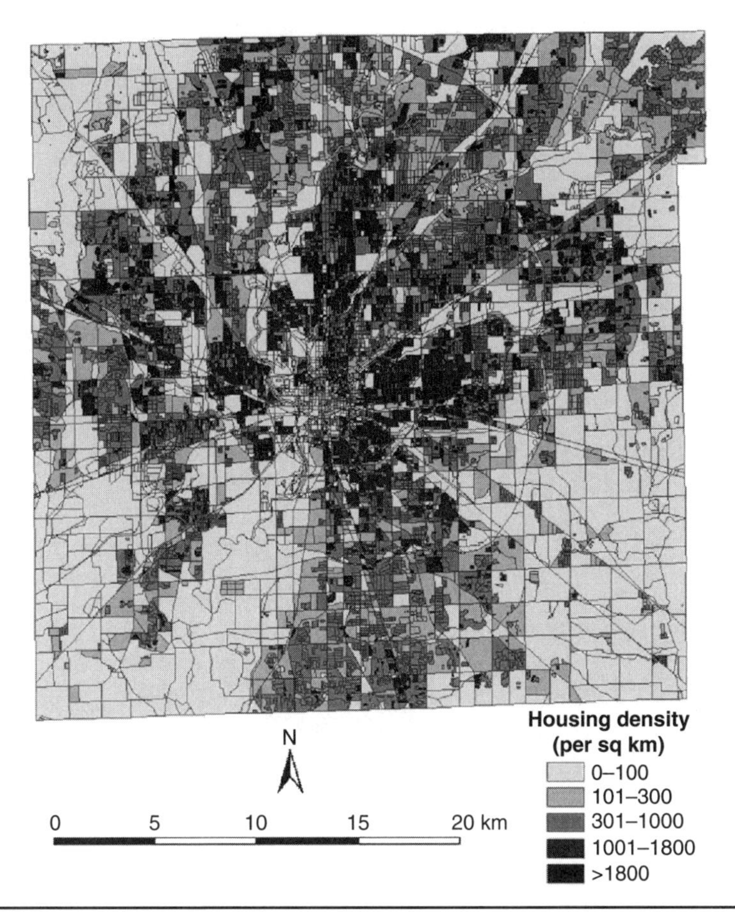

Figure 3.4 Housing density distribution at Census block, Marion County, Indiana, in 2000. See also color insert.

housing surface by using spatial interpolation algorithms. In this study, the centroids of blocks (Fig. 3.5) were identified with the aid of ArcGIS, and then inverse-distance-weighting (IDW) interpolation was used to generate a housing surface.

The smooth and continuous housing surface derived from the centroids of blocks using IDW (Fig. 3.6) enabled conversion of spatial data from irregular zonal units (blocks) into regular units (surface cells or pixels) that had a similar format as the ETM+ image. One important characteristic of this model is that it preserves the total housing units (Mesev, 1998). In contrast, the rasterized housing density (Fig. 3.7) was not continuous; cell values were the same within a block, but it preserved most of the original housing density of the blocks. Both were integrated with the ETM+ image as extra layers by stacking them on Landsat ETM+ six bands for further classification.

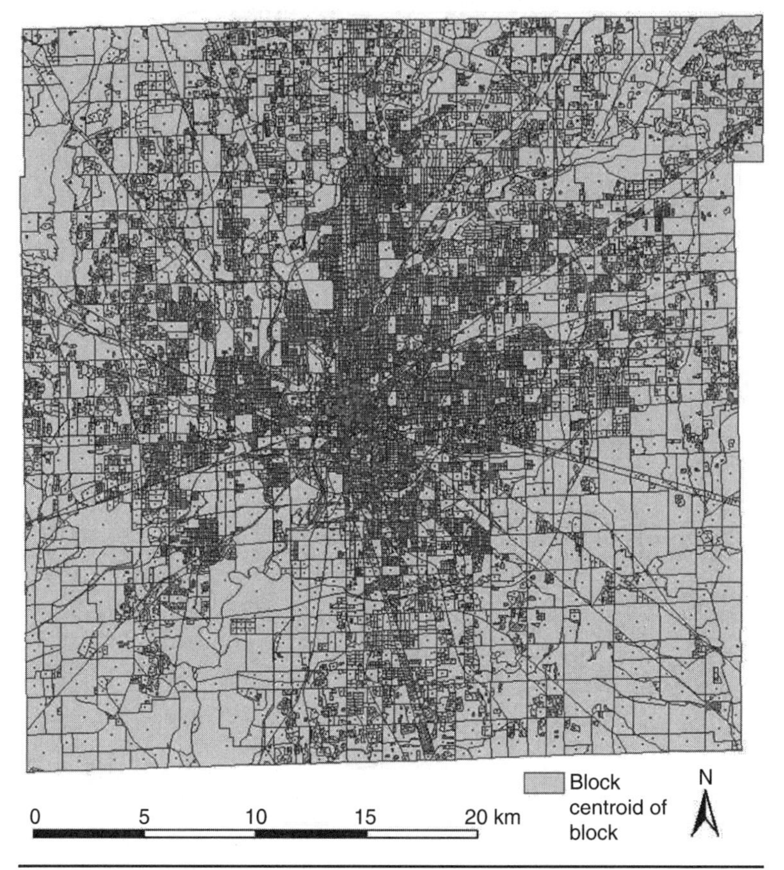

Figure 3.5 Centroids of Census block. See also color insert.

In the Pre-Classification Stage

A traditional supervised classification (specifically, the maximum-likelihood classifier) was used to classify the ETM+ image using its light-reflective bands. Before implementing image classification, selection of good-quality training samples is critical. In this research, housing density, zoning data, and high spatial-resolution aerial photographs were used to assist the identification of training samples of different urban LULC classes. The ETM+ image was classified initially into 11 classes (i.e., commercial, transportation, industrial, water, low-density residential, medium-density residential, high-density residential, grass, crop land, fallow, and forest). Figure 3.8 provides examples of typical LULC data appearing on the aerial photograph. The 11 classes then were merged to 8 classes by combining commercial, transportation, and industrial as urban and combining crop and fallow as agricultural land.

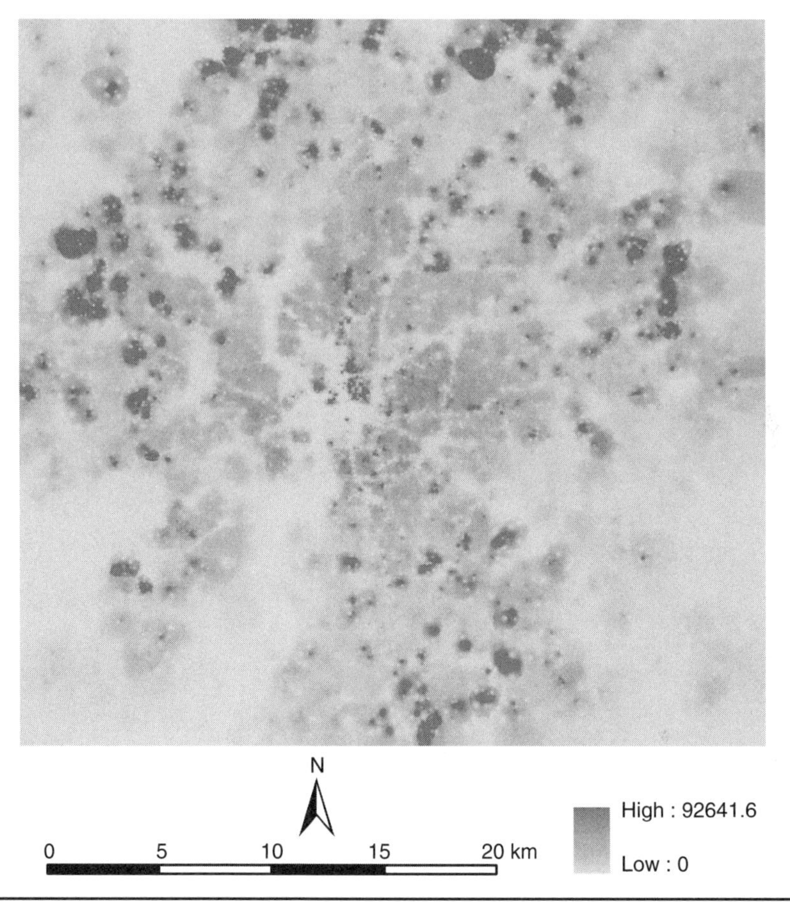

FIGURE 3.6 Housing surface generated by IDW interpolation. See also color insert.

For residential density, there are no agreed-on standards for what constitutes high, medium, and low density. A high density in Indianapolis might be medium or even low density in Shanghai. Therefore, in this study, the criteria for separating different densities of residential lands were adopted based on housing density and zoning data. For example, residential lands having housing units of more than 1300 per square kilometer were assigned as high-density residential areas, those having housing units of fewer than 400 per square kilometer were assigned as low-density residential areas, and those having housing units between 400 and 1300 per square kilometer were assigned as medium-density residential areas. Table 3.1 gives the definitions for these eight classes. This classification result was used as a base map for the post-classification sorting.

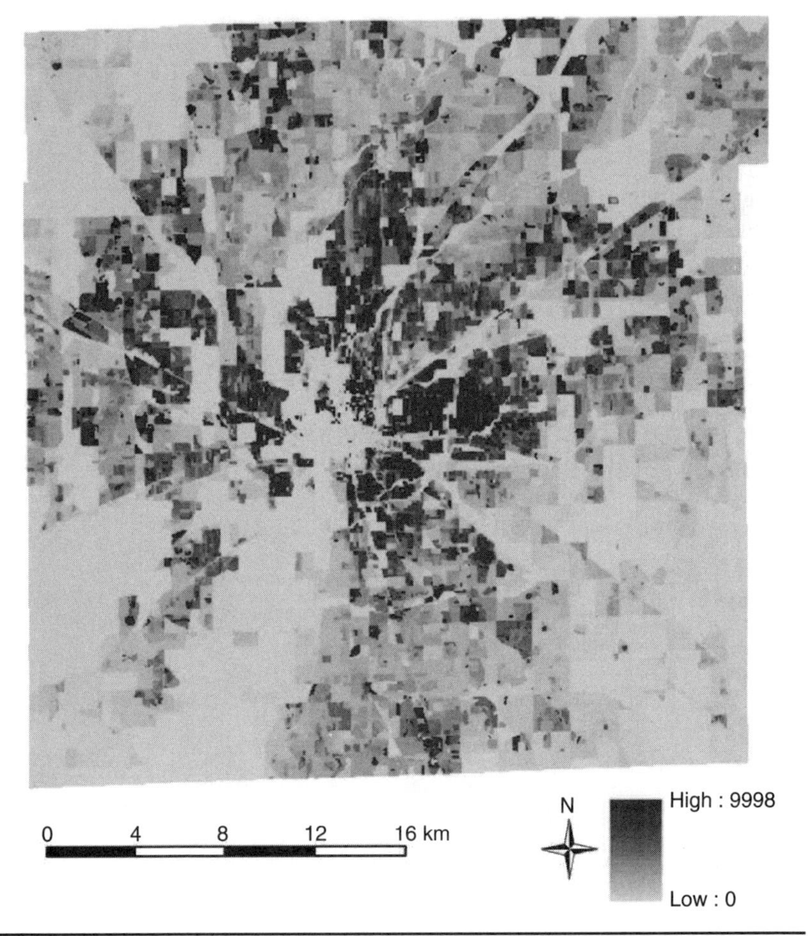

FIGURE 3.7 Rasterized housing density. See also color insert.

During the Classification

In this stage, the housing dataset was incorporated as an extra layer into the ETM+ image. Two housing data layers—rasterized housing density and housing surface—were combined with the ETM+ image as additional channels for image classification. Supervised classification with the maximum likelihood algorithm then was used to classify the combined image.

In the Post-Classification Stage

Because of similar spectral characteristics in certain LULC types, such as between high-density residential and commercial areas and between low-density residential and forest areas, urban LULC classification is often difficult based on spectral signatures (Lu and Weng, 2004). Thus, in this research, housing density was used to correct the pixels that were misclassified during the classification procedure. Based on the

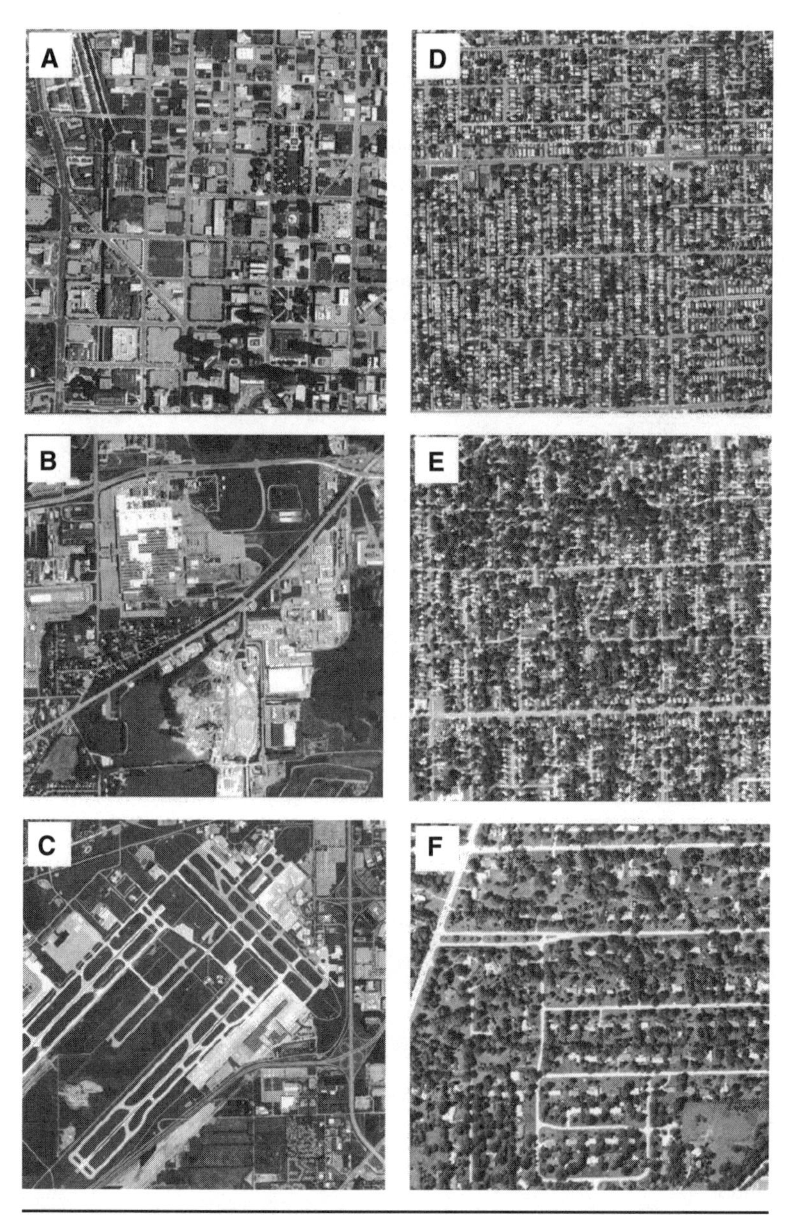

FIGURE 3.8 Examples of typical land use: (*a*) commercial, (*b*) industrial, (*c*) transportation, (*d*) high-density residential, (*e*) medium-density residential, (*f*) low-density residential. See also color insert.

housing density distribution at the block level, the following rules indicated in Table 3.2 were developed. Finally, for convenience, grass, crop land, and forest areas were merged into a new group called *vegetation*. By applying the rules shown in the table, the problematic pixels could be merged into the categories most appropriate for this research.

LULC Type		Definition
Urban		Commercial, transportation, and industrial
Residential	High-density residential	Residential areas having a housing density of greater than 1300 per square kilometer
	Medium-density residential	Residential areas having a housing density of greater than 400 and fewer than 1300 per square kilometer
	Low-density residential	Residential areas having a housing density of fewer than 400 per square kilometer
Crop land		Herbaceous vegetation that has been planted or is intensively managed for the production of food, including croplands such as corn, wheat, and soybean, as well as fallow land
Water		All areas of open water, including lakes, rivers, and streams and ponds
Forest		The areas covered by trees, including natural deciduous forest, evergreen forest, mixed forest, and shrubs
Grass		The areas covered by herbaceous vegetation, including pasture/hay planted for livestock grazing or the production of hay; also includes the urban/recreational grasses, such as parks, lawns, golf courses, airport grasses, etc

Table 3.1 Definitions of Land Use and Land Cover Types

Condition		Modification
From LULC Type	**Housing Density**	**To LULC Type**
Urban	>1300	High-density residential
High-density residential	<50	Urban
Medium-density residential	<50	Urban
High-density residential	>400 and <1300	Medium-density residential
Medium-density residential	<400	Low-density residential
Low-density residential	>400 and <1300	Medium-density residential
Low-density residential	<50	Vegetation

Table 3.2 Decision Rules Developed for Postsorting

3.2.4 Accuracy Assessment

In LULC classification, accuracy assessment of the classification results is often required (Foody, 2002). Many approaches, such as overall accuracy, producer's accuracy, user's accuracy, and kappa coefficient, have been used for evaluating classification accuracy. Their meanings and calculations have been described extensively in the literature (Congalton, 1991; Congalton and Mead, 1983; Hudson and Ramm, 1987; Smits et al., 1999). In reality, the most frequently used method for quantitatively analyzing LULC classification accuracy may be the error matrix, and thus it was used in this study. The accuracy assessment for the three classification images was conducted with a randomly sampling method. Fifty samples for each LULC type were selected. The reference data were collected from high-spatial-resolution aerial photographs. Overall accuracy, producer's accuracy, user's accuracy, and kappa statistic were calculated based on the error matrices.

3.3 Classification Result by Using Housing Data at the Pre-Classification Stage

Supervised classification using maximum likelihood classification was performed with all six ETM+ reflective bands. Training samples were selected with the assistance of housing-density and zoning GIS data, as well as aerial photographs. Owing to high variability within the same LULC types, training samples were selected to be as detailed as possible. For example, transportation, central business district (CBD), and industrial classes vary considerably in spectral response, although they were classified as one class in the final result. Therefore, training samples were selected separately based on different types of urban land use. A similar situation happened with agricultural lands (lands planted with crops and lands lying fallow). Because of the spectral confusion problem, the image was classified initially into 11 categories. The cell-array table of separabilities (Table 3.3) indicated that separability between vegetation types (crop land and forest), between residential intensity levels, between high-density residential and transportation areas, and between low-density residential and grass/crop areas was very low owing to the spectral similarity of these features, which may result in poor classification accuracy. Figure 3.9 shows the classification image, in which the 11 categories were merged to 8 LULC classes (i.e., water, urban, high-density residential, medium-density residential, low-density residential, forest, grass, and crop land). On this image, some obvious errors were found. For example, the commercial area in the northwest (*a*) was mixed with the medium-density residential area. Medium-density residential area also invades into the airport (*b*) and industrial areas (*c*).

Signature Name	2	3	4	5	6	7	8	9	10	11
Industrial (1)	2000	2000	2000	2000	2000	2000	2000	2000	2000	2000
Crop land (2)		2000	1982	2000	2000	2000	2000	2000	2000	1970
Fallow (3)			2000	2000	2000	2000	2000	2000	2000	2000
Forest (4)				2000	2000	2000	2000	2000	2000	2000
Grass (5)					2000	2000	2000	2000	2000	1985
Transport (6)						2000	2000	1482	1931	1997
Commercial (7)							2000	2000	2000	2000
Water (8)								2000	2000	2000
Residential-H (9)									1239	1763
Residential-M (10)										820
Residential-L (11)										0

TABLE 3.3 Separability Cell-Array for Six ETM+ Bands Using Transformed Divergence (Best Average Separability: 1957.1 with Combination 1, 2, 3, 4, 5, 6)

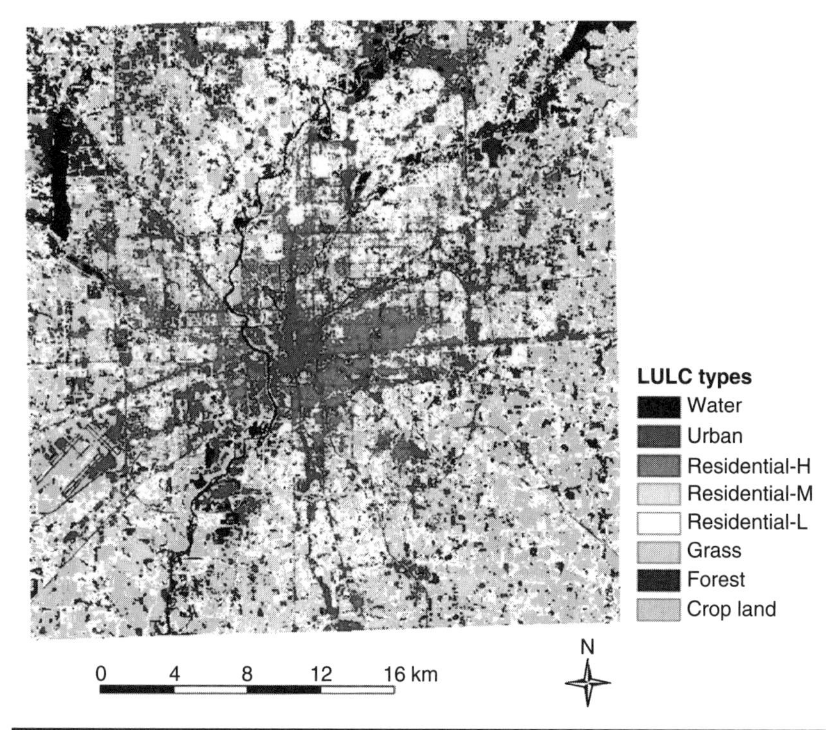

Figure 3.9 LULC classification image based on intergration of housing and the ETM+ image at the pre-classification stage. See also color insert.

However, medium-density residential areas supposed to appear in areas adjacent to high-density residential areas were classified as low-density residential areas (*d*). Table 3.4 shows the classification accuracy based on the eight LULC types. An overall accuracy of 70.5 percent and a kappa value of 0.6629 were achieved. The accuracies of the residential types were the lowest. Confusion occurred between urban and high-density residential areas; between high-density residential and medium-density residential areas; between medium-density residential and low-density residential areas; between low-density residential, grass, and forest areas; and between grass and crop land areas. These confusions were the major factors contributing to the low accuracy of image classification. Mixed pixels, such as mixtures of impervious surface, forest, and grass in the low-density residential areas, were another factor for low classification accuracy. Selecting training samples for low-density residential land was difficult even with the assistance of zoning and housing data.

	Pre-classification		During classification		Post-classification	
LULC	**PA**	**UA**	**PA**	**UA**	**PA**	**UA**
Water	100.00	96.00	98.00	98.00	97.92	95.92
Urban	61.19	82.00	75.47	80.00	86.57	79.45
Residential-H	70.97	44.00	73.47	72.00	51.61	72.73
Residential-M	46.51	40.00	56.10	46.00	81.40	74.47
Residential-L	65.79	50.00	60.00	36.00	76.32	64.44
Grass	56.96	90.00	60.53	92.00	65.82	78.79
Crop land	79.49	62.00	82.35	56.00	76.92	61.22
Forest	90.91	100.00	74.63	100.00	89.09	100.00
Overall accuracy	70.50		72.50		79.00	
Overall kappa	0.6629		0.6857		0.7574	

PA = producer's accuracy; UA = user's accuracy.

TABLE 3.4 Comparison of Different Classification Schemes

3.4 Classification Result by Integrating Housing Data during the Classification

Ancillary data also can be used as extra channels by combining them with remotely sensed data. In this study, rasterized housing density and a housing surface developed with the IDW interpolator were integrated into the ETM+ image as additional layers. A supervised classification was applied to classify the combined dataset. It was found that incorporation of rasterized housing density failed to identify some LULC classes, such as forest, crop, and grass. The integration of housing surface layer with the ETM+ image did improve the average separability to a certain degree (separability value from 1957 to 1984). A comparison of Table 3.3 with Table 3.5 indicates that the separability between any pair of LULC types mentioned in the last section was increased. Table 3.4 shows that the accuracies of high-density residential and medium-density residential were improved, but the accuracies of low-density residential, urban, and crop land decreased. Both overall accuracy and the kappa value increased, although not significantly. Figure 3.10 shows the classified image based on the during-classification approach.

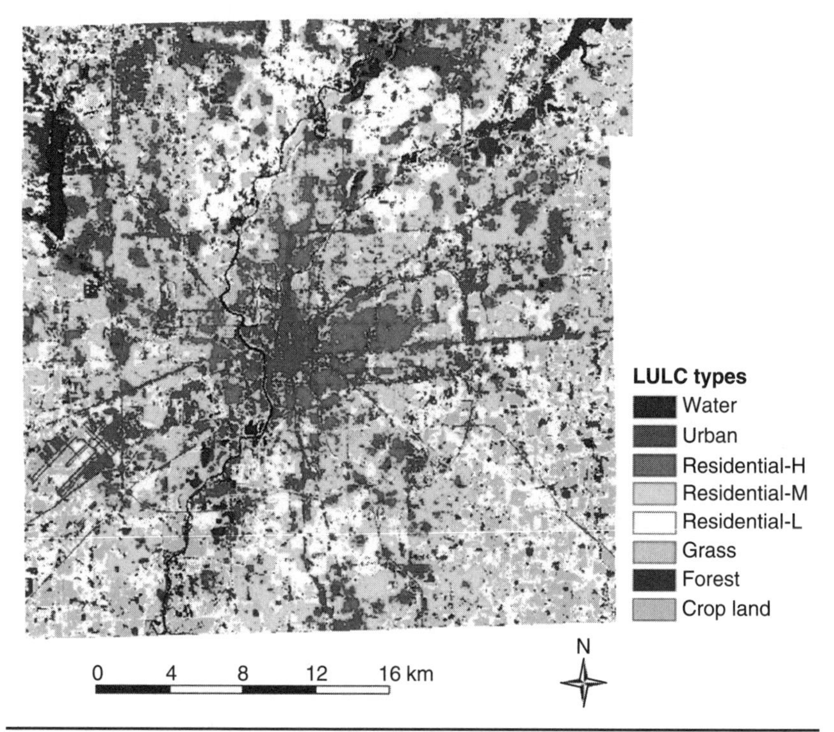

Figure 3.10 LULC map based on integration of housing surface into the ETM+ image as an additional layer (during the classification). See also color insert.

Signature Name	2	3	4	5	6	7	8	9	10	11
Industrial (1)	2000	2000	2000	2000	2000	2000	2000	2000	2000	2000
Crop land (2)		2000	1982	2000	2000	2000	2000	2000	2000	1992
Fallow (3)			2000	2000	2000	2000	2000	2000	2000	2000
Forest (4)				2000	2000	2000	2000	2000	2000	2000
Grass (5)					2000	2000	2000	2000	2000	1998
Transp. (6)						2000	2000	1770	1944	1999
Commercial (7)							2000	2000	2000	2000
Water (8)								2000	2000	2000
Residential-H (9)									1641	2000
Residential-M (10)										1578
Residential-L\ (11)										0

TABLE 3.5 Separabilty Cell-Array for ETM/Housing Surface Using Transformed Divergence (Best Average Separability: 1983.69 with Combination 1, 2, 3, 4, 5, 6, 7)

3.5 Classification Result by Using Housing Data at the Post-Classification Stage

The ancillary data can be used further for postprocessing of the classified images. In this research, the LULC image (see Fig. 3.9) developed from the ETM+ image was used as a base. Postsorting rules were applied to this image to produce a refined LULC image (Fig. 3.11). Because the objective of using these rules was to refine urban and residential land uses, they did not have much impact on water, agricultural land, and forest in terms of accuracy. However, the post-classification sorting had a great impact on the residential and urban areas, as well as grasslands. Table 3.4 indicated that the overall accuracy and kappa value were increased to 79 percent and 0.7574 respectively, compared with 70.5 percent and 0.6629 before the sorting. The accuracies of residential land areas were improved considerably, but the accuracies of urban areas and grasslands were decreased to varying degrees. Also, the obvious errors mentioned in the pre-classification stage disappeared. A comparison of the three classification schemes indicates that integration of housing data through postsorting is more effective than through adding housing surface as a layer into the dataset to be classified.

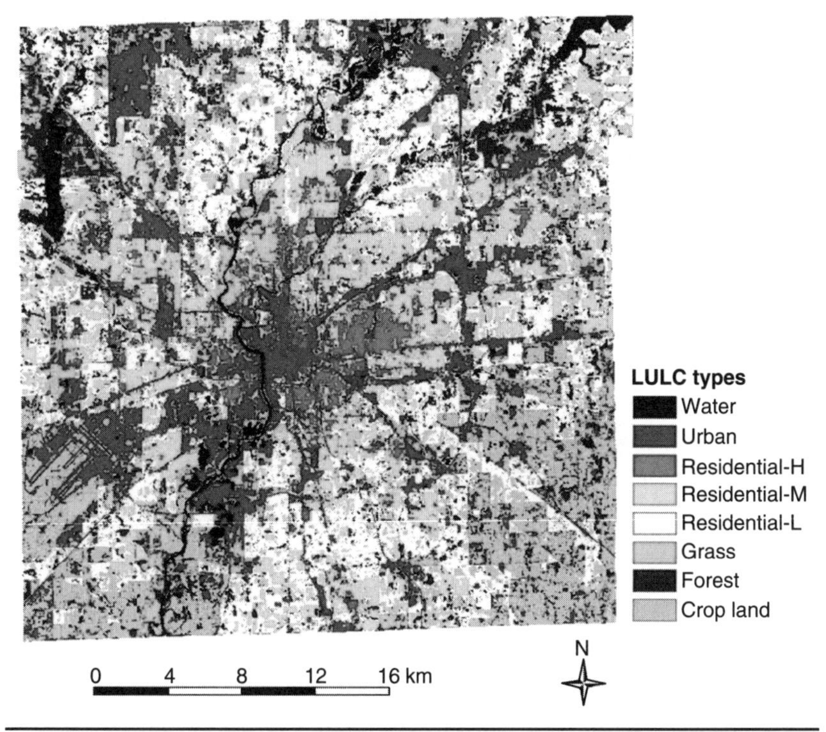

FIGURE 3.11 LULC map based on integration of housing data at the post-classification stage. See also color insert.

3.6 Summary

This chapter examined the role of Census housing data in LULC classification for an urban area through integration with the Landsat ETM+ image at different classification stages based on a case study of Indianapolis. The results indicate that use of housing data in the post-classification stage provides a better result than use of such data during the classification stage as an additional layer.

In urban areas, an important data source is housing data. However, the use of the housing data has not fully been examined in urban LULC classification. This research examined the use of housing data at different stages of image classification, that is, for assisting in the selection of training sample plots at the pre-classification stage, for combination with the remote sensing image as an extra channel at the during-classification stage, and for modifying the classified image through use of expert rules at the post-classification stage. When housing data were used as an extra channel during the classification procedure, two factors may contribute to their ineffectiveness for improving classification accuracy. One possible factor is the abnormal distribution of housing data in the training samples. Because the assumption of a normal distribution of the training samples was not satisfied, the effectiveness of the maximum likelihood classifier, if employed, may suffer. Another factor is associated with the relatively coarse spatial resolution of the interpolated housing surface image, which was developed from Census data at the block level. In this study, the control points used for generating housing surface were geometric centers, that is, centroids of the blocks, which often introduced errors when used to represent the source zones. Calculation of the centroid of an area depends on the coordinates of points defining the boundaries of a source zone. If a source zone is simple or symmetric, the centroid would be a good control point, and the estimated value for a grid cell would be more reliable. However, if a zone is irregular and complicated, the location of centroid can be significantly affected, and the interpolation result may be biased. In reality, Census blocks are often not in regular shapes. The nonuniform distribution of housing units within a block can further complicate the problem. Another important issue is how to select an appropriate spatial interpolation algorithm because different interpolators may influence the final result of LULC classification substantially.

Use of housing data in the post-classification stage provided the best results in this research. The deterministic decision rules used to refine the base LULC are critical for their effectiveness; therefore, they need to be developed carefully. In reality, developing the decision rules is a challenge because there are no standard criteria for distinguishing the different residential densities. An analyst needs to carefully check the existing information, such as Census data and historic land use, and information collected by local planning agencies. In this research, decision rules were developed based on housing density distribution with zoning data as a reference, but the break points for housing density seemed to be somewhat difficult to determine. This may have contributed to unsatisfactory improvement in separating the low-density residential class from the others.

Image classification in urban areas using medium-spatial-resolution (10 to 100 m) data is a challenge task. Although much research has been conducted with different techniques, a comprehensive comparison on which to generate some common guidelines for urban classification is lacking because of the use of different datasets, different image processing procedures, and different complexities of urban landscapes. For this same study area, we have conducted research with different approaches. It is pertinent to make a comparison and to arrive at some general observations.

Lu and Weng (2005) compared eight different image processing routines to identify suitable remote sensing variables for urban classification for the same study area by using the ETM+ image. Incorporation of spectral signature, texture, and surface temperature was examined, as well as data fusion techniques for combining a higher-spatial-resolution image with lower-spatial-resolution multispectral images. The research indicated that incorporation of texture from lower-spatial-resolution images or of a temperature image cannot improve classification accuracies. However, incorporation of textures derived from a higher-spatial-resolution panchromatic image did improve the classification accuracy. In particular, use of data fusion result and texture image yielded the best classification accuracy. Incorporation of texture and spectral signatures is conventionally assumed to provide better classification results than use of pure spectral signatures (Butusov, 2003; Shaban and Dikshit, 2001). However, the success for incorporating texture images depends on the type of texture measure, the image used to derive textures, and the selection of moving window size (Lu and Weng, 2005).

Owing to the complexity of urban landscapes and the limitations of current sensor technology, per-pixel-based classification approaches have proven to be difficult for urban LULC classification. Recently, subpixel analysis using spectral mixture models, which unmix an image into different fractions, has been demonstrated to improve classification accuracy (Lu and Weng, 2004; Phinn et al., 2002; Rashed et al., 2001). Spectral unmixing provides a more realistic representation of the true nature of a surface compared with that provided by the assignment of a single dominant class to every pixel by statistical models (Campbell, 2002) and is suitable to solve the mixture problem for medium- or low-spatial-resolution data. Lu and Weng (2004) compared a subpixel-based classification approach with traditional supervised classification in this study area and found that the subpixel-based approach can improve classification accuracy significantly (by 9 percent) compared with the maximum likelihood classifier.

Although the use of textures, data fusion of multiresolution images, and subpixel information improved urban LULC classification accuracies, the separation of different densities of residential areas is still difficult in urban LULC classification. Therefore, in another study in the same study area, Lu and Weng (2006) developed a new approach based on the combined use of impervious surface and population density for urban land use classification. Five urban land use classes (i.e., low-, medium-, high-, and very-high-density residential areas and commercial/industrial/transportation uses) were developed with an overall classification accuracy of 83.78 percent. This research indicates that setting up suitable thresholds for impervious surface and population density for each urban land use category is the key to success with this approach.

References

Butusov, O. B. 2003. Textural classification of forest types from Landsat 7 imagery. *Mapping Sciences and Remote Sensing* **40,** 91–104.

Campbell, J. B. 2002. *Introduction to Remote Sensing,* 3rd ed. New York: Guilford Press.

Catlow, D. R., Parsell, R. J., and Wyatt, B. K. 1984. The integrated use of digital cartographic data and remotely sensed imagery. *Earth-Oriented Applications of Space Technology* **4,** 255–260.

Congalton, R. G. 1991. A review of assessing the accuracy of classification of remotely sensed data. *Remote Sensing of Environment* **37,** 35–46.

Congalton, R. G., and Mead, R. A. 1983. A quantitative method to test for consistency and correctness in photo interpretation. *Photogrammetric Engineering and Remote Sensing* **49,** 69–74.

Ehlers, M., Jadkowski, M. A., Howard, R. R., and Brostuen, D. E. 1990. Application of SPOT data for regional growth analysis and local planning. *Photogrammetric Engineering and Remote Sensing* **56,** 175–180.

Fleming, M. D., and Hoffer, R. M. 1979. Machine processing of Landsat MSS and DMA topographic data for forest cover type mapping. In *Proceedings of the 1979 Symposium on Machine Processing of Remotely Sensed Data,* Purdue University, pp. 377–390.

Foody, G. M. 2002. Status of land cover classification accuracy assessment. *Remote Sensing of Environment* **80,** 185–201.

Franklin, S. E. 1987. Terrain analysis from digital patterns in geomorphometry and Landsat MSS spectral response. *Photogrammetric Engineering and Remote Sensing* **53,** 59–65.

Franklin, S. E. 1989. Ancillary data input to satellite remote sensing of complex terrain phenomena. *Computers and Geosciences* **15,** 799–808.

Frankin, S. E., and Wilson, B. A. 1991. Spatial and spectral classification of remote sensing imagery. *Computers and Geosciences* **17,** 1151–1172.

Gluch, R. 2002. Urban growth detection using texture analysis on merged Landsat TM and SPOT-P data. *Photogrammetric Engineering and Remote Sensing* **68,** 1283–1288.

Gong, P., and Howarth, P. J. 1990. The use of structure information for improving land cover classification accuracies at the rural-urban fringe. *Photogrammetric Engineering and Remote Sensing* **56,** 67–73.

Haack, B. N., Guptill, S. C., Holz, R. K., et al. 1997. Urban analysis and planning. In *Manual of Photographic Interpretation,* 2nd ed., edited by W. Philipson, pp. 517–554. Bethesda, MD: American Society for Photogrammetry and Remote Sensing.

Haack, B. N., Solomon, E. K., Bechdol, M. A., and Herold, N. D. 2002. Radar and optical data comparison/integration for urban delineation: A case study. *Photogrammetric Engineering and Remote Sensing* **68,** 1289–1296.
Harris, P. M., and Ventura, S. J. 1995. The integration of geographic data with remote sensed imagery to improve classification in an urban area. *Photogrammetric Engineering and Remote Sensing* **61,** 993–998.
Hudson, W. D., and Ramm, C. W. 1987. Correct formulation of the kappa coefficient of agreement. *Photogrammetric Engineering and Remote Sensing* **53,** 421–422.
Hung, M., and Ridd, M. K. 2002. A subpixel classifier for urban land cover mapping based on a maximum likelihood approach and expert system rules. *Photogrammetric Engineering and Remote Sensing* **68,** 1173–1180.
Hutchinson, C. F. 1982. Techniques for combining Landsat and ancillary data for digital classification improvement. *Photogrammetric Engineering and Remote Sensing* **48,** 123–130.
Janssen, L. L. F., Jaarsma, M. N., and Linden, E. T. M. 1990. Integrating topographic data with remote sensing for land cover classification. *Photogrammetric Engineering and Remote Sensing* **56,** 1503–1506.
Jensen, J. R., and Cowen, D. C. 1999. Remote sensing of urban/suburban infrastructure and socioeconomic attributes. *Photogrammetric Engineering and Remote Sensing* **65,** 611–622.
Kenk, E., Sondheim, M., and Yee, B. 1988. Methods for improving the accuracy of Thematic Mapper ground cover classifications. *Canadian Journal of Remote Sensing* **14,** 17–31.
Lu, D., and Weng, Q. 2004. Spectral mixture analysis of the urban landscapes in Indianapolis city with Landsat ETM+ imagery. *Photogrammetric Engineering and Remote Sensing* **70,** 1053–1062.
Lu, D., and Weng, Q. 2005. Urban classification using full spectral information of Landsat ETM+ imagery in Marion County, Indiana. *Photogrammetric Engineering and Remote Sensing* **71,** 1275–1284.
Lu, D., and Weng, Q. 2006. Use of impervious surface in urban land use classification. *Remote Sensing of Environment* **102,** 146–160.
Lu, D., and Weng, Q., 2007. A survey of image classification methods and techniques for improving classification performance. *International Journal of Remote Sensing* **28,** 823–870.
Markham, B. L., and Barker, J. L. 1987. Thematic Mapper bandpass solar exoatmospheric irradiances. *International Journal of Remote Sensing* **8,** 517–523.
Maselli, F., Conese, C., Filippis, T. D., and Romani, M. 1995. Integration of ancillary data into a maximum likelihood classifier with nonparametric priors, *ISPRS Journal of Photogrammetry and Remote Sensing* **50,** 2–11.
Mesev, V. 1998. The use of census data in urban image classification. *Photogrammetric Engineering and Remote Sensing* **64,** 431–438.
Myint, S. W. 2001. A robust texture analysis and classification approach for urban land use and land cover feature discrimination. *Geocarto International* **16,** 27–38.
Pedroni, L. 2003. Improved classification of Landsat TM data using modified prior probabilities in large and complex landscapes. *International Journal of Remote Sensing* **24,** 91–113.
Phinn, S., Stanford, M., Scarth, P., et al. 2002. Monitoring the composition of urban environments based on the vegetation-impervious surface-soil (VIS) model by subpixel analysis techniques. *International Journal of Remote Sensing* **23,** 4131–4153.
Rashed, T., Weeks, J. R., Gadalla, M. S., and Hill, A. G. 2001. Revealing the anatomy of cities through spectral mixture analysis of multispectral satellite imagery: A case study of the Greater Cairo region, Egypt. *Geocarto International* **16,** 5–15.
Ricchetti, E. 2000. Multispectral satellite image and ancillary data integration for geological classification. *Photogrammetric Engineering and Remote Sensing* **66,** 429–435.
Shaban, M. A., and Dikshit, O. 2001. Improvement of classification in urban areas by the use of textural features: The case study of Lucknow City, Uttar Pradesh. *International Journal of Remote Sensing* **22,** 565–593.

Smits, P. C., Dellepiane, S. G., and Schowengerdt, R. A. 1999. Quality assessment of image classification algorithms for land cover mapping: A review and a proposal for a cost-based approach. *International Journal of Remote Sensing* **20,** 1461–1486.

Stefanov, W. L., Ramsey, M. S., and Christensen, P. R. 2001. Monitoring urban land cover change: An expert system approach to land cover classification of semi-arid to arid urban centers. *Remote Sensing of Environment* **77,** 173–185.

Strahler, A. H., Logan, T. L., and Bryant, N. A. 1978. Improving forest cover classification accuracy from Landsat by incorporating topographic information. In *Proceedings of the Twelfth International Symposium on Remote Sensing of Environment,* pp. 927–942. Environmental Research Institute of Michigan.

Stuckens, J., Coppin, P. R., and Bauer, M. E. 2000. Integrating contextual information with per-pixel classification for improved land cover classification. *Remote Sensing of Environment* **71,** 282–296.

Treitz, P. M., Howard, P. J., and Gong, P. 1992. Application of satellite and GIS technologies for land cover and land use mapping at the rural-urban fringe: A case study. *Photogrammetric Engineering and Remote Sensing* **58,** 439–448.

Weng, Q. 2002. Land use change analysis in the Zhujiang Delta of China using satellite remote sensing, GIS, and stochastic modeling. *Journal of Environmental Management* **64,** 273–284.

Zhang, J., and Foody, G. M. 2001. Fully-fuzzy supervised classification of sub-urban land cover from remotely sensed imagery: Statistical neural network approaches. *International Journal of Remote Sensing* **22,** 615–628.

CHAPTER 4

Urban Landscape Characterization and Analysis

Urban landscapes are typically a complex combination of buildings, roads, parking lots, sidewalks, gardens, cemeteries, soil, water, and so on. Each of the urban component surfaces possesses unique biophysical properties and relates to their surrounding environment to create the spatial complexity of urban ecologic systems and landscape patterns. To understand the dynamics of patterns and processes and their interactions in heterogeneous landscapes such as urban areas, one must be able to quantify accurately the spatial pattern of the landscape and its temporal changes (Wu et al., 2000). In order to do so, it is necessary (1) to have a standardized method to define those component surfaces and (2) to detect and map them in repetitive and consistent ways so that a global model of urban morphology may be developed and changes to that morphology may be monitored and modeled over time (Ridd, 1995).

Remote sensing technology has been applied widely in urban land-use, land-cover (LULC) classification and change detection. However, it is rare that a classification accuracy of greater than 80 percent can be achieved using per-pixel classification (so-called hard classification) algorithms (Mather, 1999:10). Therefore, the soft/fuzzy approach to LULC classification has been applied, in which each pixel is assigned a class membership of each LULC type rather than a single label (Wang, 1990). Nevertheless, as Mather (1999) suggested, either hard or soft classification is not an appropriate tool for the analysis of heterogeneous landscapes. Mather (1999) maintained that identification/description/quantification rather than classification should be applied to provide a better understanding of the composition and processes of heterogeneous landscapes such as urban areas.

Ridd (1995) proposed a major conceptual model for remote sensing analysis of urban landscapes, namely, the vegetation–impervious surface–soil (V-I-S) model. It assumes that land cover in urban environments is a linear combination of three components: vegetation, impervious surface, and soil. Ridd believed that this model can be applied to spatial-temporal analyses of urban morphology and biophysical and human systems. Having realized that the V-I-S model may be used as a method to define standardized urban landscape components, in this chapter we employ linear spectral mixture analysis (LSMA) as a remote sensing technique to estimate and map V-I-S components to analyze urban morphology and dynamics. The case study was conducted in Indianapolis, Indiana, from 1991 to 2000 using multitemporal satellite images, that is, Landsat Thematic Mapper/Enhanced Thematic Mapper Plus (TM/ETM+) imagery from 1991, 1995, and 2000. The specific objectives of this chapter are (1) to apply LSMA to derive V-I-S components so as to characterize the urban morphology of the study area at three observation times, (2) to analyze spatial-temporal changes of the urban morphology by assessing changes in the V-I-S components, and (3) to examine intra-urban variations of landscape structures via comparative analysis of the V-I-S components and dynamics among the nine townships in the city.

4.1 Urban Landscape Analysis with Remote Sensing

4.1.1 Urban Materials, Land Cover, and Land Use

Land cover can be defined as the biophysical state of the earth's surface and immediate subsurface, including biota, soil, topography, surface water and groundwater, and human structures (Turner et al., 1995). In other words, it describes both natural and human-made coverings of the earth's surface. *Land use* can be defined as the human use of the land. Land use involves both the manner in which the biophysical attributes of the land are manipulated and the purpose for which the land is used (Turner et al., 1995). The relationship between land use and land cover is not always direct and obvious (Weng, 1999). A single class of land cover may support multiple uses, whereas a single land use may involve the maintenance of several distinct land covers.

Urban areas are composed of a variety of materials, including different types of artificial materials (e.g., concrete, asphalt, metal, plastic, glasses etc.), soils, rocks, minerals, and green and nonphotosynthetic vegetation. Remote sensing technology often has been applied to map land use or land cover instead of materials. Each type of land cover may possess unique surface properties (material), but mapping land covers and materials has different requirements. Land-cover mapping needs to consider characteristics in addition to those that come from the material (Herold et al., 2006). The surface structure (roughness) may influence the spectral response as much as the intraclass variability (Gong and Howarth, 1990; Herold et al., 2006; Myint, 2001; Shaban and Dikshit, 2001). Two different land covers, for example, asphalt roads and composite shingle/tar roofs, may have very similar materials (hydrocarbons) and thus may be difficult to discern, although from a materials perspective, these surfaces can be mapped accurately with hyperspectral remote sensing techniques (Herold et al., 2006). Therefore, land-cover mapping requires taking into account of intraclass variability and spectral separability. On the other hand, analysis of land-use classes would be nearly impossible with spectral information alone. Additional information, such as spatial, textural, and contextual information, usually is required to have a successful land-use classification in urban areas (Gong and Howarth, 1992; Herold et al., 2003; Stuckens et al., 2000). The relationship among remote sensing of urban materials, land cover, and land use is illustrated in Fig. 4.1.

Traditional classification methods of LULC based on detailed field work suffered two major common drawbacks: confusion between land use and land cover and a lack of uniformity or comparability in classification schemes, leaving behind a sheer difficulty for comparing land-use patterns over time or between areas (Mather, 1986). The use of aerial photographs and satellite images after the late 1960s did not solve this problem because these techniques are based on the

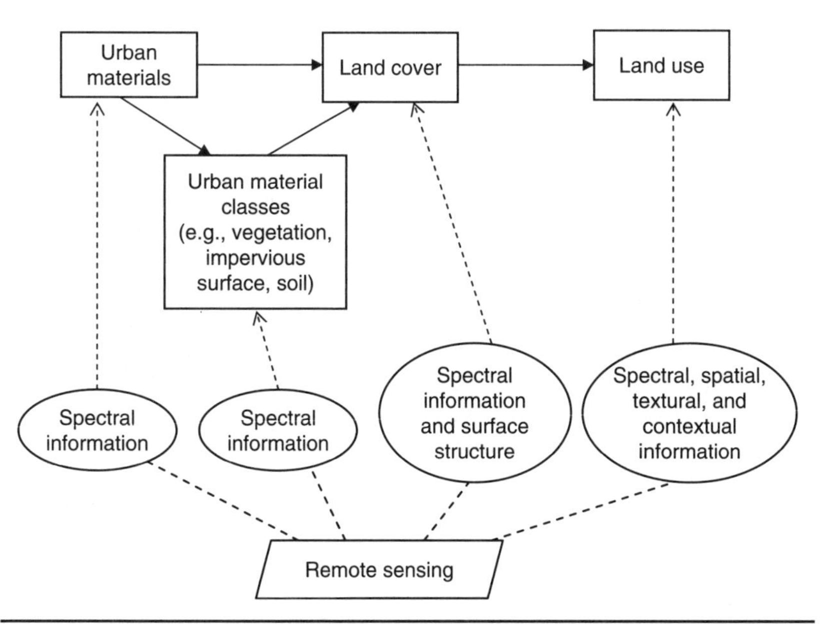

FIGURE 4.1 Relationship among remote sensing of urban materials, land cover, and land use, a conceptual framework. (*Adapted from Weng and Lu, 2009.*)

formal expression of land use rather than on the actual activity itself (Mather, 1986). In fact, many land-use types cannot be identified from the air. As a result, mapping of the earth's surface tends to present a mixture of land use and land cover data with an emphasis on the latter (Lo, 1986). This problem is reflected in the title of the classification developed in United States for the mapping of the country at a scale of 1:100,000 or 1:250,000, commencing in 1974 (Anderson et al., 1976). Moreover, the USGS Land Use/Land Cover Classification System has been designed as a resource-oriented one. Therefore, eight out of nine in the first level categories relate to nonurban areas. The 2001 National Land-Cover Database developed by the USGS reflects both problems (Homer et al., 2004). Alternative to the USGS scheme, Land-Based Classification Standard developed by the American Planning Association emphasizes on extracting urban/suburban land use information. The parcel-level land use information is obtained from in situ survey, aerial photography, and high resolution satellite imagery, based on the characteristics of activity, function, site development, structure, and ownership (American Planning Association, 2004). Generally speaking, success in most land-use or land-cover mapping typically is measured by the ability to match remote sensing spectral signatures with the Anderson classification scheme (Anderson et al., 1976), which, in urban areas, involves mainly land use (Ridd, 1995). The confusion between land use and land cover contributes to the low classification accuracy (Foody, 2002), whereas less emphasis on land cover in urban areas weakens the ability of digital remote sensing as a research tool for characterizing and quantifying the urban ecologic structure and process (Ridd, 1995).

4.1.2 The Scale Issue

The Anderson classification scheme was designed for use with remote sensing data, originally with aerial photography, at various scales and resolutions (Campbell, 1983). However, the spatial scale and categorical scale are not explicitly linked in the classification scheme. The former refers to the manner in which image information content is determined by spatial resolution and the way the spatial resolution is handled in the image processing, whereas the latter refers to the level of detail in classification categories (Ju et al., 2005). This disconnection leads to the problem of simply lumping classes into more general classes in multiscale LULC classifications, which may cause a great lost of categorical information. Since most classifications are conducted on a single spatial and categorical scale, there remains an important issue of matching an appropriate categorical scale of the Anderson scheme with the spatial resolution of the satellite image used (Jensen and Cowen, 1999; Welch, 1982). However, the nature of some applications requires LULC classification to be conducted on multiple spatial and/or categorical scales because a single scale cannot delineate all classes owing to contrasting sizes, shapes, and internal variations of different landscape patches (Raptis et al., 2003; Wu and David, 2002). This is especially true for complex, heterogeneous landscapes, such as urban areas. Moreover, when statistical clusters are grouped into LULC classes, in which smaller areas (e.g., pixels) are combined into larger ones (e.g., patches), both spatial resolution and statistical information are lost (Clapham, 2003).

4.1.3 The Image "Scene Models"

Strahler and colleagues (1986) defined H- and L-resolution scene models based on the relationship between the size of the scene elements and the resolution cell of the sensor. The scene elements in the L-resolution model are smaller than the resolution cells and thus are not detectable. When the objects in the scene become increasingly smaller than the resolution-cell size, they may no longer be regarded as objects individually. Hence the reflectance measured by the sensor may be treated as a sum of interactions among various types of scene elements as weighted by their relative proportions (Strahler et al., 1986). This is what happens with medium-resolution satellite imagery, such as those of Landsat TM or ETM+, applied for urban mapping. As the spatial resolution interacts with the fabric of urban landscapes, the problem of mixed pixels is created. Such a mixture becomes especially prevalent in residential areas, where buildings, trees, lawns, concrete, and asphalt all can occur within a pixel (Epstein et al., 2002). Mixed pixels have been recognized as a major problem in the effective use of remotely sensed data in LULC classification and change detection (Cracknell, 1998; Fisher, 1997). The low accuracy of LULC classification in urban areas reflects, to certain degree, the inability of traditional per-pixel classifiers such as the maximum likelihood classifier to handle composite signatures.

4.1.4 The Continuum Model of Urban Landscape

The prevalence of mixed pixels in urban areas implies that the instantaneous field of the view of medium-resolution sensors does not match with the operational scale of the landscapes. Such landscapes are better viewed as a continuum formed by continuously varying proportions of idealized materials, just as soils may be described in terms of the proportions of sand, silt, and clay (Mather, 1999). Agricultural land in the Midwest, residential areas, and semiarid areas are typical examples of continuum-type landscapes.

Ridd's V-I-S model (Fig. 4.2) is an example of continuum model for urban areas. The V-I-S model was developed for Salt Lake City, Utah, a semiarid area, but has been tested in other cities. Ward and colleagues (2000) applied a hierarchical unsupervised classification approach to a Landsat TM image in southeast Queensland, Australia,

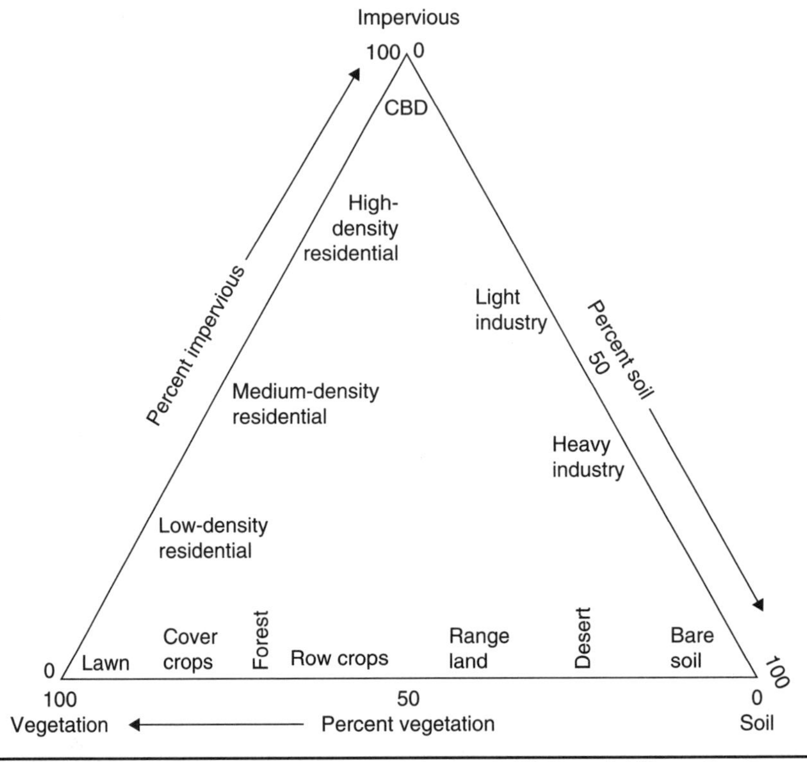

FIGURE 4.2 The vegetation–impervious surface–soil (V-I-S) model illustrating the characteristics of urban landscapes. (*Adapted from Ridd, 1995.*)

based on the V-I-S model. An adjusted overall accuracy of 83 percent was achieved. Madhavan and colleagues (2001) used an unsupervised classifier to classify Landsat TM images in Bangkok, Thailand, with the V-I-S model and found it to be useful for improving classification and analysis of change trends. Similar to the approach of Ward and colleagues, Setiawan and colleagues (2006) applied a V-I-S model–based hieratical procedure to classify a Landsat TM image of Yogyakarta, Indonesia, and found that its accuracy was 27 percent better than the maximum likelihood algorithm. All these studies employed the V-I-S model as the conceptual framework to relate urban morphology to medium-resolution satellite imagery, but hard-classification algorithms were applied. Therefore, the problem of mixed pixels cannot be solved, and the analysis of urban landscapes still was based on pixels or pixel groups.

The V-I-S model has demonstrated usefulness for characterizing and quantifying urban landscape patterns, but its use in practice is still constrained owing to the following factors: First, the V-I-S model cannot explain all land-cover types such as water and wetlands. Second, impervious surface is difficult to identify as a single surface material or material class (Lu and Weng, 2006a, 2006b; Wu and Murray, 2003). Finally, the distinction between soils and impervious surfaces may not be easy in medium-resolution remote sensing imagery. In addition, Ridd's statistical sampling method for deriving the vegetation, impervious surface, and soil components has a high subjectivity and may not be representative of the whole urban landscape. The data used was a color infrared photograph at the scale of 1:30,000. In order to apply this conceptual model to digital remote sensing data, it is necessary to use digital image processing algorithms. Previous studies have attempted to relate the V-I-S model to various remote sensing data, but its linkage with advanced computational algorithms has been less than successful. This chapter applies the V-I-S model concept to a spatial-temporal analysis of the urban morphology in Indianapolis, Indiana, with LSMA, and in so doing, the potentials and limitations of this model for characterizing and quantifying urban landscape components can further be examined.

4.1.5 Linear Spectral Mixture Analysis (LSMA)

LSMA is another method that has been employed to handle the mixed-pixel problem besides the fuzzy classification. Instead of using statistical methods, LSMA is based on physically deterministic modeling to unmix the signal measured at a given pixel into its component parts, called *endmembers* (Adams et al., 1986; Boardman, 1993; Boardman et al., 1995). Endmembers are recognizable surface materials that have homogeneous spectral properties all over the image. LSMA assumes that the spectrum measured by a sensor is a linear combination of the spectra of all components within the pixel (Boardman, 1993). Because of its effectiveness in handling the spectral mixture problem and its ability to provide continuum-based biophysical variables, LSMA has been used widely in (1) estimation of vegetation cover (Asner and Lobell, 2000; Lee and Lathrop, 2005; McGwire et al., 2000; Small, 2001; Weng et al., 2004), (2) impervious surface estimation and/or urban morphology analysis (Lu and Weng, 2006a, 2006b; Phinn et al., 2002; Rashed et al., 2003; Wu and Murray 2003; Wu et al., 2005), (3) vegetation or land-cover classification (Adams et al., 1995; Aguiar et al., 1999; Cochrane and Souza, 1998; Lu and Weng, 2004), and (4) change detection (Powell et al., 2007; Rashed et al., 2005). However, with a few exceptions, these studies have focused on technical specifics and examination of the effectiveness of LSMA. Only a few studies have explicitly adopted the V-I-S model as the conceptual model to explain urban land-cover patterns (Lu and Weng, 2006a, 2006b; Phinn et al., 2002; Powell et al., 2007; Wu and Murray, 2003; Wu et al., 2005), whereas others do so implicitly (Rashed et al., 2003, 2005). Rashed and colleagues (2005) and Powell and colleagues (2007) are the only research attempts to examine urban land-cover "change" with the V-I-S model. Rashed and colleagues (2005) assessed changes between landscape components aggregated to census tracts in Cairo, Egypt, but determination of the thresholds of change may be problematic because they may vary from image to image. Powell and colleagues (2007) identified the stages of urban development by selecting four neighborhoods from an image of Manaus, Brazil, that did not involve change detection from multitemporal satellite images.

The mathematical model of LSMA can be expressed as

$$R_i = \sum_{k=1}^{n} f_k R_{ik} + E_i \tag{4-1}$$

where i = number of spectral bands
k = number of endmembers
R_i = the spectral reflectance of band i of a pixel that contains one or more endmembers
f_k = the proportion of endmember k within the pixel
R_{ik} = the known spectral reflectance of endmember k within the pixel in band i
E_i = the error for band i

To solve f_k, the following conditions must be satisfied: (1) Selected endmembers should be independent of each other, (2) the number of endmembers should not be larger than the spectral bands used, and (3) selected spectral bands should not be highly correlated. A constrained least-squares solution assumes that the following two conditions are satisfied simultaneously:

$$\sum_{k=1}^{n} f_k = 1 \quad \text{and} \quad 0 \le f_k \le 1 \tag{4-2}$$

$$\text{Root-mean-square error (RMSE)} = \sqrt{\left(\sum_{i=1}^{m} ER_i^2\right) \Big/ m} \tag{4-3}$$

Estimation of endmember fraction images involves four steps: image processing, endmember selection, unmixing solution, and evaluation of fraction images. Of these steps, selecting suitable endmembers is the most critical one in the development of high-quality fraction images. Two types of endmembers may be applied: image endmembers and reference endmembers. The former are derived directly from the image itself, whereas the latter are derived from field measurements or the laboratory spectra of known materials (Roberts et al., 1998). Many remote sensing applications have employed image endmembers because they can be obtained easily and are capable of representing the spectra measured on the same scale as the image data themselves (Roberts et al., 1998). Image endmembers may be derived from the extremes of the image feature space based on the assumption that they represent the purest pixels in the image (Boardman, 1993).

4.2 Case Study: Urban Landscape Patterns and Dynamics in Indianapolis

4.2.1 Image Preprocessing

A description of the study area can be found in Sec. 3.2.1 of Chapter 3. Landsat TM images of June 6, 1991 (acquisition time approximately 10:45 a.m.) and July 3, 1995 (approximately 10:28 a.m.) and a Landsat ETM+ image of June 22, 2000 (approximately 11:14 a.m.) were used in this study. Although the images purchased were corrected geometrically, the geometric accuracy was determined not to be high enough to combining them with other high-resolution datasets. The images therefore were further rectified to a common Universal Transverse Mercator (UTM) coordinate system based on 1:24,000 scale topographic maps and were resampled to a pixel size of 30 m for all bands using the nearest-neighbor algorithm. A root-mean-square error (RMSE) of less than 0.2 pixel was obtained for all the rectifications. These Landsat images were acquired under clear-sky conditions, and an improved image-based dark-object subtraction model was applied to implement atmospheric corrections (Chavez, 1996; Lu et al., 2002).

4.2.2 Image Endmember Development

In order to identify image endmembers effectively and to achieve high-quality endmembers, different image-transform approaches, such as principal components analysis (PCA) and minimum noise fraction (MNF), may be applied to transform the multispectral images into a new dataset (Boardman and Kruse, 1994; Green et al., 1988). In this research, image endmembers were selected from the feature spaces formed by the MNF components (Cochrane and Souza 1998; Garcia-Haro et al., 1996; Small, 2001, 2002, 2004; van der Meer and de Jong, 2000). The MNF transform contains two steps: (1) decorrelation and rescaling of the noise in the data based on an estimated noise covariance matrix, producing transformed data in which the noise has unit variance and no band-to-band correlations, and (2) implementation of a standard PCA of the noise-whitened data. The result of MNF transformation is a two-part dataset, one part associated with large eigenvalues and coherent eigenimages and a complementary part with near-unity eigenvalues and noise-dominated images (ENVI, 2000). By performing an MNF transform, noise can be separated from the data by saving only the coherent portions, thus improving spectral processing results. In this research, the MNF procedure was applied to transform the Landsat ETM+ (the 2000 image) six reflective bands into a new coordinate set. The first three MNF components accounted for the majority of the information (99 percent) and were used for selection of endmembers. The scatterplots between the MNF components are shown in Fig. 4.3*a*, revealing the potential endmembers. Four endmembers, namely, green vegetation, high albedo, low albedo, and soil, were finally selected. Figure 4.3*b* shows spectral reflectance characteristics of the selected endmembers. Next, a constrained least-squares solution was applied to unmix the six Landsat ETM+ reflective bands into four fraction images. The same procedures were employed for derivation of fraction images from the Landsat TM 1991 and 1995 images. The first three MNF components computed from the 1991 and 1995 images also accounted for more than 99 percent of the scene

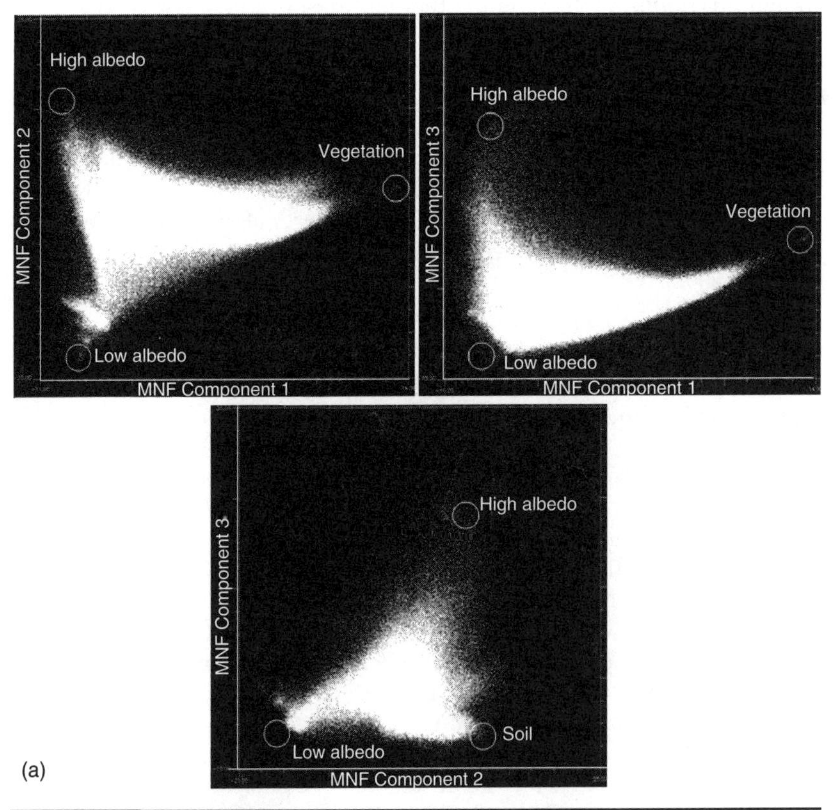

FIGURE 4.3 (*a*) Feature spaces between the MNF components illustrating potential endmembers of Landsat ETM+ image. (*b*) Spectral reflectance characteristics of the selected endmembers. (*Adapted from Weng and Lu, 2009.*)

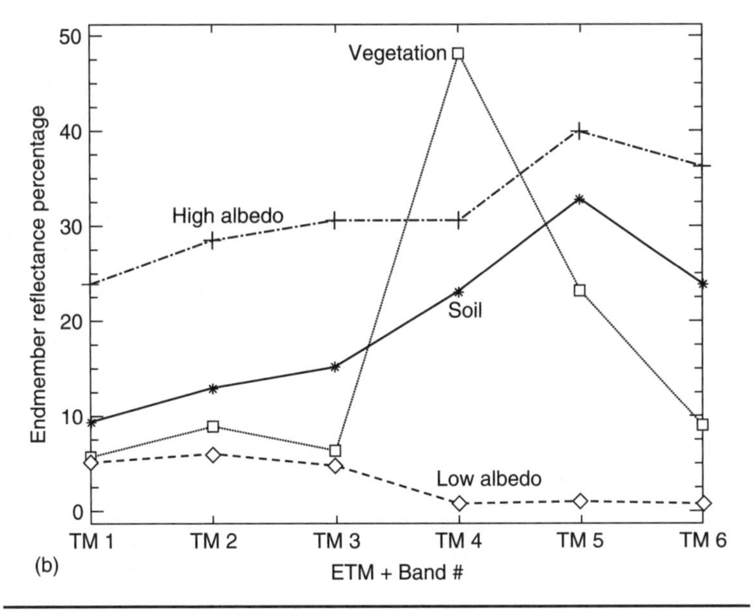

FIGURE 4.3 (*Continued*)

variance, and the topologies of the triangular mixing space were consistent with that shown in Fig. 4.3*a*. Figure 4.4 shows four fraction images for the three years.

4.2.3 Extraction of Impervious Surfaces

Previous research indicated that impervious surface can be computed by adding the high- and low-albedo fractions (Wu and Murray, 2003), but this method did not consider the impact of pervious surfaces on the low- and high-albedo fraction images, which often resulted in overestimation of impervious surface. Our experiment with Landsat ETM+ imagery indicates that although the high-albedo fraction image related mainly to impervious surface information such as buildings and roads, it also related to other covers such as dry soils. On the other hand, the low-albedo fraction image was found to associate with water and shadows, such as water bodies, shadows from forest canopy and tall buildings, and moistures in crops or pastures. However, some impervious surfaces, especially dark impervious surfaces, also were linked to the low-albedo fraction image. Therefore, it is important to develop a suitable analytical procedure for removal of nonimpervious information from the fraction images. In this study, we developed a procedure using land surface temperature data to isolate nonimpervious from impervious surfaces and using soil fraction images as the thresholds to purify the high-albedo fraction images.

For the high-albedo fraction images, impervious surface was predominantly confused with dry soils. Therefore, the soil fraction

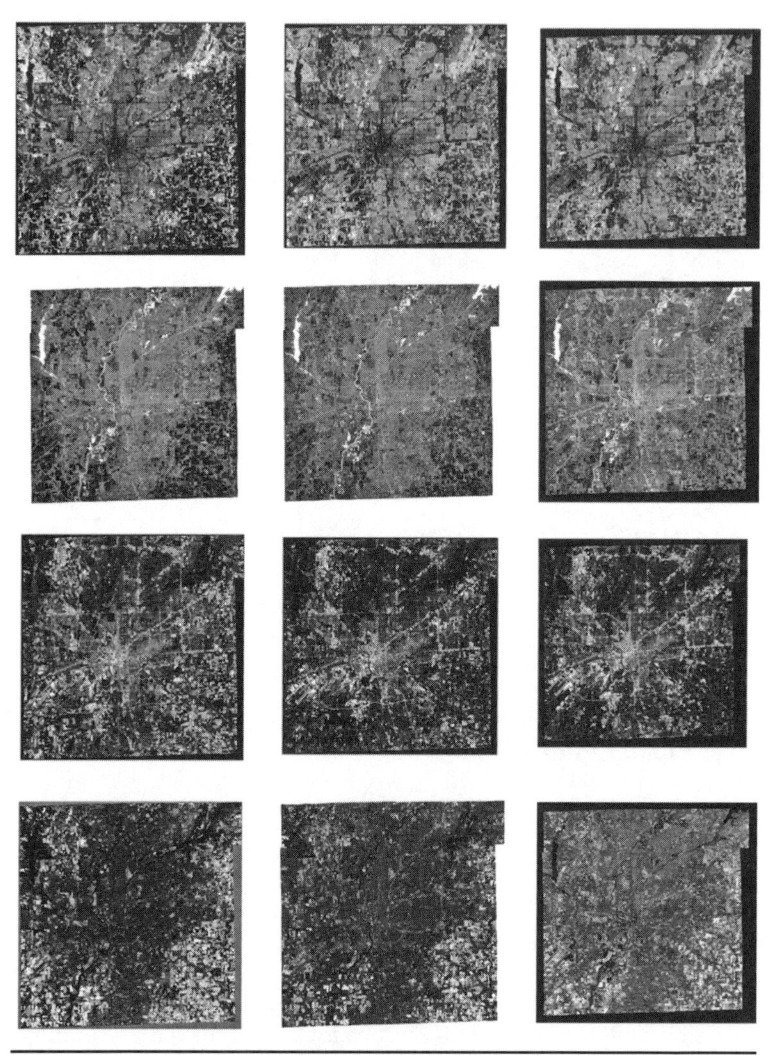

FIGURE 4.4 Fraction images from spectral mixture analysis of each year (first row: green vegetation; second row: low albedo; third row: high albedo; and fourth row: soil). (*Adapted from Weng and Lu, 2006.*)

images may be used to remove soils from the high-albedo fraction images. For the low-albedo fraction images, dark impervious surface was confused with water and shadows. Therefore, the critical step was to separate impervious surface from pervious pixels, including water, vegetation (e.g., forest, pasture, grass, and crops), and soils. In this study, we developed some expert rules to remove pervious pixels. The impervious surface image then was developed by adding the adjusted low- and high-albedo fraction images. Figure 4.5 provided a comparison of the impervious surface images before and after the adjustment. Our accuracy assessment of the Landsat ETM+ image

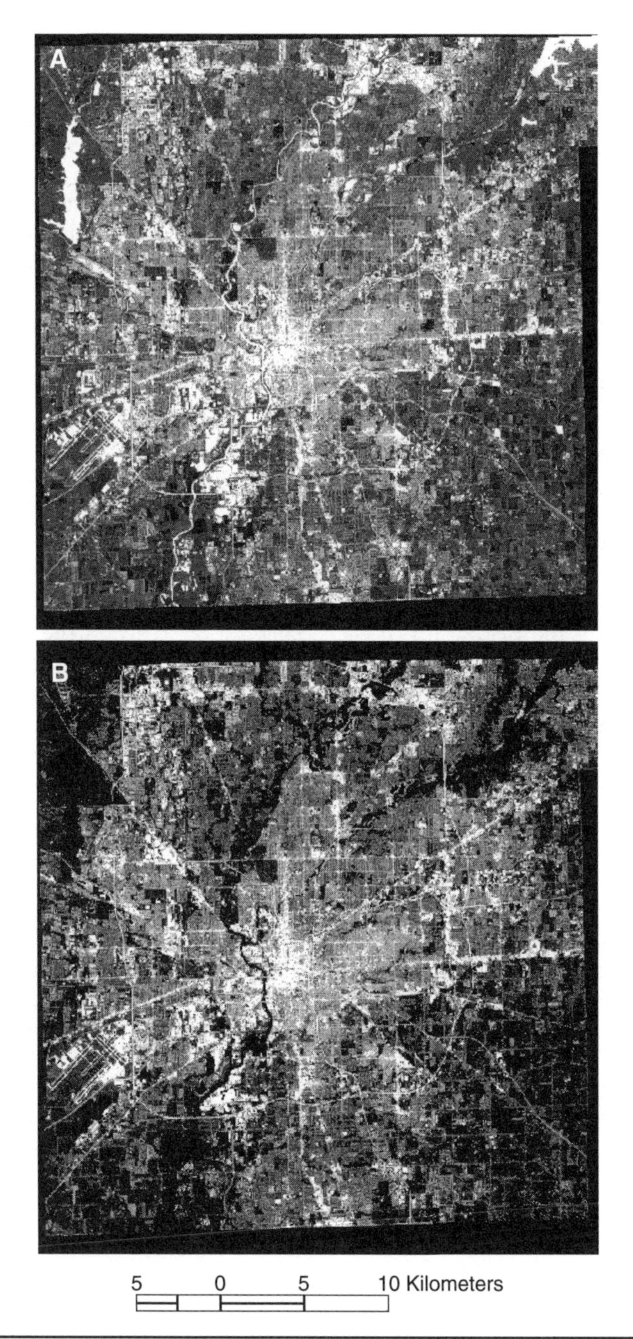

FIGURE 4.5 Comparison of impervious surface images developed from different methods. (*Adapted from Lu and Weng, 2006a.*)

indicated that an overall RMSE of 9.22 percent and a system error of 5.68 percent were obtained (Lu and Weng, 2006a).

4.2.4 Image Classification

Fraction images were used for thematic land classification via a hybrid procedure that combined maximum likelihood and decision-tree classifiers (Lu and Weng, 2004). Sample plots were identified from high-resolution aerial photographs covering initially 10 LULC types: commercial and industrial, high-density residential, low-density residential, bare soil, crop, grass, pasture, forest, wetland, and water. On average, 10 to 16 sample plots for each class were selected. A window size of 3 × 3 was applied to extract the fraction value for each plot. The mean and standard deviation values were calculated for each LULC class. The characteristics of fractional composition for selected LULC types then were examined. Next, the maximum likelihood classification algorithm was applied to classify the fraction images into 10 classes, generating a classified image and a distance image. A distance threshold was selected for each class to screen out the pixels that probably did not belong to that class, which was determined by examining the histogram of each class interactively in the distance image. Pixels with a distance value greater than the threshold were assigned a class value of zero. A decision-tree classifier then was applied to reclassify these pixels. The parameters required by the decision-tree classifier were identified based on the mean and standard deviation values of the sample plots for each class. Finally, the accuracy of the classified image was checked with a stratified random-sampling method (Jensen, 2005) against the reference data of 150 samples collected from large-scale aerial photographs. To simplify urban landscape analysis, 10 classes were merged into 6 LULC types, including (1) commercial and industrial urban land, (2) residential land, (3) agricultural and pasture land, (4) grassland, (5) forest, and (6) water (Lu and Weng, 2004). Figure 4.6 shows the classified LULC maps in the three years.

The overall accuracy, producer's accuracy, and user's accuracy were calculated based on the error matrix for each classified map, as well as the KHAT statistic, kappa variance, and Z statistic. The overall accuracies of the LULC maps for 1991, 1995, and 2000 were determined to be 90, 88, and 89 percent, respectively. Apparently, LULC data derived from the LSMA procedure have a reasonably high accuracy and are sufficient for urban landscape analysis.

4.2.5 Urban Morphologic Analysis Based on the V-I-S Model

The three images in the first row of Fig. 4.4 show the geographic patterns of green vegetation (GV) fractions. These images display a large dark area (low fraction values) at the center of the study area corresponding to the central business district of the city of Indianapolis. Bright areas of high GV values were found in the surrounding areas.

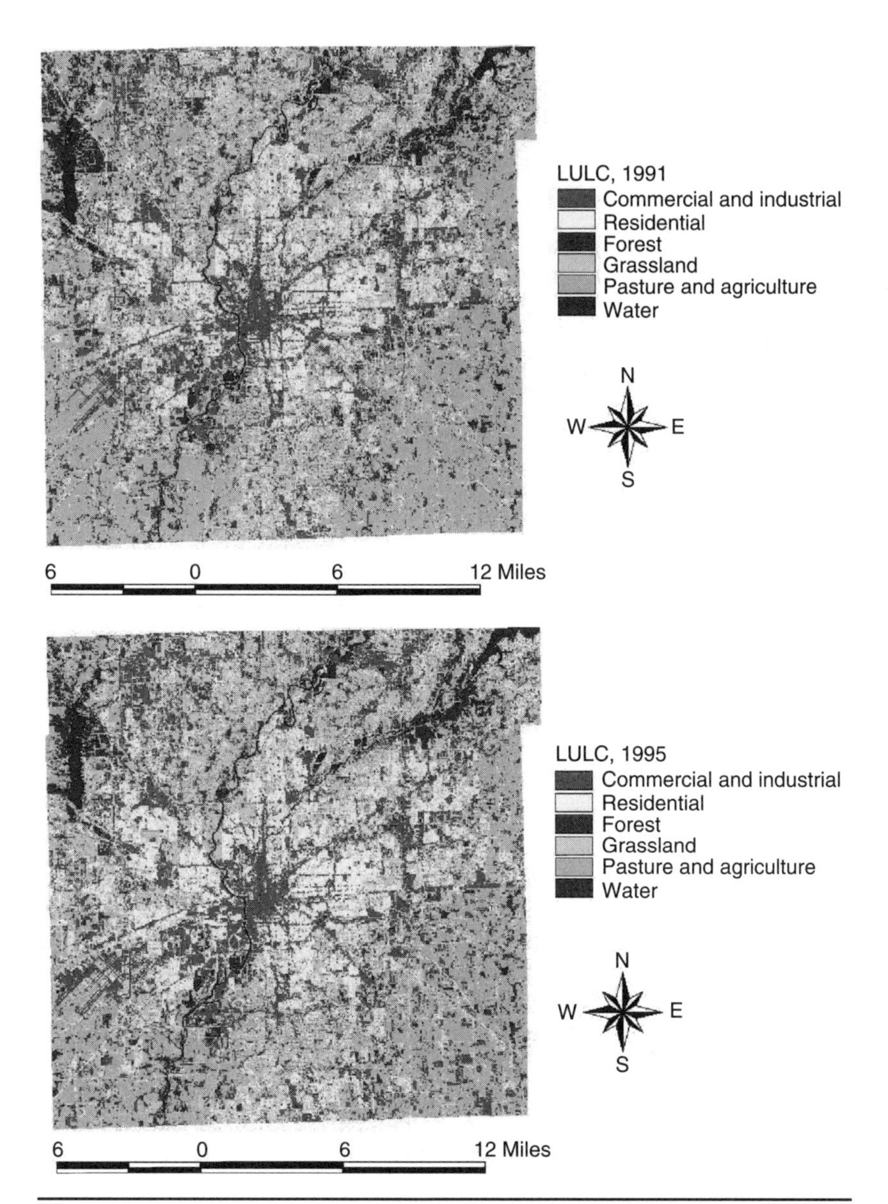

FIGURE 4.6 LULC maps of 1991, 1995, and 2000. (*Adapted from Weng and Lu, 2006.*) See also color insert.

Various types of crops were still at the early stage of growth or had not emerged, as indicated by medium-gray to dark tone of the GV fraction images in the southeastern and southwestern parts of the city. Table 4.1 indicates that forest had the highest GV fraction values, followed by grassland. In contrast, commercial and industrial land displayed the lowest GV values. Very little vegetation was found in

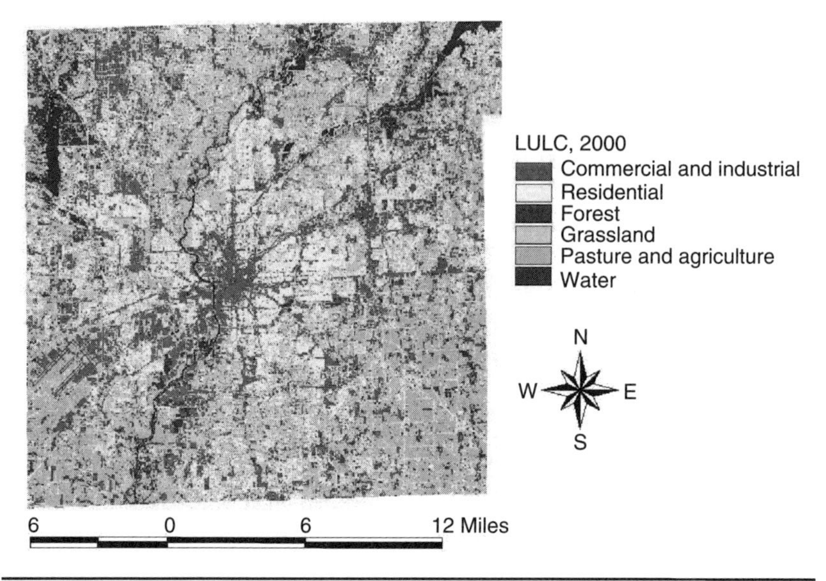

Figure 4.6 (*Continued*)

water bodies, as indicated by the GV fraction values. Both residential land and pasture/agricultural land yielded a mediate GV fraction value, subject to the impact of the dates on which the images were acquired. In all the years observed, pasture/agricultural land exhibited a large standard deviation value, suggesting that pasture and agricultural land may hold various amount of vegetation.

The percentage of land covered by impervious surface may vary significantly with LULC categories and subcategories (Soil Conservation Service, 1975). This study shows a substantially different estimate for each LULC type because this study applied a spectral unmixing model to the remote sensing images, and the modeling had introduced some errors, as expected. For example, a high impervious surface fraction value was found in water because water related to the low-albedo fraction, and the latter was included in the computation of impervious surface. Generally speaking, an LULC type with a higher GV fraction appeared to have a lower impervious surface fraction. Commercial and industrial land detected very high impervious surface fraction values around 0.7 in all years. Residential land came after with fraction values around 0.5. Grassland, agricultural/pasture land, and forest land detected lower values of impervious surface owing largely to their exposed bare soil, confusion with commercial/industrial and residential land, and modeling errors.

Soil fraction values generally were low in most of the urban area but high in the surrounding areas. Especially in agricultural fields located in the southeastern and southwestern parts of the city, soil fraction images appeared very bright because various types of crops

Land-Cover Type	1991 TM Image			1995 TM Image			2000 ETM+ Image		
	Mean Vegetation (SD)	Mean Impervious Surface (SD)	Mean Soil (SD)	Mean Vegetation (SD)	Mean Impervious Surface (SD)	Mean Soil (SD)	Mean Vegetation (SD)	Mean Impervious Surface (SD)	Mean Soil (SD)
Commercial and Industrial	0.167 (0.128)	0.709 (0.190)	0.251 (0.193)	0.127 (0.097)	0.679 (0.178)	0.273 (0.177)	0.125 (0.092)	0.681 (0.205)	0.276 (0.191)
Residential	0.314 (0.132)	0.558 (0.138)	0.198 (0.152)	0.371 (0.115)	0.508 (0.108)	0.149 (0.092)	0.298 (0.095)	0.467 (0.124)	0.247 (0.137)
Grassland	0.433 (0.176)	0.451 (0.135)	0.268 (0.208)	0.553 (0.145)	0.366 (0.096)	0.155 (0.131)	0.442 (0.099)	0.276 (0.083)	0.305 (0.119)
Agriculture and Pasture	0.304 (0.213)	0.374 (0.112)	0.602 (0.285)	0.388 (0.191)	0.291 (0.091)	0.378 (0.236)	0.371 (0.168)	0.275 (0.072)	0.407 (0.222)
Forest	0.654 (0.162)	0.436 (0.128)	0.182 (0.166)	0.716 (0.085)	0.388 (0.065)	0.046 (0.052)	0.584 (0.075)	0.327 (0.074)	0.175 (0.055)
Water	0.226 (0.186)	0.730 (0.197)	0.188 (0.178)	0.176 (0.210)	0.805 (0.167)	0.094 (0.068)	0.111 (0.120)	0.891 (0.136)	0.078 (0.071)

TABLE 4.1 V-I-S Compositions of LULC Types in Indianapolis in 1991, 1995, and 2000

were still at the early stage of growth. Table 4.1 shows that agricultural/pasture land observed a fraction value close to 0.4 at all times. Grassland had medium fraction values averaging 0.25, substantially higher than the fraction values of forest land and residential land. Commercial and industrial land displayed similar fraction values as grassland, which had much to do with its confusion with dry soils in the high-albedo images. Water generally had a minimal impervious surface fraction value. Like the GV fraction, the soil fraction displayed the highest standard deviation values in agricultural/pasture land owing to various amounts of emerging vegetation.

The V-I-S composition may be examined by taking samples along transects. Figure 4.7 shows ternary plots of four transects across the geometric center of the city sampled from the 2000 Landsat ETM+ image. Sample 1 runs from west to east, sample 2 from north to south, sample 3 from southwest to northeast, and sample 4 from southeast to northwest. Errors from the spectral unmixing modeling are not included in these diagrams because their low values clustered near zero. Along the east-west transect, nearly all pixels sampled showed a GV fraction of less than 0.6, whereas the soil fraction values ranged from 0.1 to 0.7. A clustering pattern was apparent if impervious surface fraction values were observed in the range from 0.2 to 0.7 and GV fraction values were observed in the range of 0.5 to 0.8. A more clustered pattern can be observed in the ternary diagrams based on the north-south and the southwest-northeast transects. However, the southeast-northwest transect clearly exhibited a more dispersed pattern of pixel distribution, suggesting a variety of V-I-S composition types. GV along this transect yielded fraction values from 0.3 to 1, whereas the impervious surface fraction might have any value between 0 and 1. Soil fraction values continued to increase up to 0.8. When mean signature values of the fractions for each LULC type are plotted, quantitative relationships among the thematic LULC types in terms of V-I-S composition can be examined. Figure 4.8 shows the V-I-S composition by LULC in each year, with an area delineating one standard deviation from the mean fraction value.

4.2.6 Landscape Change and the V-I-S Dynamics

The fraction images were classified into three thematic maps (as shown in Fig. 4.6). Table 4.2 shows the composition of LULC by year and changes that occurred between the two intervals. In 1991, residential use and pasture/agriculture accounted equally for 27 percent of the total land, whereas grassland shared another 20 percent. The combination of commercial and industrial land used 13 percent of the total area, and forest land had a close match, yielding another 10 percent. Water bodies occupied the remaining 3 percent, and this percentage kept unchanged from 1991 to 2000. However, LULC dynamics occurred in all other categories, as seen in the last three columns of Table 4.2. The most notable increment was observed in residential use, which grew

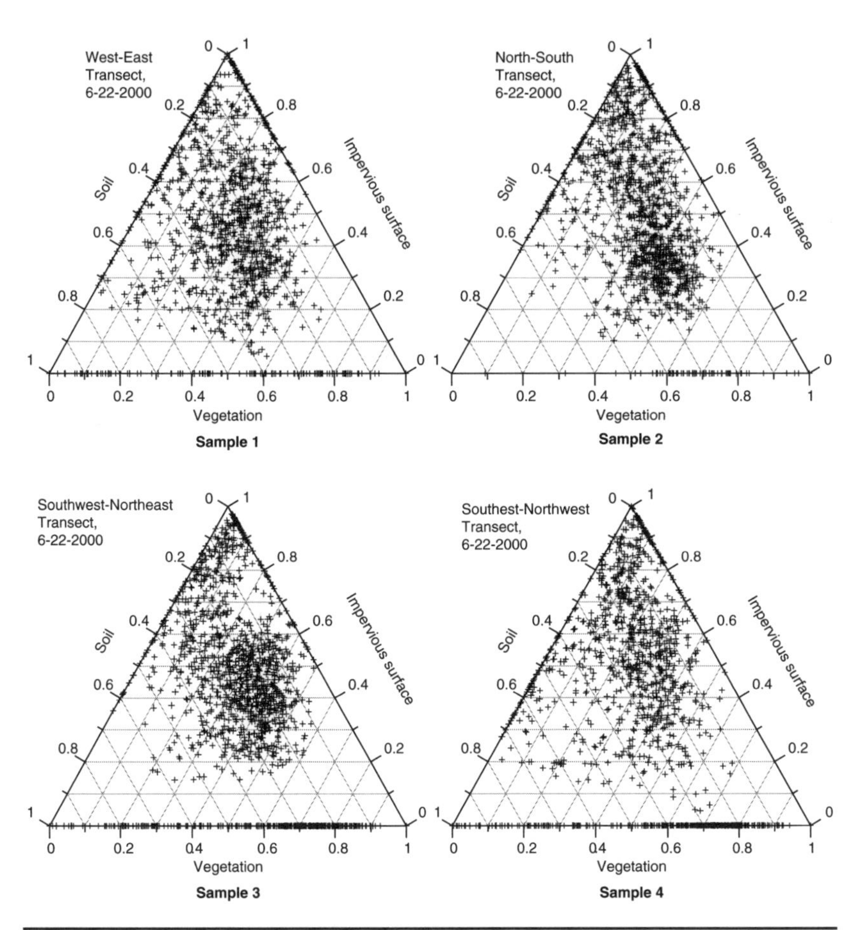

FIGURE 4.7 V-I-S composition along four sampled transects. Sample 1: west-east transect; Sample 2: north-south transect; Sample 3: southwest-northeast transect; and Sample 4: southeast-northwest transect.

from 27 percent in 1991 to 33 percent in 1995, reaching 38 percent in 2000. Associated with this change, grassland increased from 20 to 23 percent. Highly developed land, mainly for commercial, industrial, transportation, and utilities uses, continued to expand. In 2000, it accounted for over 15,000 hectares, or 15 percent, generating a 2 percent increase over the 9 years. These results suggest that urban land dispersal in Indianapolis was related both to population increases and to economic growth. In contrast, a pronounced decrease in pasture and agricultural land was discovered from 1991 (27 percent) to 1995 (20 percent). This decrease also was evident between 1995 and 2000, when pasture and agricultural land shrank further by 6581.30 hectares (31.56 percent). Forest land in a city like Indianapolis was understandably limited in size. Our remote sensing–GIS analysis indicates, however, that forest land continued to disappear at a

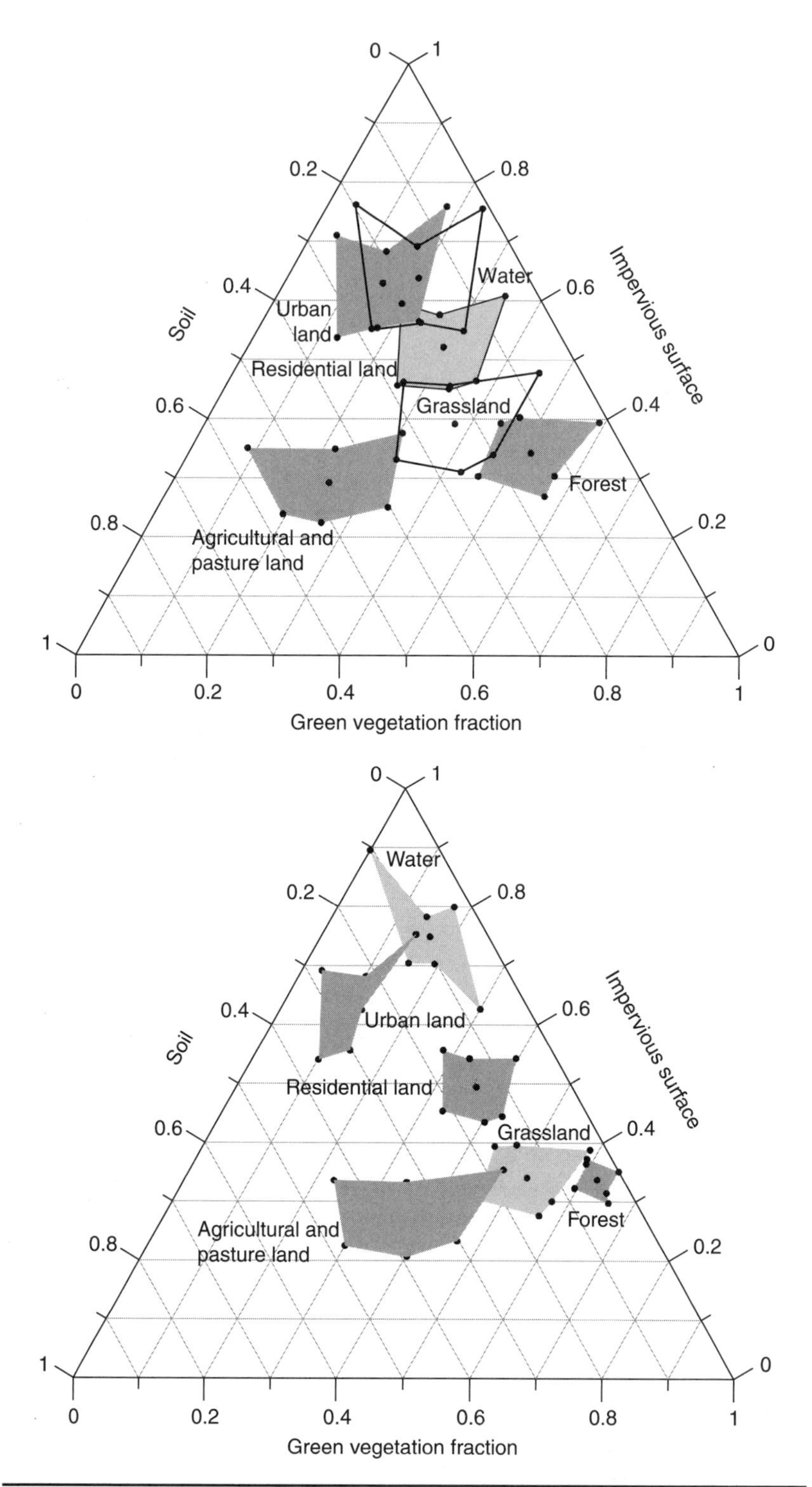

Figure 4.8 Quantitative relationships among the LULC types in respect to the V-I-S model. (*Adapted from Weng and Lu, 2009.*)

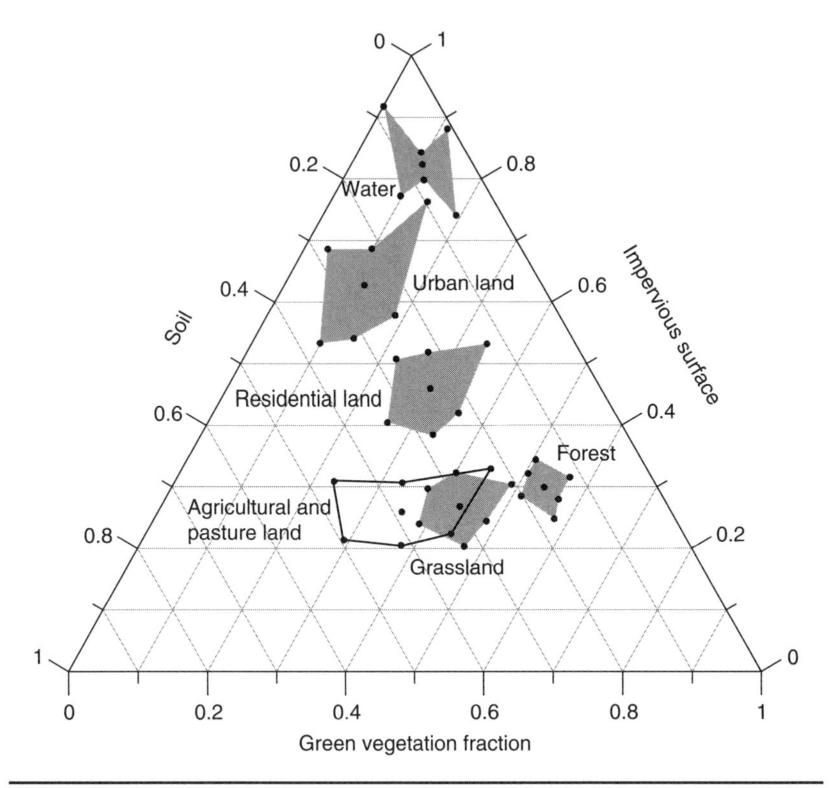

Figure 4.8 (*Continued*)

markedly stable rate. Between 1991 and 2000, forest land was reduced by 2864.81 hectares (i.e., 28.75 percent), leveling down to approximately 7100 hectares. Cross-tabulation of the 1991 and 2000 LULC maps reveals that most of the losses in pasture, agricultural, and forest land were the result of conversion to residential and other urban uses owing to the continued process of urbanization and suburbanization. GIS overlay of the two maps further shows the spatial occurrence of urban expansion to be mostly on the edges of the city. These changes in LULC have led to changes in the composition of image fractions.

Figure 4.9 shows the pattern of changes in the V-I-S components of the LULC classes. Impervious surface, as an important urban land-cover feature, indicates not only the degree of urbanization but also a major contribution to the environmental impacts of urbanization (Arnold and Gibbons, 1996). To examine how impervious surfaces in Indianapolis changed from 1991 to 2000, Fig. 4.10 was created to show the distribution of impervious surfaces for the observed years in four categories. Table 4.3 shows that the number of pixels with values of greater than zero increased from 42,501 in 1991 to 45,804 in 1995 and further increased to 46,560 in 2000. A comparison between 1991 and 2000 indicates that this increase in the number of pixels took place

LULC Type	Area, 1991 (ha)	Area, 1995 (ha)	Area, 2000 (ha)	Change, 1991–1995	Change, 1995–2000	Change, 1991–2000
Commercial and industrial	13,322.10	16,706.50	15,489.00	3,384.40 (25.40%)	–1,217.50 (–7.29%)	2,166.90 (16.27%)
Residential	28,708.90	34,123.70	40,771.70	5,414.80 (18.86%)	6,648 (19.48%)	12,062.80 (42.02%)
Grassland	21,132.50	21,356.60	23,976.40	224.10 (1.06%)	2,619.80 (12.27%)	2,843.90 (13.46%)
Pasture and agriculture	28,466.00	20,853.80	14,272.50	–7,612.20 (–26.74%)	–6,581.30 (–31.56%)	–14,193.50 (–49.86%)
Forest	9,965.58	8,547.71	7,100.77	–1,417.87 (–14.23%)	–1,446.94 (–16.93%)	–2,864.81 (–28.75%)
Water	2,894.21	2,903.96	2,903.06	9.75 (0.34%)	–0.90 (–0.03%)	8.85 (0.31%)

TABLE 4.2 Changes in Land Use and Land Cover in Indianapolis, 1991–2000

over the entire range of impervious surface fraction values (except for the category of 0.1 to 0.2). This analysis further substantiates the preceding findings; that is, Indianapolis underwent an extensive urbanization process during which impervious or impenetrable surfaces, such as rooftops, roads, parking lots, driveways, and sidewalks, were widely generated. In other words, a significant number of nonurban pixels became urbanized during the study period. Furthermore, in 1991, only 7.03 percent of the urbanized pixels (pixels that contain some impervious surface areas) had a value of impervious surface fraction greater than 0.6. In 1995 and 2000, they were 7.41 and 9.24 percent, respectively. This increase in pixel counts in the higher percentage categories of the impervious surface fraction suggests that even more construction had taken place in previously urbanized pixels; that is, there existed an infill type of urban development (Wilson et al., 2003).

4.2.7 Intra-Urban Variations and the V-I-S Compositions

The intra-urban variations of landscape structures can be examined by a comparative analysis of the V-I-S compositions of each township in the city. Tables 4.4, 4.5, and 4.6 show the mean values (with standard

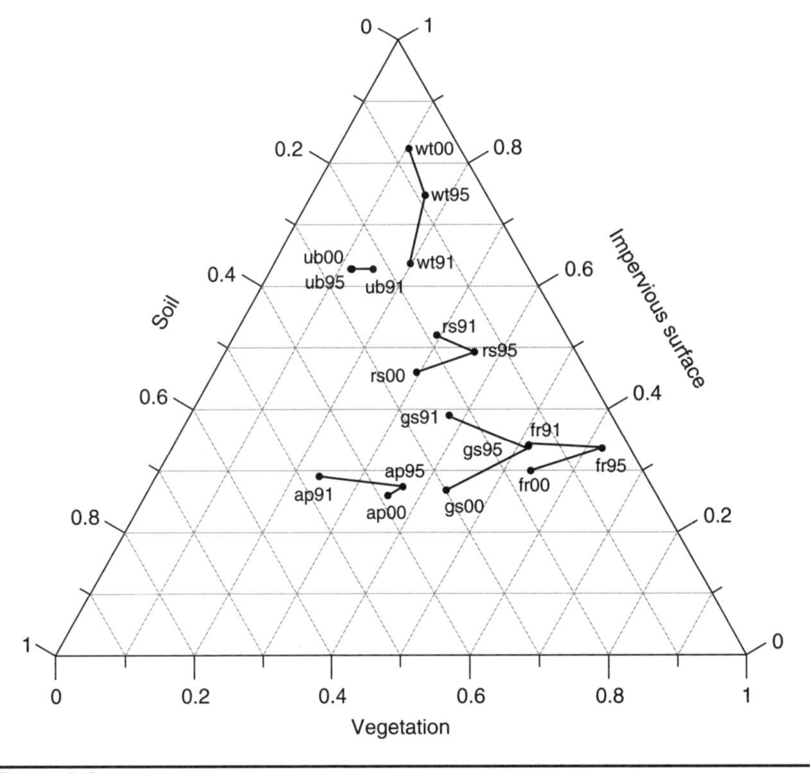

FIGURE 4.9 V-I-S Dynamics.

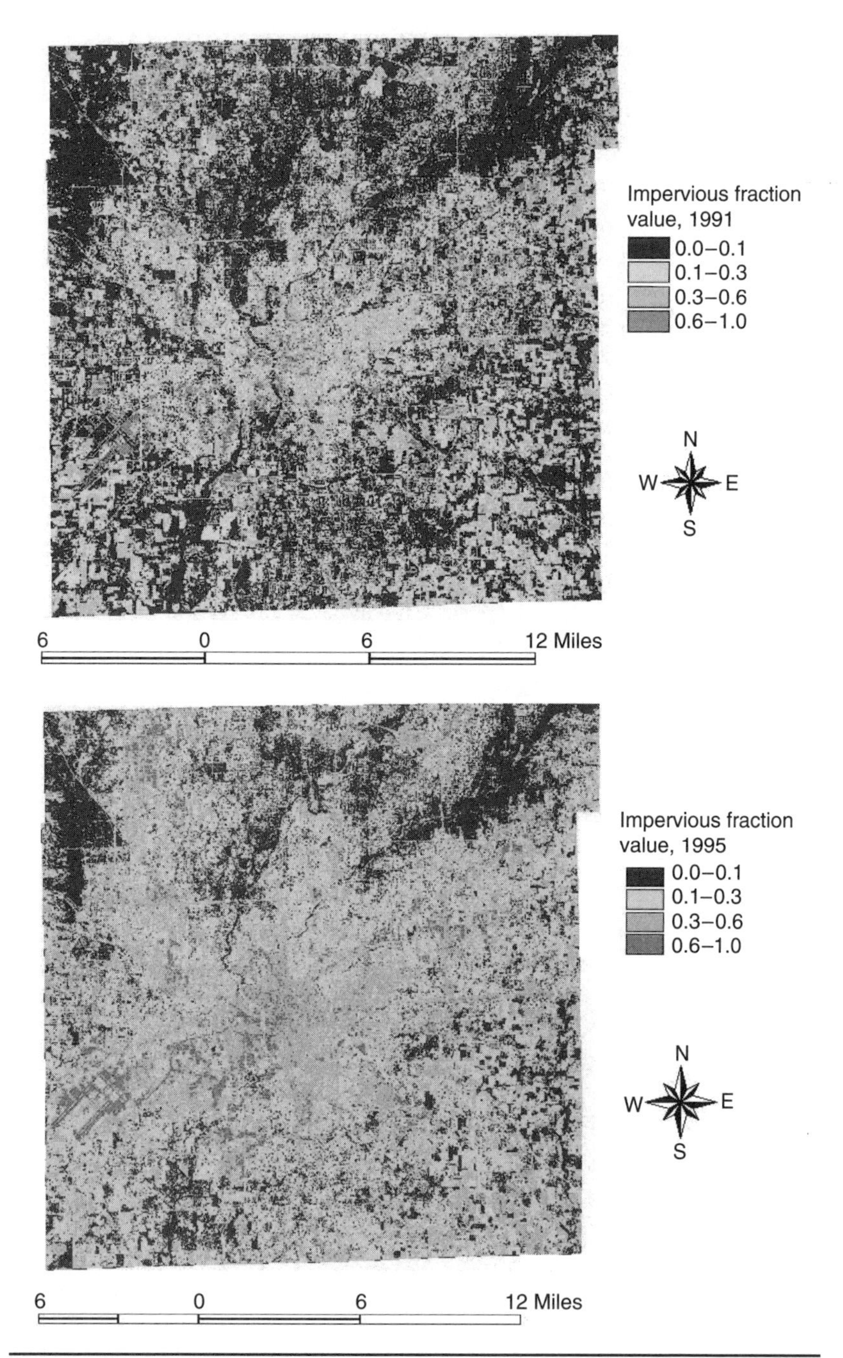

FIGURE 4.10 Distribution of impervious coverage by year. (*Adapted from Weng and Lu, 2009.*) See also color insert.

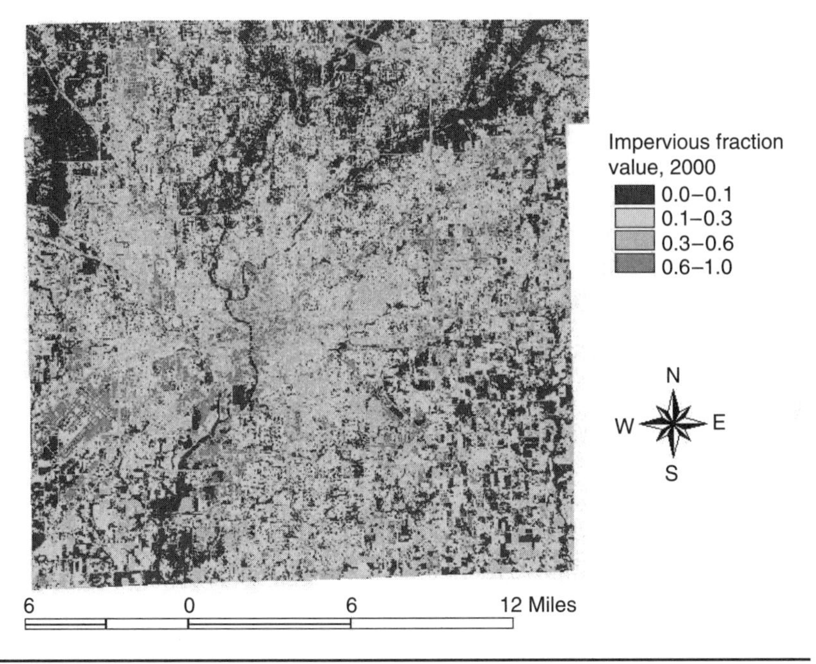

Figure 4.10 (*Continued*)

Impervious Surface Fraction Value	Pixel Counts, 1991	Pixel Counts, 1995	Pixel Counts, 2000
≤0.1	1,355	1,373	1,823
≤0.2	17,880	17,350	17,300
≤0.3	9,630	12,080	12,010
≤0.4	6,904	7,485	7,276
≤0.5	3,743	4,121	3,850
≤0.6	1,459	1,751	1,961
≤0.7	781	810	1,106
≤0.8	406	414	651
≤0.9	215	269	347
≤1.0	128	151	236
Total	42,501	45,804	46,560

Table 4.3 Number of Pixels (Those Containing Some Impervious Surface) in Each Percentage Category of Impervious Surface

	Commercial and Industrial Land			Residential Land			Agricultural and Pasture Land		
Township	GV	IM	SL	GV	IM	SL	GV	IM	SL
Pike	0.177 (0.144)	0.664 (0.200)	0.290 (0.204)	0.312 (0.174)	0.543 (0.162)	0.251 (0.180)	0.360 (0.230)	0.416 (0.148)	0.530 (0.298)
Washington	0.179 (0.127)	0.711 (0.175)	0.213 (0.165)	0.342 (0.131)	0.553 (0.122)	0.170 (0.129)	0.258 (0.181)	0.425 (0.145)	0.532 (0.276)
Lawrence	0.147 (0.100)	0.686 (0.180)	0.263 (0.188)	0.295 (0.129)	0.539 (0.126)	0.221 (0.151)	0.354 (0.220)	0.385 (0.105)	0.542 (0.289)
Wayne	0.193 (0.143)	0.698 (0.201)	0.247 (0.190)	0.327 (0.136)	0.568 (0.146)	0.183 (0.150)	0.296 (0.201)	0.408 (0.158)	0.549 (0.286)
Center	0.156 (0.119)	0.772 (0.157)	0.186 (0.148)	0.302 (0.113)	0.621 (0.123)	0.142 (0.115)	0.290 (0.187)	0.473 (0.138)	0.457 (0.213)
Warren	0.143 (0.115)	0.714 (0.185)	0.259 (0.191)	0.304 (0.124)	0.546 (0.127)	0.203 (0.147)	0.274 (0.206)	0.367 (0.095)	0.593 (0.267)
Decatur	0.213 (0.164)	0.614 (0.236)	0.358 (0.246)	0.342 (0.155)	0.494 (0.146)	0.257 (0.195)	0.315 (0.221)	0.369 (0.117)	0.641 (0.301)
Perry	0.163 (0.117)	0.695 (0.190)	0.267 (0.192)	0.313 (0.125)	0.519 (0.132)	0.216 (0.149)	0.278 (0.194)	0.367 (0.116)	0.639 (0.285)
Franklin	0.153 (0.111)	0.596 (0.204)	0.403 (0.248)	0.272 (0.132)	0.460 (0.131)	0.311 (0.177)	0.273 (0.199)	0.357 (0.087)	0.634 (0.268)

	Forest Land			Grass Land			Water		
Township	**GV**	**IM**	**SL**	**GV**	**IM**	**SL**	**GV**	**IM**	**SL**
Pike	0.667 (0.171)	0.486 (0.157)	0.198 (0.183)	0.198 (0.212)	0.469 (0.154)	0.289 (0.216)	0.189 (0.180)	0.666 (0.201)	0.247 (0.228)
Washington	0.620 (0.160)	0.466 (0.117)	0.165 (0.142)	0.165 (0.155)	0.479 (0.116)	0.208 (0.186)	0.255 (0.188)	0.748 (0.177)	0.141 (0.130)
Lawrence	0.696 (0.130)	0.422 (0.098)	0.150 (0.131)	0.150 (0.181)	0.438 (0.122)	0.301 (0.219)	0.180 (0.159)	0.767 (0.177)	0.146 (0.127)
Wayne	0.604 (0.191)	0.467 (0.152)	0.200 (0.165)	0.200 (0.181)	0.471 (0.163)	0.300 (0.219)	0.277 (0.201)	0.693 (0.210)	0.203 (0.174)
Center	0.611 (0.178)	0.453 (0.130)	0.185 (0.153)	0.185 (0.163)	0.477 (0.141)	0.264 (0.198)	0.260 (0.172)	0.763 (0.186)	0.160 (0.143)
Warren	0.669 (0.138)	0.415 (0.100)	0.153 (0.129)	0.153 (0.175)	0.433 (0.125)	0.271 (0.198)	0.207 (0.169)	0.762 (0.195)	0.168 (0.138)
Decatur	0.619 (0.184)	0.427 (0.141)	0.246 (0.227)	0.246 (0.203)	0.436 (0.153)	0.318 (0.244)	0.266 (0.196)	0.633 (0.251)	0.268 (0.231)
Perry	0.646 (0.158)	0.403 (0.112)	0.169 (0.149)	0.169 (0.163)	0.415 (0.118)	0.269 (0.198)	0.274 (0.204)	0.718 (0.209)	0.211 (0.200)
Franklin	0.671 (0.139)	0.383 (0.087)	0.152 (0.130)	0.152 (0.200)	0.379 (0.112)	0.274 (0.185)	0.217 (0.180)	0.733 (0.210)	0.194 (0.164)

TABLE 4.4 V-I-S Composition by Township, 1991

	Commercial and Industrial Land			Residential Land			Agricultural and Pasture Land		
Township	**GV**	**IM**	**SL**	**GV**	**IM**	**SL**	**GV**	**IM**	**SL**
Pike	0.151 (0.116)	0.638 (0.184)	0.276 (0.165)	0.343 (0.142)	0.490 (0.131)	0.197 (0.105)	0.484 (0.165)	0.301 (0.110)	0.269 (0.198)
Washington	0.140 (0.101)	0.665 (0.143)	0.235 (0.132)	0.377 (0.109)	0.488 (0.093)	0.157 (0.085)	0.437 (0.160)	0.347 (0.122)	0.294 (0.219)
Lawrence	0.125 (0.075)	0.649 (0.151)	0.274 (0.157)	0.361 (0.093)	0.495 (0.088)	0.161 (0.086)	0.418 (0.168)	0.311 (0.094)	0.313 (0.213)
Wayne	0.150 (0.121)	0.703 (0.192)	0.227 (0.153)	0.372 (0.130)	0.513 (0.125)	0.145 (0.096)	0.377 (0.161)	0.322 (0.112)	0.369 (0.187)
Center	0.100 (0.079)	0.749 (0.139)	0.208 (0.119)	0.347 (0.105)	0.551 (0.102)	0.124 (0.079)	0.530 (0.110)	0.383 (0.174)	0.132 (0.111)
Warren	0.110 (0.074)	0.684 (0.166)	0.282 (0.167)	0.379 (0.101)	0.506 (0.091)	0.141 (0.086)	0.366 (0.183)	0.271 (0.077)	0.399 (0.229)
Decatur	0.160 (0.121)	0.644 (0.241)	0.345 (0.208)	0.414 (0.142)	0.488 (0.136)	0.160 (0.112)	0.435 (0.191)	0.300 (0.109)	0.343 (0.229)
Perry	0.119 (0.083)	0.691 (0.176)	0.269 (0.184)	0.397 (0.106)	0.496 (0.097)	0.134 (0.084)	0.383 (0.197)	0.276 (0.081)	0.429 (0.253)
Franklin	0.111 (0.075)	0.524 (0.177)	0.480 (0.213)	0.389 (0.101)	0.477 (0.092)	0.162 (0.087)	0.336 (0.194)	0.282 (0.078)	0.428 (0.243)

	Forest Land			Grass Land			Water		
Township	GV	IM	SL	GV	IM	SL	GV	IM	SL
Pike	0.711 (0.090)	0.392 (0.075)	0.053 (0.054)	0.507 (0.173)	0.399 (0.091)	0.219 (0.168)	0.115 (0.190)	0.761 (0.171)	0.096 (0.053)
Washington	0.692 (0.096)	0.396 (0.057)	0.052 (0.047)	0.523 (0.125)	0.358 (0.088)	0.144 (0.117)	0.214 (0.207)	0.807 (0.167)	0.097 (0.062)
Lawrence	0.720 (0.066)	0.385 (0.047)	0.032 (0.025)	0.538 (0.142)	0.354 (0.117)	0.168 (0.130)	0.123 (0.176)	0.857 (0.120)	0.091 (0.086)
Wayne	0.695 (0.107)	0.401 (0.088)	0.063 (0.061)	0.535 (0.151)	0.350 (0.080)	0.172 (0.137)	0.220 (0.221)	0.775 (0.185)	0.094 (0.072)
Center	0.717 (0.082)	0.387 (0.064)	0.053 (0.058)	0.553 (0.132)	0.356 (0.099)	0.148 (0.122)	0.215 (0.218)	0.778 (0.183)	0.088 (0.056)
Warren	0.727 (0.063)	0.373 (0.046)	0.027 (0.022)	0.559 (0.140)	0.362 (0.111)	0.153 (0.138)	0.240 (0.237)	0.855 (0.144)	0.098 (0.064)
Decatur	0.735 (0.100)	0.386 (0.082)	0.063 (0.082)	0.665 (0.131)	0.337 (0.072)	0.109 (0.092)	0.235 (0.237)	0.755 (0.188)	0.079 (0.092)
Perry	0.729 (0.080)	0.373 (0.063)	0.040 (0.055)	0.592 (0.125)	0.353 (0.073)	0.124 (0.105)	0.246 (0.224)	0.821 (0.158)	0.093 (0.069)
Franklin	0.731 (0.067)	0.374 (0.045)	0.026 (0.020)	0.628 (0.142)	0.386 (0.104)	0.127 (0.132)	0.191 (0.207)	0.880 (0.119)	0.095 (0.061)

Table 4.5 V-I-S Composition by Township, 1995

	Commercial and Industrial Land			Residential Land			Agricultural and Pasture Land		
Township	GV	IM	SL	GV	IM	SL	GV	IM	SL
Pike	0.126 (0.088)	0.668 (0.197)	0.285 (0.194)	0.286 (0.093)	0.466 (0.128)	0.258 (0.128)	0.469 (0.136)	0.284 (0.070)	0.295 (0.161)
Washington	0.144 (0.099)	0.674 (0.179)	0.229 (0.149)	0.321 (0.082)	0.458 (0.110)	0.225 (0.089)	0.513 (0.080)	0.311 (0.058)	0.239 (0.076)
Lawrence	0.142 (0.093)	0.621 (0.197)	0.293 (0.183)	0.301 (0.086)	0.455 (0.114)	0.254 (0.127)	0.450 (0.141)	0.307 (0.062)	0.299 (0.165)
Wayne	0.122 (0.090)	0.719 (0.210)	0.259 (0.186)	0.301 (0.091)	0.488 (0.125)	0.220 (0.119)	0.419 (0.160)	0.296 (0.073)	0.319 (0.182)
Center	0.110 (0.091)	0.727 (0.181)	0.234 (0.161)	0.279 (0.084)	0.520 (0.120)	0.206 (0.100)	0.458 (0.121)	0.316 (0.059)	0.245 (0.113)
Warren	0.126 (0.097)	0.672 (0.199)	0.276 (0.189)	0.297 (0.096)	0.453 (0.118)	0.263 (0.148)	0.373 (0.165)	0.276 (0.070)	0.403 (0.211)
Decatur	0.130 (0.090)	0.670 (0.250)	0.323 (0.205)	0.287 (0.116)	0.434 (0.128)	0.302 (0.174)	0.339 (0.164)	0.260 (0.068)	0.456 (0.222)
Perry	0.119 (0.090)	0.662 (0.221)	0.328 (0.227)	0.313 (0.098)	0.457 (0.124)	0.247 (0.150)	0.363 (0.190)	0.277 (0.074)	0.445 (0.276)
Franklin	0.100 (0.066)	0.608 (0.214)	0.384 (0.232)	0.282 (0.116)	0.423 (0.134)	0.325 (0.184)	0.330 (0.164)	0.260 (0.070)	0.456 (0.225)

	Forest Land			Grass Land			Water		
Township	**GV**	**IM**	**SL**	**GV**	**IM**	**SL**	**GV**	**IM**	**SL**
Pike	0.611 (0.065)	0.306 (0.073)	0.178 (0.049)	0.445 (0.103)	0.271 (0.083)	0.311 (0.116)	0.080 (0.109)	0.819 (0.133)	0.047 (0.052)
Washington	0.548 (0.084)	0.348 (0.067)	0.180 (0.053)	0.460 (0.082)	0.301 (0.081)	0.267 (0.091)	0.137 (0.125)	0.765 (0.150)	0.082 (0.064)
Lawrence	0.569 (0.067)	0.331 (0.069)	0.193 (0.055)	0.443 (0.097)	0.277 (0.085)	0.306 (0.107)	0.087 (0.114)	0.806 (0.136)	0.088 (0.066)
Wayne	0.600 (0.073)	0.301 (0.080)	0.157 (0.052)	0.436 (0.105)	0.263 (0.084)	0.320 (0.127)	0.124 (0.119)	0.854 (0.123)	0.082 (0.066)
Center	0.556 (0.084)	0.342 (0.081)	0.148 (0.060)	0.452 (0.096)	0.271 (0.093)	0.294 (0.110)	0.124 (0.123)	0.827 (0.125)	0.110 (0.100)
Warren	0.587 (0.072)	0.318 (0.075)	0.179 (0.057)	0.435 (0.097)	0.272 (0.081)	0.312 (0.119)	0.128 (0.108)	0.838 (0.125)	0.107 (0.089)
Decatur	0.614 (0.064)	0.286 (0.071)	0.163 (0.054)	0.395 (0.127)	0.262 (0.074)	0.363 (0.146)	0.118 (0.114)	0.834 (0.110)	0.094 (0.068)
Perry	0.594 (0.075)	0.313 (0.076)	0.153 (0.056)	0.450 (0.095)	0.271 (0.077)	0.303 (0.128)	0.118 (0.122)	0.849 (0.129)	0.086 (0.069)
Franklin	0.584 (0.067)	0.305 (0.077)	0.170 (0.054)	0.420 (0.103)	0.264 (0.076)	0.337 (0.126)	0.161 (0.115)	0.815 (0.139)	0.086 (0.063)

Table 4.6 V-I-S Composition by Township, 2000

deviations in parentheses) of vegetation, impervious surface, and soil by LULC type for each township. The mean values are plotted further in the bar diagrams of Fig. 4.11. It can be observed from these tables and the figure that each LULC had a distinct V-I-S composition pattern but was quite consistent over the three dates of observation, implying relatively small changes in the V-I-S composition for each land class. A further comparison of the V-I-S mean values of six LULC class for the nine townships shows that the composition was generally stable, with slight changes, for the three observed years. For commercial and industrial land of any township, the mean value of vegetation was around 0.15, the impervious surface close to 0.7, and soil a little over 0.25. For residential land, the mean values of V-I-S components for each township were detected to be approximately 0.3, 0.5, and 0.2 respectively. These patterns of V-I-S composition appear similar to the characteristic of the V-I-S composition pattern for the whole city, as indicated by Table 4.1.

The disparity in the V-I-S composition in the townships may be revealed by plotting and examining in detail the mean values and changes in the vegetation, impervious surface, and soil components with ternary diagrams. Figure 4.12 shows the changes of V-I-S compositions (mean fraction values) by LULC type by township from 1991 to 2000. By focusing our analysis on the V-I-S composition of commercial/industrial and residential lands, some interesting observations can be made. Two townships, that is, Pike and Decatur, experienced continued development in commercial/industrial land during the study period. This development had resulted in an increase in the impervious surface component but a decrease in the vegetation component, whereas the soil component largely remained constant over the time. The trend in commercial development in these townships was heading along the interstate highways, such as I-65, I-465, and I-70, converting large areas of forest and agricultural land into impervious surfaces. Figure 4.12*a* and 4.12*g* further shows that commercial/industrial land was moving away from residential land. This implies that commercial activities did not occur within (or near) the residential areas. Urban development in these areas may be characterized as urban fringe expansion and linear and leapfrog clustered development. In the Center and Washington townships, no such change occurred in the impervious surface component, as shown in Fig. 4.12*b* and 4.12*e* because these townships had been fully developed in the past. The decrease in the vegetation component and the increase in the soil component probably were associated with infill type of development or redevelopment activities. This urbanization pattern was characterized by the development of vacant land in already built-up areas and/or redevelopment of abandoned industries, warehouses, and other types of old buildings, which usually occurred in places where public facilities already existed (Forman, 1995). In contrast to the preceding two types of urban development, Franklin township in the

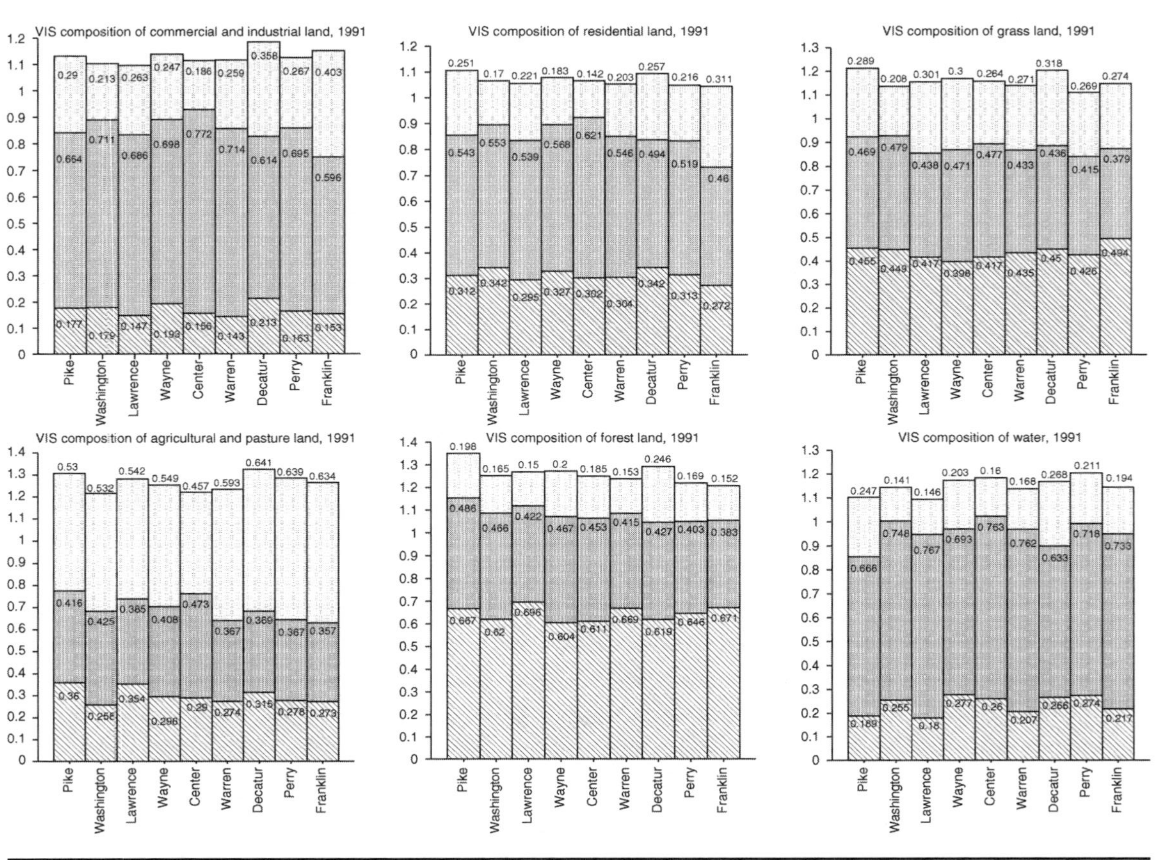

FIGURE 4.11 The values of vegetation, impervious surface, and soil by LULC type by township

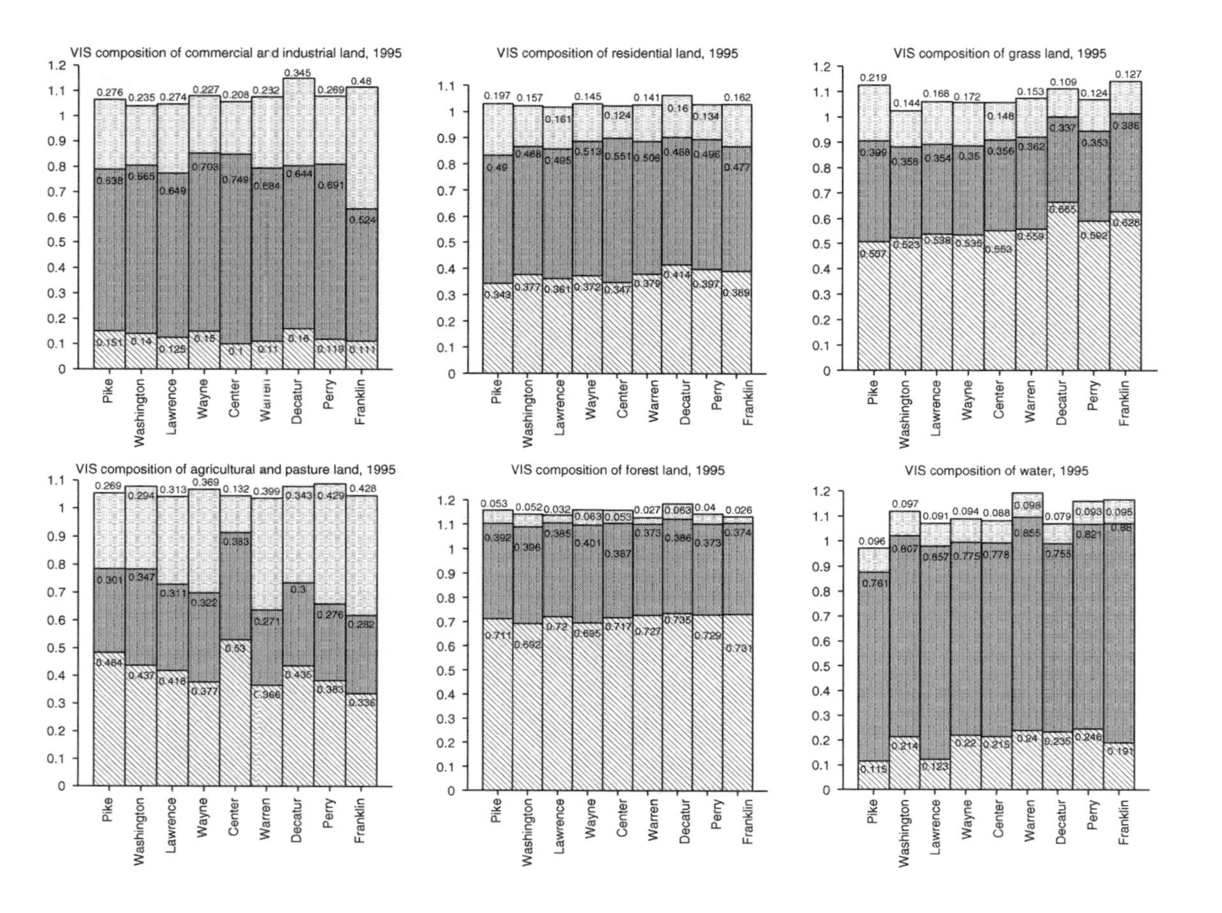

VIS composition of commercial and industrial land, 1995
VIS composition of residential land, 1995
VIS composition of grass land, 1995
VIS composition of agricultural and pasture land, 1995
VIS composition of forest land, 1995
VIS composition of water, 1995
Pike
Washington
Lawrence
Wayne
Center
Warren
Decatur
Perry
Franklin

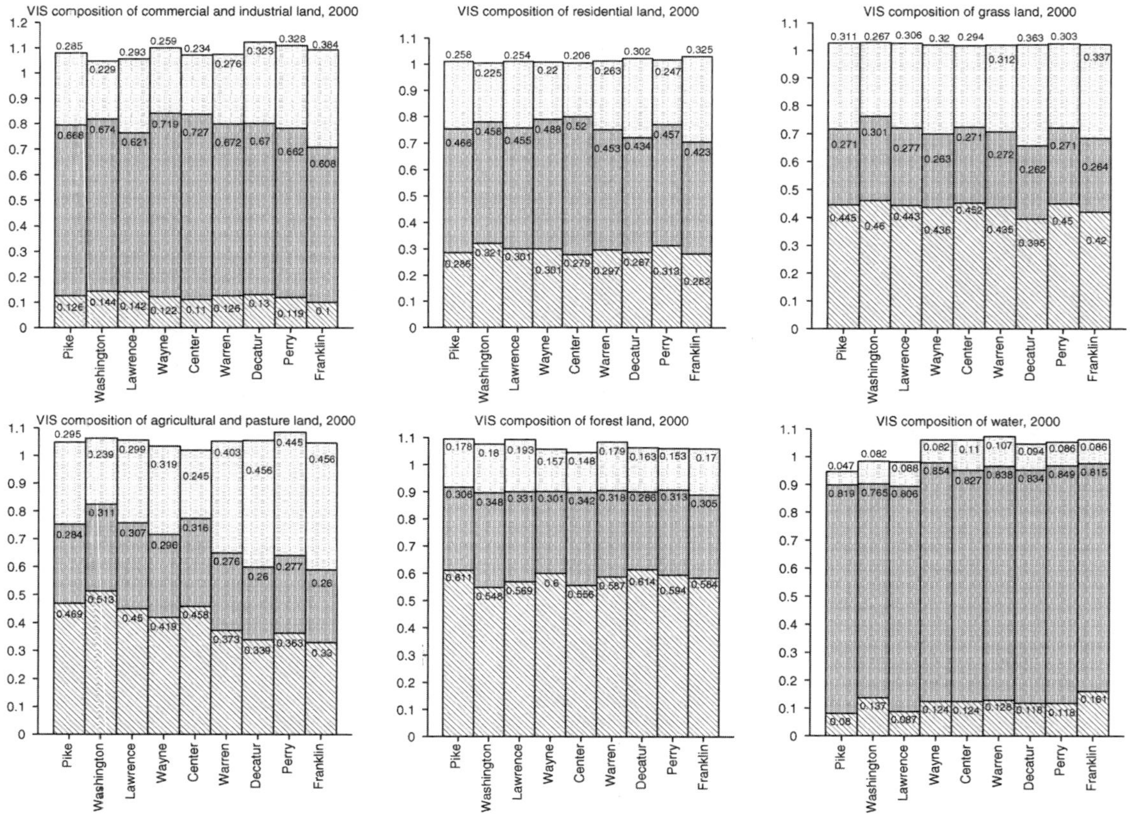

VIS composition of commercial and industrial land, 2000

VIS composition of residential land, 2000

VIS composition of grass land, 2000

VIS composition of agricultural and pasture land, 2000

VIS composition of forest land, 2000

VIS composition of water, 2000

Figure 4.11 *(Continued)*

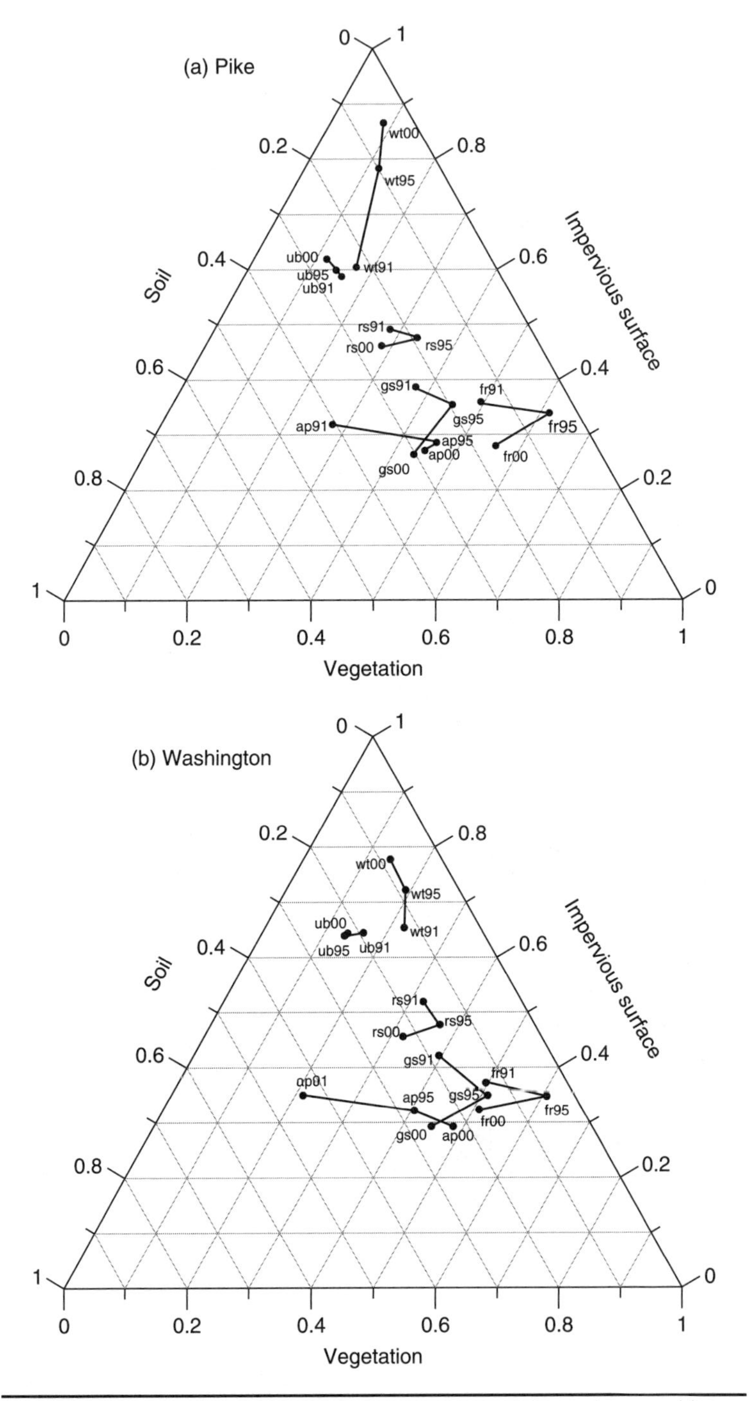

FIGURE 4.12 Changes in the V-I-S composition by LULC type by township. (*Adapted from Weng and Lu, 2009.*)

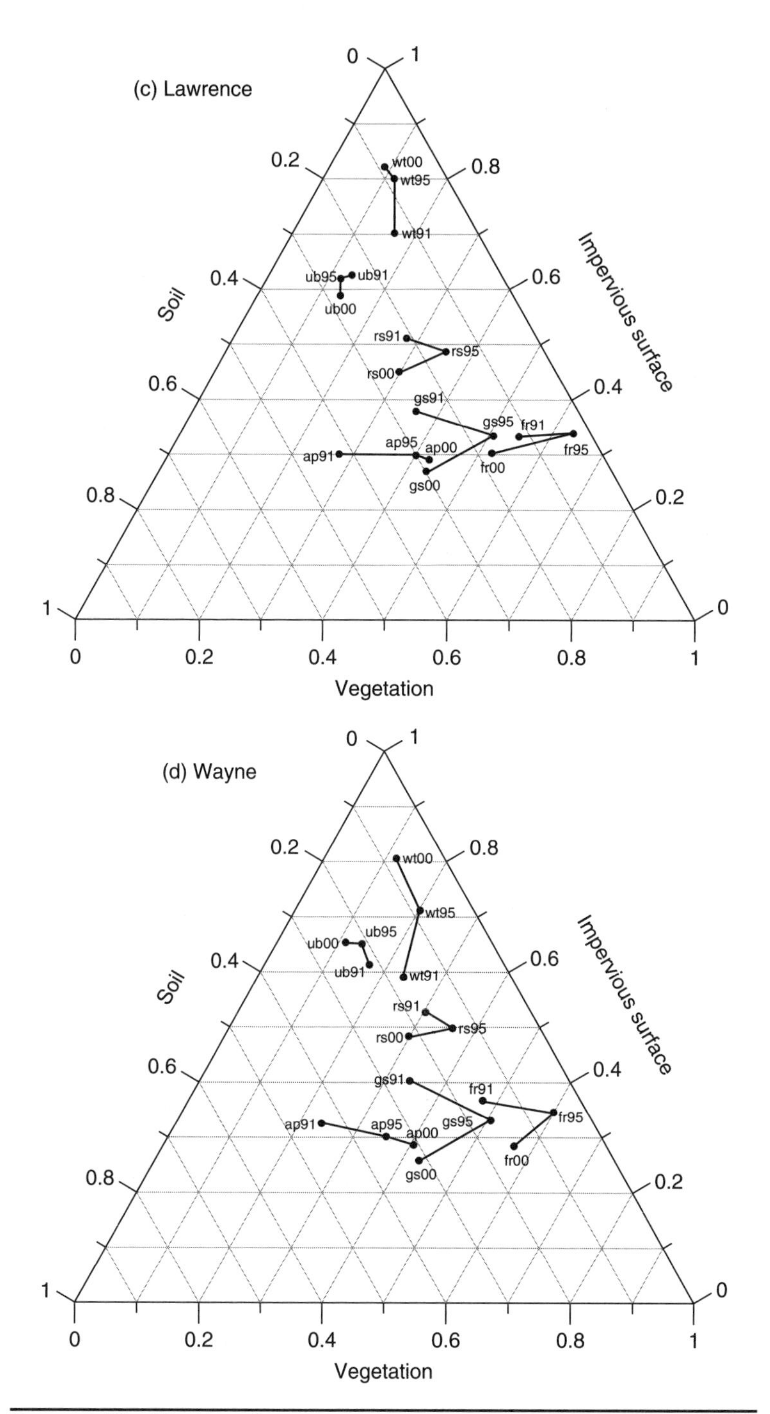

FIGURE 4.12 *(Continued)*

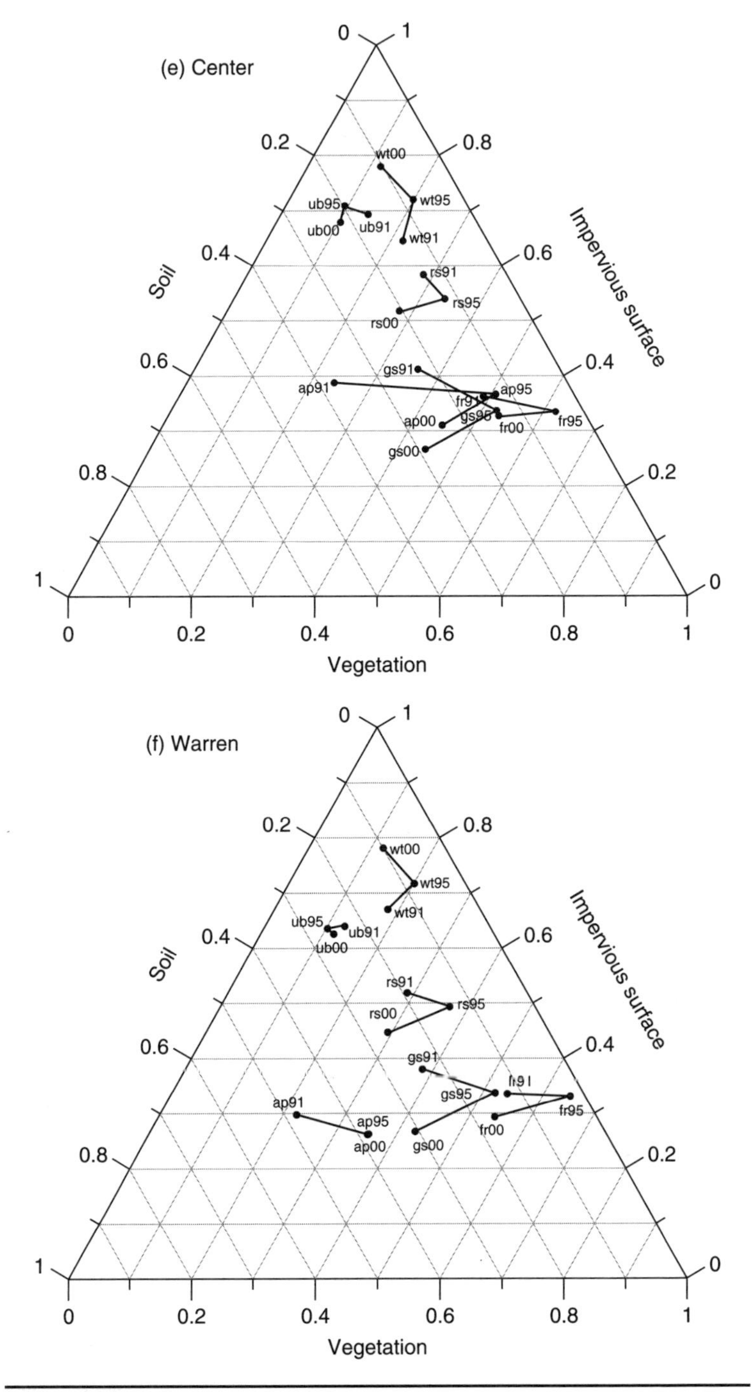

FIGURE 4.12 *(Continued)*

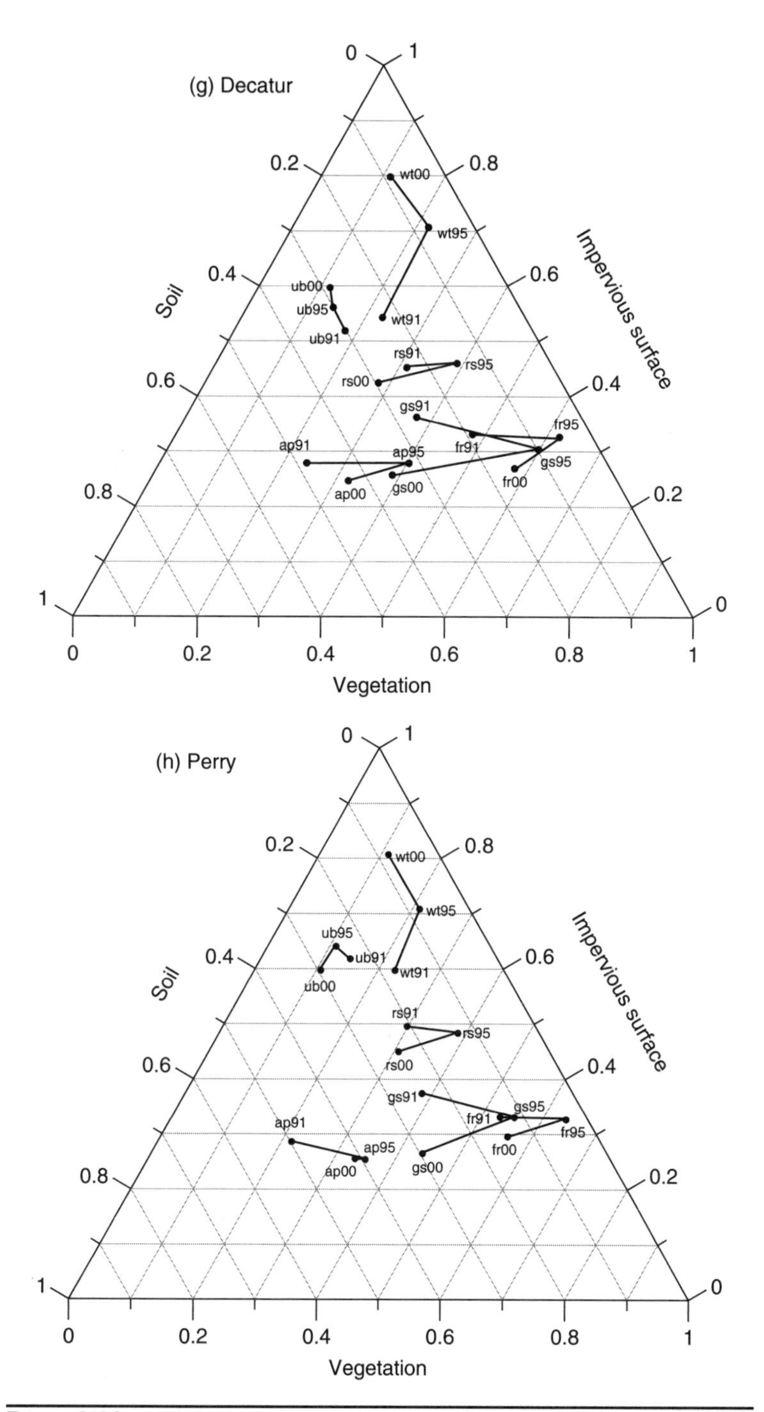

FIGURE 4.12 *(Continued)*

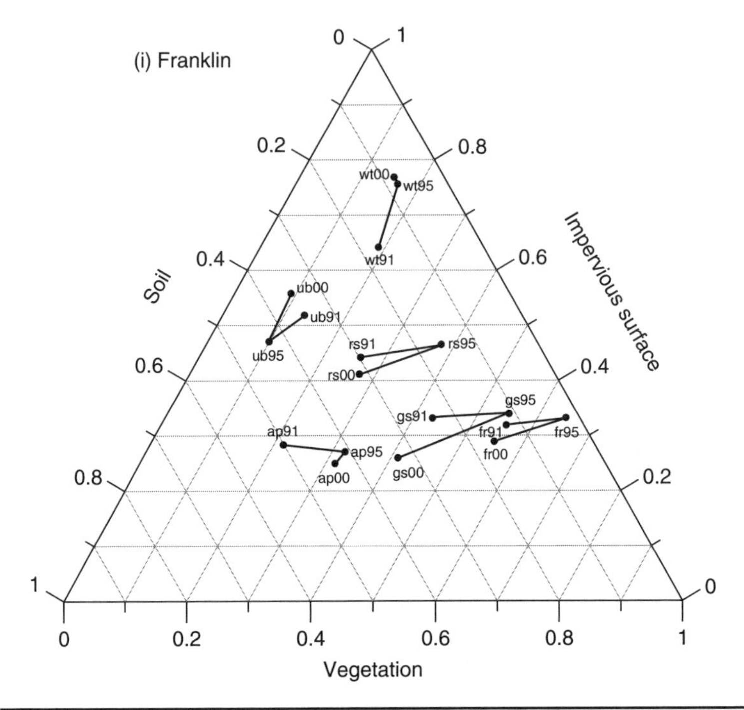

FIGURE 4.12 (*Continued*)

southeastern part of the city exhibited another type of development, which may be called *clustered* or *leap-frog development*. The urban development in Franklin typically involved compact, high-density development of functionally separated clusters. The ternary plot in Fig 4.12*i* shows that the pattern of V-I-S composition has not become stabilized for the periods observed. Commercial development was largely introduced into the residential areas, especially after 1995. Residential and commercial lands thus appeared increasingly intermingled. The remaining four townships, that is, Wayne, Perry, Lawrence, and Warren, did not discover urban growth as distinct as the preceding townships. Instead, urban development in these areas displayed some sorts of combination among fringe development, infill development, and outlying growth. The latter can be defined as development beyond existing developed areas and may be broken down into isolated, linear branch, and clustered branch (Wilson et al., 2003). Ternary plots (c), (d), (f), and (h) in Fig. 4.12*c*, 4.12*d*, 4.12*f*, and 4.12*h* show that that the impervious surface component in commercial land altered over the time but may have experienced the opposite direction of changes between the period 1991 to 1995 and the period 1995 to 2000. The changes in impervious surfaces inevitably triggered changes in the vegetation and soil components, leading to the V-I-S dynamics. These dynamics also can be observed clearly in residential land.

4.3 Discussion and Conclusions

Remote sensing of urban material, land use, and land cover has different requirements. This study intended to approach their relationship by using a continuum-field model. This model was developed based on reconciliation of the V-I-S model (Ridd, 1995) and LSMA of Landsat imagery. The case study demonstrated successfully that the continuum model is effective for characterizing and quantifying the spatial and temporal changes in the landscape compositions of Indianapolis between 1991 and 2000.

The linkage between the Ridd's model and LSMA lies in the urban material. Ridd (1995) suggested that the urban landscape can be decomposed into three classes of urban materials (i.e., vegetation, impervious surfaces, and soil) but did not implement the model using digital image processing algorithms. LSMA has been used widely to analyze spectrally heterogeneous urban reflectances based on endmembers, which are recognizable surface materials that have homogeneous spectral properties over the entire image. The number of endmembers that can be extracted from a satellite image is determined by the data dimensionality reflected in the image mixing space, which, in turn, is subject to the number of spectral bands available. Theoretically speaking, Landsat TM/ETM+ images, with six spectral bands (excluding the thermal infrared band), can derive as many as seven endmembers. However, our study indicated that the three lower-order principal components accounted for more than 99 percent of the image variance, implying that the mixing space was perfectly three-dimensional. The RMSE was small (<0.1 of mean pixel reflectance) for all three unmixings. Therefore, the endmembers selected for this study (i.e., high albedo, low albedo, vegetation, and soil) were regarded as proper to account for the observed radiance. When high- and low-albedo endmembers were used to model impervious surfaces, the RMSE also was found to be reasonably small for the LSMA model (<0.02). As such, LSMA with the four-endmember model offered a simple, robust, physically based solution to quantify urban reflectance. Our study further indicated that although the spectral endmembers were different over time because of changes in urban materials and land cover/use, the topology of the triangular mixing space (see Fig. 4.3) was consistent. It became apparent that LSMA can provide a repetitive way to derive consistent image endmembers from Landsat images. Therefore, by representing Ridd's V-I-S components as image fractions, the continuum model developed in this study provides an effective approach for quantifying urban landscape patterns as standardized component surfaces.

The V-I-S image fractions can be used not only for characterizing urban morphology and biophysical conditions but also for land-use/cover classification. The classifications based on the image fractions would be able to accommodate the spectral heterogeneity that characterizes the urban landscape in medium-resolution imagery. In contrast, with the hard classifiers, each thematic class is assumed to be both spatially and spectral homogeneous at the pixel scale (Small, 2005). Moreover, the continuum variables that LSMA provided can more accurately characterize the fuzzy nature of the urban landscapes. Fuzzy classification intends to address this nature but does not provide a physical solution for urban reflectance (Small, 2004). This study further shows that statistically-derived LSMA fractions may be used to generate better classification results. Stable and reliable fraction estimates derived from the multitemporal image data were found to be straightforward for an analysis of urban landscape dynamics because the fractional characteristics of LULC types on one date were comparable with those on other dates. Many digital change-detection methods for urban growth analysis employ pixel-by-pixel comparison algorithms, which may cause the problem of error propagation in the overlay process. Compared with other urban growth models based on multitemporal remote sensing imagery (e.g., Auch et al., 2004; Betty and Longley, 1994; Clarke and Gaydos, 1998), this continuum model offers a more flexible way for urban time-space modeling and comparison of urban morphologies in different geographic settings because it took into account of the physical process responsible for the observed radiance and provided a more accurate representation of urban landscapes.

The V-I-S model has been applied successfully to both temperate (Salk Lake City, Utah, Columbus, Ohio, Indianapolis, Indiana, Los Angels, California, Queensland, Australia, Cairo, Egypt, and Shanghai, China) and tropical cities (Bangkok, Thailand, Yogyakarta, Indonesia, and Manaus, Brazil) under different climate conditions. A main strength of the V-I-S model lies in its generalization and simplification of urban land cover into the components of vegetation, impervious surface, and soil. The soil in the model may include various types of soils, bare ground, and sparsely vegetated areas. The vegetation refers to green vegetation and may include trees, grasses, pastures, crops, and so on. Both vegetation and soil may have distinct seasonal variability, whereas impervious surface is relatively stable. Urban mapping must consider the daily, seasonal, and annual phonologic cycles of vegetation, which affect the temporal pattern of soil. In addition to the temporal variability, the spectral variability of the V-I-S components must be addressed. For instance, because of the complexity of impervious surfaces, different urban areas may have substantially different types of impervious surfaces. Identifying a single endmember to represent all types of impervious surfaces may not be realistic. This study shows that impervious surface was overestimated in the less developed areas but underestimated in well-developed areas. Impervious surface may have highly similar spectral responses with nonphotosynthesis vegetation and dry soils. On the other hand, shadows from tall buildings, large tree crowns, and dark objects may cause impervious surface to be underestimated. The limitation in the spectral resolution of Landsat imagery makes it impractical to derive and depict various types of impervious surfaces with several endmembers. To overcome this problem, a multiple-endmember LSMA model has been developed and found to be more effective when a large number of endmembers are required to be modeled across an image scene (Okin et al., 2001; Painter et al., 1998; Powell et al., 2007; Roberts et al., 1998). In addition, calibration with reference endmembers from a spectral library or in situ measurements also would be desirable.

The spectral variability of the vegetation fraction is another significant issue in LSMA of Landsat imagery for urban landscapes (Song, 2005). Our study indicates that a vegetation endmember usually corresponded to grass or dense crop or pasture. Because forest possessed an internal shade component, it was located along the mixing line between the vegetation and low-albedo endmembers. Shade is an important component captured by optical remote sensors and therefore was included in the Lu-Weng model (Lu and Weng, 2004:1053–1062). Impervious surface, shade, and vegetation were considered three basic components in urban areas, whereas in surrounding areas, soil, vegetation, and shade were used to account for the spectral variability. However, the Lu-Weng model was not able to separate soil and impervious surfaces. In this study, low- and high-albedo endmembers were derived to depict impervious surfaces better. Low-albedo endmembers were found mainly to correspond to water, canopy shadow, tall-building shadow, or dark impervious surface materials. Future study should refine this endmember and examine how it is related to shade, water, and other materials.

The spatial variability of the V-I-S model needs to be examined as well. The intra-urban variability is a consequence of the wide variety of urban materials, land cover, and land use in different parts of the city. Our investigation of the V-I-S dynamics in the nine townships of Indianapolis owing to changes in the urban landscape found an urban-to-rural gradient in these components. This explains the effectiveness of the continuum model for such an urban growth analysis. The concepts of "urban" and "city" were not differentiated well in Ridd's study. We suggest that based on previous studies, the V-I-S model may be extended to the land cover of cities (including urban, suburban, and surrounding rural areas). Wu and colleagues (2005) suggested that different V-I-S models should be constructed for different functional districts. The interurban variability results from the diversity of biophysical, geological, economic, social, and urban development history among cities. Small (2005) compared 28 cities in the world and found that these cities had a similar mixing topology in Landsat ETM+ imagery and could be represented by three-component (i.e., high-albedo, dark, and vegetation endmembers) linear mixture models in both scene-specific and global composite mixing spaces.

This current research could be extended in the future by testing the hypothesis of whether the Indianapolis model can be applied only to cities in developed countries and not to cities in developing countries. The model can be parameterized and validated for certain time periods of selected cities using available medium-resolution satellite data. The results of our model applied to and parameterized for developed countries with those for developing countries then can be compared, with the goal of developing a global model for remotely sensed analysis of urban landscapes.

References

Adams, J. B., Smith, M. O., and Johnson, P. E. 1986. Spectral mixture modeling: A new analysis of roack and soil types at the Viking Lander site. *Journal of Geophysical Research* **91,** 8098–8112.

Adams, J. B., Sabol, D. E., Kapos, V., et al. 1995. Classification of multispectral images based on fractions of endmembers: Application to land cover change in the Brazilian Amazon. *Remote Sensing of Environment* **52,** 137–154.

Aguiar, A. P. D., Shimabukuro, Y. E., and Mascarenhas, N. D. A. 1999. Use of synthetic bands derived from mixing models in the multispectral classification of remote sensing images. *International Journal of Remote Sensing* **20,** 647–657.

American Planning Association. 2004. *Land-Based Classification System.* Washington: American Planning Association, *www.planning.org/lbcs/* (last accessed January 20, 2007).

Anderson, J. R., Hardy, E. E., Roach, J. T., and Witmer, R. E. 1976. A land use and land cover classification systems for use with remote sensing data. USGS Professional Paper 964. Washington: U.S. Geological Survey.

Arnold, C. L., Jr., and Gibbons, C. J. 1996. Impervious surface coverage: The emergence of a key environmental indicator. *Journal of the American Planning Association* **62,** 243–258.

Asner, G. P., and Lobell, D. B. 2000. A biogeophysical approach for automated SWIR unmixing of soils and vegetation. *Remote Sensing of Environment* **74,** 99–112.

Auch, R., Taylor, J., and Acevedo, W. 2004. *Urban Growth in American Cities: Glimpses of U.S. Urbanization.* USGS Circular 1252. Washington: U.S. Geological Survey.

Betty, M., and Longley, P. 1994. *Fractal Cities.* San Diego: Academic Press.

Boardman, J. W. 1993. Automated spectral unmixing of AVIRIS data using convex geometry concepts. In *Summaries of the Fourth JPL Airborne Geoscience Workshop,* pp. 11–14. JPL Publication 93-26. Pasadena, CA: NASA Jet Propulsion Laboratory.

Boardman, J. W., and Kruse, F. A. 1994. Automated spectral analysis: A geological example using AVIRIS data, north Grapevine Mountains, Nevada. In *Proceedings, ERIM Tenth Thematic Conference on Geologic Remote Sensing,* Ann Arbor, MI, pp. 407–418.

Boardman, J. M., Kruse, F. A., and Green, R. O. 1995. Mapping target signature via partial unmixing of AVIRIS data. In *Summaries of the Fifth JPL Airborne Earth Science Workshop,* pp. 23–26. JPL Publication 95-1. Pasadena, CA: NASA Jet Propulsion Laboratory.

Campell, J. B. 1983. *Mapping the Land: Aerial Imagery for Land Use Information.* Washington: Association of American Geographers.

Chavez, P. S., Jr. 1996. Image-based atmospheric corrections—revisited and improved. *Photogrammetric Engineering and Remote Sensing* **62,** 1025–1036.

Clapham, W. B., Jr. 2003. Continuum-based classification of remotely sensed imagery to describe urban sprawl on a watershed scale. *Remote Sensing of Environment* **86,** 322–340.

Clarke, K. C., and Gaydos, L. J. 1998. Loose-coupling a cellular automaton model and GIS: Long-term urban growth prediction for San Francisco and Washing/Baltimore. *International Journal of geographic Information Science* **12,** 699–714.
Cochrane, M. A., and Souza, C. M., Jr. 1998. Linear mixture model classification of burned forests in the eastern Amazon. *International Journal of Remote Sensing* **19,** 3433–3440.
Cracknell, A. P. 1998. Synergy in remote sensing: What's in a pixel? *International Journal of Remote Sensing* **19,** 2025–2047.
ENVI. 2000. *ENVI User's Guide*. Boulder, CO: Research Systems, Inc.
Epstein, J., Payne, K., and Kramer, E. 2002. Techniques for mapping suburban sprawl. *Photogrammetric Engineering and Remote Sensing* **63,** 913–918.
Fisher, P. 1997. The pixel: A snare and a delusion. *International Journal of Remote Sensing* **18,** 679–685.
Foody, G. M. 2002. Status of land cover classification accuracy assessment. *Remote Sensing of Environment* **80,** 185–201.
Forman, R. T. T. 1995. *Land Mosaics: The Ecology of Landscapes and Regions*. Cambridge, England: Cambridge University Press.
Garcia-Haro, F. J., Gilabert, M. A., and Melia, J. 1996. Linear spectral mixture modeling to estimate vegetation amount from optical spectral data. *International Journal of Remote Sensing* **17,** 3373–3400.
Gong, P., and Howarth, P. J. 1990. The use of structure information for improving land-cover classification accuracies at the rural-urban fringe. *Photogrammetric Engineering and Remote Sensing* **56,** 67–73.
Gong, P., and Howarth, P. J. 1992. Frequency-based contextual classification and gray-level vector reduction for land-use identification. *Photogrammetric Engineering and Remote Sensing* **58,** 423–437.
Green, A. A., Berman, M., Switzer, P., and Craig, M. D. 1988. A transformation for ordering multispectral data in terms of image quality with implications for noise removal. *IEEE Transactions on Geoscience and Remote Sensing* **26,** 65–74.
Herold, M., Liu, X., and Clark, K. C. 2003. Spatial metrics and image texture for mapping urban land use. *Photogrammetric Engineering and Remote Sensing* **69,** 991–1001.
Herold, M., Schiefer, S., Hostert, P., and Roberts, D. A. 2006. Applying imaging spectrometry in urban areas. In *Urban Remote Sensing,* edited by Q. Weng and D. Quattrochi, pp. 137–161. Boca Raton, FL: CRC/Taylor & Francis.
Homer, C., Huang, C., Yang, L., et al. 2004. Development of a 2001 national land-cover database for the United States. *Photogrammetric Engineering and Remote Sensing* **70,** 829–840.
Jensen, J. R. 2005. *Introductory Digital Image Processing: A Remote Sensing Perspective*. Upper Saddle River, NJ: Pearson Prentice Hall.
Jensen, J. R., and Cowen, D. C. 1999. Remote sensing of urban/suburban infrastructure and socio-economic attributes. *Photogrammetric Engineering and Remote Sensing* **65,** 611–622.
Ju, J., Gopal, S., and Kolaczyk, E. D. 2005. On the choice of spatial and categorical scale in remote sensing land cover classification. *Remote Sensing of Environment* **96,** 62–77.
Lee, S., and Lathrop, R.G., Jr. 2005. Sub-pixel estimation of urban land cover components with linear mixture model analysis and Landsat Thematic Mapper imagery. *International Journal of Remote Sensing* **26,** 4885–4905.
Lo, C. P. 1986. *Applied Remote Sensing*. New York: Longman.
Lu. D., and Weng, Q. 2004. Spectral mixture analysis of the urban landscapes in Indianapolis with Landsat ETM+ imagery. *Photogrammetric Engineering and Remote Sensing* **70,** 1053–1062.
Lu. D., and Weng, Q. 2006a. Use of impervious surface in urban land use classification. *Remote Sensing of Environment* **102,** 146–160.
Lu. D., and Weng, Q. 2006b. Spectral mixture analysis of ASTER imagery for examining the relationship between thermal features and biophysical descriptors in Indianapolis, Indiana. *Remote Sensing of Environment* **104,** 157–167.

Lu, D., Mausel, P., Brondizio, E., and Moran, E. 2002. Assessment of atmospheric correction methods for Landsat TM data applicable to Amazon basin LBA research. *International Journal of Remote Sensing* **23,** 2651–2671.

Madhavan, B. B., Kubo, S., Kurisaki, N., and Sivakumar, T. V. L. N. 2001. Appraising the anatomy and spatial growth of the Bangkok metropolitan area using a vegetation-impervious-soil model through remote sensing. *International Journal of Remote Sensing* **22,** 789–806.

Mather, A. S. 1986. *Land Use.* London: Longman.

Mather, P. M. 1999. Land cover classification revisited. In *Advances in Remote Sensing and GIS,* edited by P. M. Atkinson and N. J. Tate, pp. 7–16. New York: Wiley.

McGwire, K., Minor, T., and Fenstermaker, L. 2000. Hyperspectral mixture modeling for quantifying sparse vegetation cover in arid environments. *Remote Sensing of Environment* **72,** 360–374.

Myint, S. W. 2001. A robust texture analysis and classification approach for urban land-use and land-cover feature discrimination. *Geocarto International* **16,** 27–38.

Okin, G. S., Roberts, D. A., Murray, B., and Okin, W. J. 2001. Practical limits on hyperspectral vegetation discrimination in arid and semiarid environments. *Remote Sensing of Environment* **77,** 212–225.

Painter, T. H., D. A. Roberts, R. O. Green, and J. Dozier, 1998. The effects of grain size on spectral mixture analysis of snow-covered area from AVIRIS data, *Remote Sensing of Environment,* 65:320–332.

Phinn, S., Stanford, M., Scarth, P., et al. 2002. Monitoring the composition of urban environments based on the vegetation–impervious surface–soil (VIS) model by subpixel analysis techniques. *International Journal of Remote Sensing* **23,** 4131–4153.

Powell, R. L., Roberts, D. A., Dennison, P. E., and Hess, L. L. 2007. Sub-pixel mapping of urban land cover using multiple endmember spectral mixture analysis: Manaus, Brazil. *Remote Sensing of Environment* **106,** 253–267.

Raptis, V. S., Vaughan, R. A., and Wright, G. G. 2003. The effect of scaling on land cover classification from satellite data. *Computers & Geosciences* **29,** 705–714.

Rashed, T., Weeks, J. R., Roberts, D. A., et al. 2003. Measuring the physical composition of urban morphology using multiple endmember spectral mixture models. *Photogrammetric Engineering and Remote Sensing* **69,** 1011–1020.

Rashed, T., Weeks, J. R., Stow, D., and Fugate, D. 2005. Measuring temporal compositions of urban morphology through spectral mixture analysis: Toward a soft approach to change analysis in crowded cities. *International Journal of Remote Sensing* **26,** 699–718.

Ridd, M. K. 1995. Exploring a V-I-S (vegetation-impervious surface-soil) model for urban ecosystem analysis through remote sensing: Comparative anatomy for cities. *International Journal of Remote Sensing* **16,** 2165–2185.

Roberts, D. A., Batista, G. T., Pereira, J. L. G., et al. 1998. Change identification using multitemporal spectral mixture analysis: Applications in eastern Amazonia. In *Remote Sensing Change Detection: Environmental Monitoring Methods and Applications,* edited by R. S. Lunetta and C. D. Elvidge, pp. 137–161. Chelsea, MI: Ann Arbor Press.

Setiawan, H., Mathieu, R., and Thompson-Fawcett, M. 2006. Assessing the applicability of the V-I-S model to map urban land use in the developing world: Case study of Yogyakarta, Indonesia. *Computers, Environment and Urban Systems* **30,** 503–522.

Shaban, M. A., and Dikshit, O. 2001. Improvement of classification in urban areas by the use of textural features: The case study of Lucknow city, Uttar Pradesh. *International Journal of Remote Sensing* **22,** 565–593.

Small, C. 2001. Estimation of urban vegetation abundance by spectral mixture analysis. *International Journal of Remote Sensing* **22,** 1305–1334.

Small, C. 2002. Multitemporal analysis of urban reflectance. *Remote Sensing of Environment* **81,** 427–442.

Small, C. 2004. The Landsat ETM+ spectral mixing space. *Remote Sensing of Environment* **93,** 1–17.

Small, C. 2005. A global analysis of urban reflectance. *International Journal of Remote Sensing* **26,** 661–681.
Soil Conservation Service. 1975. *Urban Hydrology for Small Watersheds.* USDA Soil Conservation Service Technical Release No. 55. Washington: U.S. Department of Agriculture.
Song, C. 2005. Spectral mixture analysis for subpixel vegetation fractions in the urban environment: How to incorporate endmember variability? *Remote Sensing of Environment* **95,** 248–263.
Strahler, A. H., Woodcock, C. E., and Smith, J. A. 1986. On the nature of models in remote sensing. *Remote Sensing of Environment* **70,** 121–139.
Stuckens, J., Coppin, P. R., and Bauer, M .E. 2000. Integrating contextual information with per-pixel classification for improved land cover classification. *Remote Sensing of Environment* **71,** 282–296.
Turner, B. L., II., Skole, D., Sanderson, S., et al. 1995. *Land-Use and Land-Cover Change: Science and Research Plan. Stockhdm and Geneva.* International Geosphere-Bioshere Program and the Human Dimensions of Global Environmental Change Program (IGBP Report No. 35 and HDP Report No. 7). Geneva: International Geosphere-Bioshere Program.
van der Meer, F., and de Jong, S. M. 2000. Improving the results of spectral unmixing of Landsat Thematic Mapper imagery by enhancing the orthogonality of endmembers. *International Journal of Remote Sensing* **21,** 2781–2797.
Wang, F. 1990. Fuzzy supervised classification of remote sensing images. *IEEE Transactions on Geoscience and Remote Sensing* **28,** 194–201.
Ward, D., Phinn, S. R., and Murray, A. T. 2000. Monitoring growth in rapidly urbanizing areas using remotely sensed data. *Professional Geographer* **53,** 371–386.
Welch, R. A. 1982. Spatial resolution requirements for urban studies. *International Journal of Remote Sensing* **3,** 139–146.
Weng, Q. 1999. Environmental impacts of land use and land cover change in the Zhujiang Delta, China: An analysis using an integrated approach of gis, remote sensing, and spatial modeling, Ph.D. dissertation, University of Georgia, Athens, GA.
Weng, Q. and Lu, D. 2006. Sub-pixel analysis of urban landscapes. In *Urban Remote Sensing,* edited by Q. Weng and D. Quattrochi, pp. 71–90. Boca Raton, FL: CRC/ Taylor & Francis.
Weng, Q., and Lu, D. 2009. Landscape as a continuum: An examination of the urban landscape structures and dynamics of Indianapolis city, 1991–2000. *International Journal of Remote Sensing,* **30,** 2547–2577.
Weng, Q., Lu, D., and Schubring, J. 2004. Estimation of land surface temperature-vegetation abundance relationship for urban heat island studies. *Remote Sensing of Environment* **89,** 467–483.
Wilson, E. H., Hurd, J. D., Civco, D. L., et al. 2003. Development of a geospatial model to quantify, describe and map urban growth. *Remote Sensing of Environment* **86,** 275–285.
Wu, C., and Murray, A. T. 2003. Estimating impervious surface distribution by spectral mixture analysis. *Remote Sensing of Environment* **84,** 93–505.
Wu, J. G., and David, J. L. 2002. A spatially explicit hierarchical approach to modeling complex ecological systems: Theory and application. *Ecological Modelling* **153,** 7–26.
Wu, J. G., Jelinski, D. E., Luck, M., and Tueller, P. T. 2000. Multiscale analysis of landscape heterogeneity: Scale variance and pattern metrics. *Geographic Information Sciences* **6,** 6–19.
Wu, J. W., Xu, J. H., and Yue, W. Z. 2005. V-I-S model for cities that are experiencing rapid urbanization and development. In *Geoscience and Remote Sensing Symposium, IGARSS'05. Proceedings* **3,** 1503–1506.

CHAPTER 5

Urban Feature Extraction

To extract urban features such as roads and buildings, high-spatial-resolution imagery is preferred over moderate-resolution imagery because of the ability of high-resolution imagery to provide more detailed information. As the spatial resolution increases, the proportion of pure pixels most likely increases, and number of mixed pixels decreases (Hsieh et al., 2001). Therefore, subpixel classifiers may not be appropriate. Moreover, traditional image classification methods are based largely on the color and tone of the pixels. Other important information that may be derived from high-resolution imagery, such as texture, shape, and context, are completely neglected (Sharma and Sarkar, 1998). As a result, it is often not suitable to employ traditional or subpixel classifiers for feature extraction from high-spatial-resolution imagery.

In recent studies, many image segmentation techniques have been developed and applied for feature extraction with a fair amount of success (Cao and Jin, 2007; Guo et al., 2007; Karimi and Liu, 2004; Mayer et al., 1997, 1998; Wei et al., 2004; Yun and Uchimura, 2007). However, most of the image segmentation techniques are not robust for a spectrally complex environment (Pal and Pal, 1993), which makes them less suitable for image classification within urban areas. Object-based classification operates on objects that are composed of many pixels grouped together by image segmentation (Shackelford and Davis, 2003). Object-based approaches have many advantages over traditional classifiers because they use not only spectral properties but also the characteristics of shape, texture, context, relationship with neighbors, superpixels, and subpixels. Successful results have been obtained using this approach (Voode et al., 2004). Rule-based classification is another approach to classifying image objects. Traditional rule-based classification is based on strict binary rules. Objects are assigned to a class if the objects meet the rules of that class. Traditional rules of crisp boundaries may not be suitable for classifying objects because the attributes of different features may overlap (Jin and

Paswaters, 2007). Fuzzy logic can better cope with the uncertainties inherent in the data and vagaries in human knowledge (Jin and Paswaters, 2007).

In this chapter, an object-oriented classification technique using fuzzy rules will be implemented for extracting roads and buildings from high spatial-resolution imagery, that is, IKONOS imagery. This approach takes full consideration of the important image characteristics, such as spectral, spatial, and texture information, during the classification. Based on a case study in central Indianapolis, Indiana, this chapter will further discuss problems that may be encountered with the object-oriented approach and probable solutions.

5.1 Landscape Heterogeneity and Per-Field and Object-Based Image Classifications

In recent years, many advanced classification approaches, such as artificial neural networks, fuzzy sets, and expert systems, have been applied for image classification. Depending on what criterion is used for grouping, image classification approaches can be categorized as supervised or unsupervised; parametric or nonparametric; hard or soft (fuzzy) classification; per-pixel, subpixel, or per-field; and so on. Table 5.1 provides a brief description of per-pixel, subpixel, per-field, and object-oriented classifiers by listing their characteristics, exemplary classifiers, and recent references in the literature. The heterogeneity in complex landscapes, especially in urban areas, results in high spectral variation within the same land-cover class. With per-pixel classifiers, each pixel is grouped individually into a certain category, and the results may be noisy owing to high spatial frequency in the landscape. The per-field classifier is designed to deal with the problem of environmental heterogeneity and has been shown to be effective in improving classification accuracy (Aplin and Atkinson, 2001; Aplin et al., 1999a, 1999b; Dean and Smith, 2003; Lloyd et al., 2004). The per-field-based classifier averages out the noise by using land parcels (called *fields*) as individual units (Aplin et al., 1999a, 1999b; Dean and Smith 2003; Lobo et al., 1996; Pedley and Curran, 1991). Geographic information systems (GIS) provide a means of implementing per-field classification through integration of vector and raster data (Dean and Smith, 2003; Harris and Ventura, 1995; Janssen and Molenaar, 1995). The vector data are used to subdivide an image into parcels, and classification then is conducted based on the parcels, thus avoiding intraclass spectral variations. However, per-field classifications often are affected by such factors as the spectral and spatial properties of remotely sensed data, the size and shape of the fields, the definition of field boundaries, and the land-cover classes chosen (Janssen and Molenaar, 1995). The difficulty in handling the dichotomy between vector and raster data models affects extensive use of the per-field classification approach.

Category	Characteristics	Example Classifiers	References
Per-pixel-based classifiers	Traditional classifiers typically develop a signature by combining the spectra of all training-set pixels from a given feature. The resulting signature contains the contributions of all materials present in the training-set pixels, ignoring the mixed-pixel problem.	Most of the classifiers, such as maximum likelihood, minimum distance, artificial neural network, decision tree, and support vector machine	See Lu and Weng (2007) for a list of references.
Subpixel-based classifiers	The spectral value of each pixel is assumed to be a linear or nonlinear combination of defined pure materials (or endmembers), providing proportional membership of each pixel to each endmember.	Fuzzy set classifiers, subpixel classifiers, spectral mixture analysis	See Lu and Weng (2007) for a list of references.
Object-oriented classifiers	Image segmentation merges pixels into objects, and classification is conducted based on the objects instead of individual pixels. No GIS vector data are used.	eCognition, ENVI feature-extraction module	Benz et al., 2004; Geneletti and Gorte, 2003; Gitas et al., 2004; Herold et al., 2003; Thomas et al., 2003; van der Sande et al., 2003; Walter, 2004
Per-field-based classifiers	GIS plays an important role in per-field classification, integrating raster data and vector data in classification. The vector data are often used to subdivide an image into parcels, and classification is based on the parcels, avoiding the spectral variation inherent in the same class.	GIS-based classification approaches	Aplin and Atkinson, 2001; Aplin et al., 1999a; Barnsley and Barr, 1997; Carlotto, 1998; Chalifoux et al., 1998; Dean and Smith, 2003; Lobo et al., 1996; Smith and Fuller, 2001

Table 5.1 Major Categories of Image Classification Methods

An alternative approach to heterogeneous landscapes is to use an object-oriented classification (Benz et al., 2004; Gitas et al., 2004; Thomas et al., 2003; Walter, 2004). Two stages are involved in an object-oriented classification. Image segmentation merges pixels into objects, and a classification then is implemented based on objects instead of individual pixels. In the process of creating objects, scale determines the presence or absence of an object class, and the size of an object affects the classification result. This approach has proven to be able to provide better classification results than per-pixel-based classification approaches, especially for fine spatial-resolution data. Object-based classification methods have been developed and applied with varying degrees of success (Antonarakis et al., 2008; Blaschke et al., 2000; Laliberte et al., 2004; Liu et al., 2005; Mallinis et al., 2008; Miliaresis and Kokkas, 2007; Mitri and Gitas, 2008; Shackelford and Davis, 2003; van Coillie et al., 2007; Walker and Briggs, 2007; Zhou and Troy, 2008; Zhou and Wang, 2007, 2008). For example, van Coillie et al. (2007) developed a three-step object-based classification that included image segmentation, feature selection by generic algorithm (GAs), and joint neural-network-based object classification. Zhou and Wang (2007, 2008) developed an algorithm of multiple agent segmentation and classification (MASC) that included four steps: image segmentation, shadow effect, MANOVA-based classification, and post-classification. This algorithm was applied to impervious surface extraction in the state of Rhode Island.

As fine spatial-resolution data (mostly better than 5 m spatial resolution), such as IKONOS and QuickBird data, have become easily available in recent years, they are employed increasingly for different applications (Goetz et al., 2003; Herold et al., 2003; Hurtt et al., 2003; Sugumaran et al., 2002; van der Sande et al., 2003; Wang et al., 2004; Xu et al., 2003; Zhang and Wang, 2003). A major advantage of these fine spatial-resolution images is that such data contain large amounts of spatial information, providing a great potential to extract much more detailed information on land-cover structures than medium or coarse spatial-resolution data. However, some new problems associated with the fine spatial-resolution image data emerge, notably the shades caused by topography, tall buildings, or trees and the high spectral variation within the same land-cover class. These disadvantages may lower classification accuracy if the classifiers used cannot handle them effectively (Cushnie, 1987; Irons et al., 1985). Increased spectral variation is common with the high degree of spectral heterogeneity in complex landscapes such as urban environments. In order to make full use of the rich spatial information inherent in fine spatial-resolution data, it is necessary to minimize the negative impact of high intraspectral variation. The combination of spectral and spatial classification is especially valuable for fine land use and land cover (LULC) classifications in urban areas or in areas with complex landscapes.

5.2 Case Study: Urban Feature Extraction from High Spatial-Resolution Satellite Imagery

5.2.1 Data Used

A description of the study area can be found in Sec. 3.2.1 of Chap. 3. Two IKONOS images were used for this study. One image covering the residential area was acquired on August 1, 2003, and another image covering the central business district (CBD) area was acquired on October 6, 2003. IKONOS offers both multispectral and panchromatic channels. The multispectral IKONOS image is composed of four bands [i.e., blue, green, red, and near-infrared (NIR)] with a spatial resolution of 4 m, and the panchromatic band has a spatial resolution of 1 m. A U.S. Geological Survey (USGS) topographic map in digital raster graphics (DRG) format was used to geocorrect the IKONOS images. Fifty ground control points were chosen over an image, and a root-mean-square error (RMSE) of less than 0.5 was achieved for both images. The IKONOS images were reprojected to the Universal Transverse Mercator (UTM) projection (zone 16, datum WGS84). Given that the image was cloud-free and covered a relatively small area, atmospheric corrections were not applied. To use both spatial and spectral information for feature extraction, a data fusion technique was performed. Four multispectral bands were pan-sharpened using the panchromatic band.

An aerial photo of Marion County with a spatial resolution of 0.14 m was used for classification refinement. The aerial photograph was provided by the Indianapolis Mapping and Geographic Infrastructure System and was acquired in April 2003. The coordinate system is Indiana State Plane East, zone 1301, with North American datum of 1983. The aerial photograph was reprojected into the same coordinate system as the IKONOS images and resampled to 1-m spatial resolution for better comparison with the IKONOS images.

5.2.2 Image Segmentation

An object-oriented classification technique based on image segmentation incorporating classification rules was applied in this study. The analytical procedures can be categorized into three steps: image segmentation, rule-based image classification, and post-classification refinement. The basic idea of object-oriented methods is to handle image primitives as objects that were derived from image segmentation. Image segmentation is the process of partitioning images into segments by grouping neighboring pixels with similar characteristics (e.g., spectral, spatial, or textural). Image segmentation is crucial to the performance of object-based classification (Baatz and Schäpe, 2000). The ideal segments should correspond to the real-world objects.

However, over- and undersegmentation occur constantly during the process; therefore, appropriate segmentation techniques need to be selected and executed carefully.

An edge-based segmentation algorithm provided in ENVI software was used in this research. This algorithm can generate multiscale segmentation results from finer to coarser segmentation by suppressing weak edges to different degrees. Only one input parameter, *scale level,* is required to be set up, which makes the algorithm easy to use. The parameter needs to be defined carefully because higher scale level would lead to fewer segments, whereas lower scale level yields more segments. Too few segments (undersegmentation) results in a mixture of different features, whereas too many segments (oversegmentation) result in divided features. Ideally, the highest scale level that can delineate the boundaries of features properly should be selected. If oversegmentation cannot be avoided, the results should be refined further by merging smaller segments. The parameter *merge level,* which represents the threshold lambda value, also needs to be defined properly. The value ranges from 0 to 100, and the highest merge level that can delineate the boundaries of features precisely should be established.

5.2.3 Rule-Based Classification

This method assumes that the features can be defined by creating rules based on the object's characteristics (e.g., spatial, spectral, and textural). Features are classified based on human knowledge about different feature types; for example, roads are elongated, vegetation has a low normalized difference vegetation index (NDVI) value, asphalt and concrete have different spectral appearances, and trees are more textured than grass. In general, to extract a feature accurately, a number of rules, instead of one rule, need to be established. For instance, the characteristics of rooftops can be summarized as (1) rooftops have little or no NVDI values, (2) the shapes of rooftops are close to rectangular, and (3) the areas of rooftops are within a certain range. In terms of the attributes of rooftops, three classification rules can be built accordingly: (1) lower NDVI value, (2) rectangular shape, and (3) area within a certain range. These rules are sufficient to differentiate rooftops from other features. For instance, rooftops can be differentiated easily from vegetation owing to the different NDVI values, and shape is another important factor for extracting rooftops, which can help to distinguish features such as roads and driveways from rooftops. Area can help to distinguish rooftops from parking lots. Although parking lots and rooftops share some similar attributes, such as spectral reflectance and shape, parking lots usually cover a larger area than rooftops. Therefore, by applying the classification rules, rooftops can be differentiated successfully from other features and precisely extracted from imagery.

	Residential		
Attribute	**Impervious Surface**	**Road**	**Building**
Band ratio	<0.4118	(0.2902, 0.4068)	<0.3409
Average Band 1	>408.4838	>408.4838	>408.4838
Area	>80.3239	>80.3239	>80.3239
Length		>114.5334	
Average Band 4			<600.2772

TABLE 5.2 Rules for Feature Extraction in the Residential Area

In this research, two IKONOS images covering residential and CBD areas, respectively, were used to extract three different features, impervious surfaces, roads, and buildings, by conducting the rule-based classification. One of the difficulties in feature extraction from high spatial-resolution imagery is within-class variation, especially in CBD areas. For example, rooftops made of different materials have different spectral appearances. Some are dark, whereas some are bright. Roads made of concrete or asphalt also appear in different tones. Therefore, to extract the features precisely and improve the accuracy of the results, roads and buildings were roughly categorized into three levels before image classification. The levels include bright roads and bright buildings, dark roads and dark buildings, and roads and buildings in between. The feature in each category was extracted separately by building and applying separate classification rules. The final map was created by overlaying three separately generated results. The classification rules established for extracting each feature in this study are shown in Tables 5.2 and 5.3.

5.2.4 Post-Classification Refinement and Accuracy Assessment

A classification refinement was performed with the panchromatic bands and aerial photographs as reference. Misclassifications were corrected, and missing segments caused by shadows were recovered using GIS tools.

An accuracy assessment was conducted on all the six final maps using the panchromatic bands and aerial photographs as reference. Four parameters were calculated to quantify the results, including producer's accuracy, user's accuracy, overall accuracy, and kappa statistic. Two hundreds sample points were randomly selected within the image mosaic for each map. One hundred points were selected from the feature class, and the other hundred points were selected from the nonfeature class. The strategy was to make sure that the feature class could be sampled sufficiently and be well inspected.

	CBD						
		Road			Building		
Attribute	**Impervious Surface**	**Dark Road**	**Bright Road**	**Road in Between**	**Dark Building**	**Bright Building**	**Building in Between**
Band ratio	<0.2235	<0.1843	<0.1216	<0.2235	<0.0824	<0.1373	<0.2235
Average Band 3	>136.5882		(355.1294, 728.4706)	>136.5882	(136.5882, 282.2823)	>236.7529	>136.5882
Area		>730.5338	>327.9876	>209.0821	>245.4749	>314.4511	>174.5188
Elongation		>1.9501	>1.5681		<3.7989	<2.1060	<2.9774
Average Band 1		(228.4902, 289.4209)	(449.3640, 594.0745)	(281.8046, 403.6660)			
Length		>109.8408		23.9004			
Rect_fit		<0.5		<0.6517		>0.3713	>0.4292
Roundness				<0.4142			
Stdband_3				<73.1534	<65.2906		
Texture_mean					<805.2845		
Texture_variance							<2939.4448

TABLE 5.3 Rules for Feature Extraction in the CBD Area

5.2.5 Results of Feature Extraction

Impervious surfaces, buildings, and roads were extracted, and three maps were created for each IKONOS image (Figs. 5.1 and 5.2). Owing to the significant within-class variations in the CBD image, bright roads, dark roads, and roads in between were extracted separately. The separate resulting maps then were overlaid to create a final road map. The same strategy was applied to building extraction in the CBD area. Visual inspection indicated that the distribution patterns of extracted features were in accordance with those in the original images and that the boundaries of features were delineated appropriately and precisely.

The results of accuracy assessment indicated that all maps achieved a relatively high accuracy. Owing to less within-class variation as well as fewer shadows existing in the residential area, features were extracted with higher accuracy than those extracted from the CBD area. To be specific, impervious surface obtained an overall accuracy of 94.5 percent and a kappa value of 0.89 in the residential area. An overall accuracy of 91.5 percent and a kappa value of 0.83 were achieved for impervious surfaces in the CBD area. Roads were extracted with an overall accuracy of 95 percent and a kappa value of 0.9 in the residential area, whereas in the CBD area an overall accuracy of 92 percent and a kappa value of 0.84 were obtained for road extraction. Buildings obtained an overall accuracy of 95.5 percent and a kappa value of 0.91 in the residential area and an overall accuracy of 93 percent and a kappa value of 0.86 in the CBD area. Detailed information about the accuracy assessment is shown in the Tables 5.4 and 5.5.

5.3 Discussion

A rule-based classification was applied to feature extraction from IKONOS images in this study. Image segmentation is the first and also crucial step in the process. The accuracy of the final maps relied tremendously on the segmentation results. The parameter *scale level* needs to be defined carefully. Higher scale level created fewer segments, whereas lower scale level yielded more segments. Ideally, the highest scale level that can delineate the boundaries of features appropriately should be selected. Nevertheless, for real practice, the conditions were not ideal most of the time because it is almost impossible to make sure that the boundaries of all the features are delineated precisely. On the other hand, to extract different features, the scale level of image segmentation also should be different. Therefore, scale level needs to be defined every time a new feature is being extracted. The strategy employed in this study was to use smaller scale levels to segment the images, which would keep most of the detailed information in the next step, and then to test different merge levels to merge the small segments until the boundaries of most of the features were

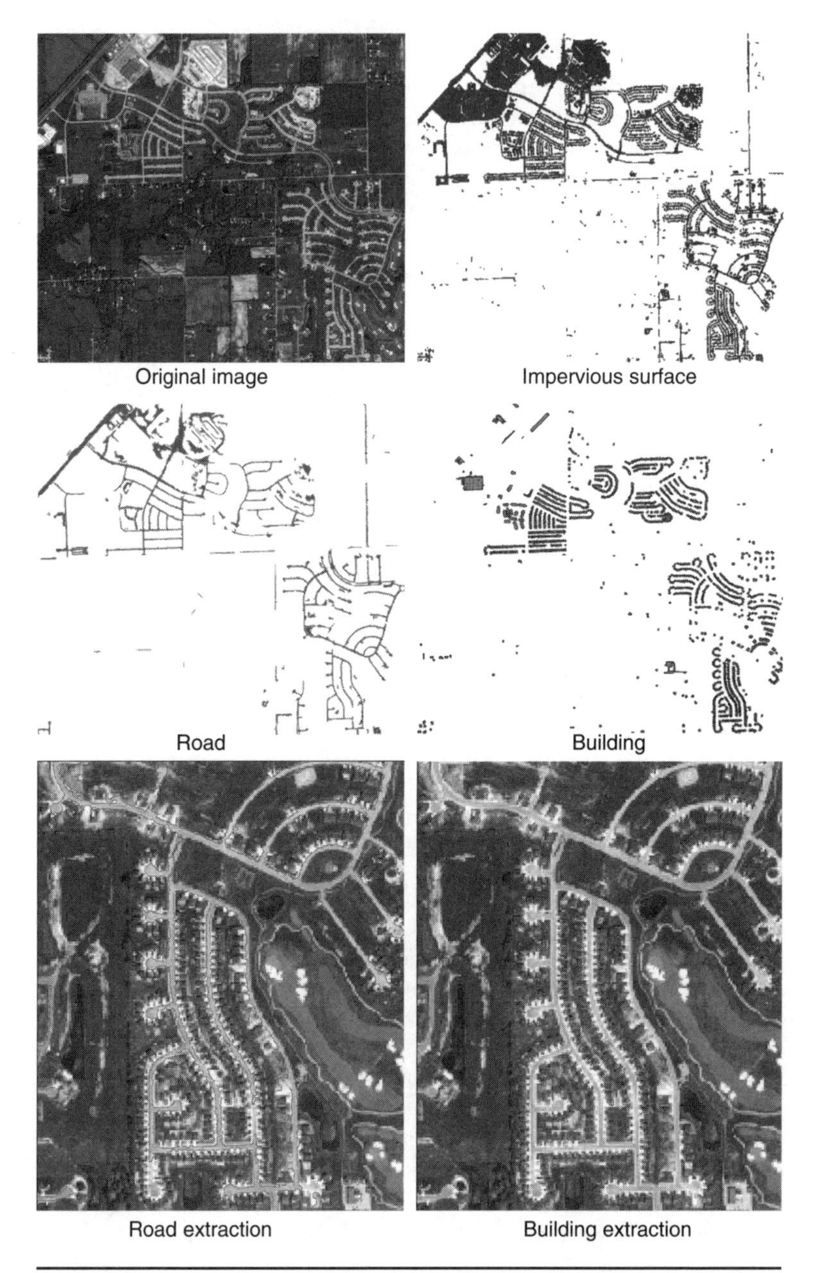

FIGURE 5.1 Feature extraction from residential area. See also color insert.

delineated appropriately. Refinement should be carried out as well to improve the accuracy of the segmentation results further.

Classification rules need to be built carefully. Each rule had to be established appropriately to make sure that the desired features remain and unwanted features are removed as much as possible. In many cases,

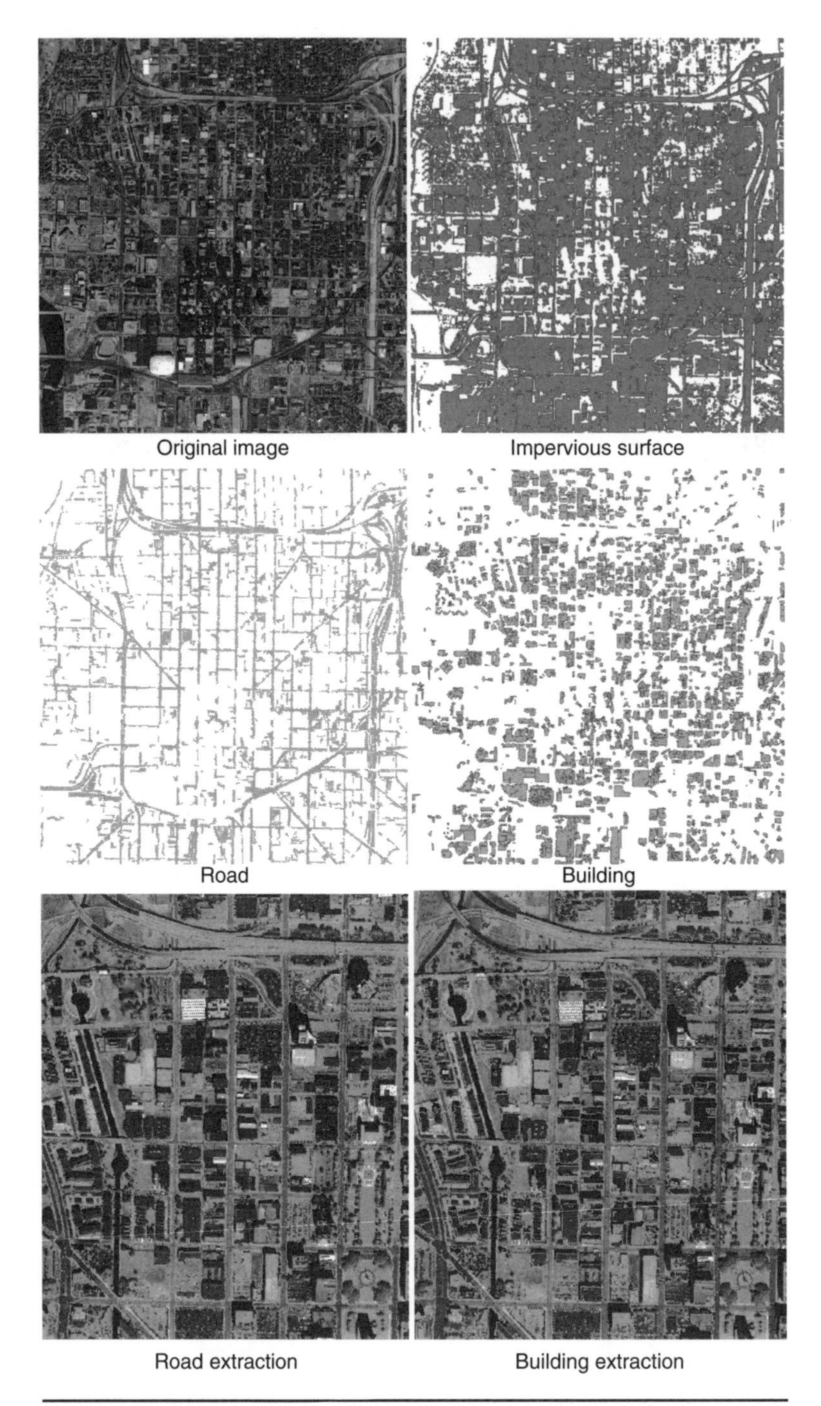

FIGURE 5.2 Feature extraction from CBD area. See also color insert.

	Reference Total	Classified Total	Number Correct	Producer's Accuracy (%)	User's Accuracy (%)
IS	101	100	95	94%	95%
Non-IS	99	100	94	95%	94%
Total	200	200	189		
Overall accuracy = 94.5%; kappa = 0.89					
Road	96	100	93	97%	93%
Nonroad	104	100	97	93%	97%
Total	200	200	190		
Overall accuracy = 95%; kappa = 0.9					
Building	97	100	94	97%	94%
Nonbuilding	103	100	97	94%	97%
Total	200	200	191		
Overall accuracy = 95.5%; kappa = 0.91					

IS = impervious surface.

TABLE 5.4 Accuracy Assessment for Feature Extraction in the Residential Area

	Reference Total	Classified Total	Number Correct	Producer's Accuracy (%)	User's Accuracy (%)
IS	101	100	92	91%	92%
Non-IS	99	100	91	92%	91%
Total	200	200	183		
Overall accuracy = 91.5%; kappa = 0.83					
Road	96	100	90	94%	90%
Nonroad	104	100	94	90%	94%
Total	200	200	182		
Overall accuracy = 92%; kappa = 0.84					
Building	92	100	89	97%	89%
Nonbuilding	108	100	97	90%	97%
Total	200	200			
Overall accuracy = 93%; kappa = 0.86					

IS = impervious surface.

TABLE 5.5 Accuracy Assessment for Feature Extraction in the CBD Area

the rules affected both sides. Therefore, building rules successfully is still a challenge and needs to be conducted carefully. Instead of using one rule, numerous rules need to be defined for extracting each feature type. However, too many rules also bring difficulties. For example, when a mistake occurs, it is difficult to find out which rule(s) caused the mistake. Thus a sensitivity test for each rule would be beneficial.

Another problem encountered in this study is within-class variation, especially in the CBD area. The same types of features, such as roads, buildings, and parking lots, are made of different materials; for example, roads can be made of concrete, asphalt, or bricks, and rooftops could be made of plastic, concrete, or glass. Different materials have different spectral characteristics. As a result, the differences in the spectral appearances of the same feature type in the remote sensing images can be significant. Therefore, classification errors may be made when spectral attributes are used for feature extraction. To tackle this problem, features (e.g., buildings and roads) were roughly categorized into three levels in this study: bright roads and bright buildings, dark roads and dark buildings, and gray roads and gray buildings. Features in different category were extracted separately. The final map was created by adding those separately extracted results. With such stratification, the accuracy can be increased significantly.

Shadow caused a lot of problems in feature extraction from high-spatial-resolution imagery, especially in the CBD area. Higher buildings usually cause larger shadows than lower buildings. In this study, a certain amount of shadow coverage was contained in the IKONOS images, and the closer to the center of the city one was, the great was the shadow coverage. To reduce the impact of shadow on feature extraction, a refinement process was carried out after the rule-based classification. The aerial photographs were used as references to recover the features covered by shadows in the original imagery.

5.4 Conclusions

A rule-based classification was presented and applied for feature extraction from high spatial-resolution images in this chapter. Three maps corresponding to impervious surfaces, buildings, and roads, respectively, were yielded from each IKONOS image of Indianapolis. The results indicate that the boundaries of features were delineated appropriately and precisely. A high accuracy of feature extraction was obtained for each map. Impervious surface achieved an overall accuracy of 94.5 percent in the residential area and an overall accuracy of 91.5 percent in the CBD area. Roads were extracted with an overall accuracy of 95 percent in the residential area and an overall accuracy of 92 percent in the CBD area. The overall accuracy of building extraction was 95.5 percent for the residential area and 93 percent for the CBD area. Overall, feature extracted from the residential area achieved a higher accuracy than that from the CBD area owing to the spectral complexity and the shadow problem in the CBD area.

The rule-based classification technique had four advantages. First, many urban land cover types, such as roads, buildings, and parking lots, share similar spectral characteristics. To differentiate them, both spatial and texture information needs to be used. Unlike pixel-based classification, rule-based classification used not only spectral information but also spatial (e.g. shape, elongation, and area) and textural characteristics and neighborhood relationship to extract features. Therefore, this method was more likely to produce maps with high accuracy. Second, image segmentation was easy to perform and required only two parameters, scale level and merge level, to be established. Third, the classification process was straightforward because of use of classification rules. Finally, fuzzy logic was incorporated into this study to overcome a problem created by traditional rule-based classifications based on strict binary rules. Fuzzy logic uses membership to represent the degree rather than that an object belonged to a feature type. The output of each fuzzy rule was a confidence map. An object was assigned to a feature type that had the maximum confidence value.

Although features can be extracted with relatively high accuracy with this method, there are some restrictions. For example, image segmentation was the first and also crucial step of an object-based classification. For this method, an edge-based segmentation technique was used. The drawback of edge-based segmentation technique is that it is easily affected by noise compared with other methods, such as region-based segmentation. Therefore, this method may not work well in a noisy environment such as the CBD area. To tackle this problem, a region-based segmentation technique can be used rather than the edge-based approach. Our approach also was based on fuzzy rules. In general, numerous rules need to be established to extract features instead of using one rule. However, too many rules would bring difficulties for classification. When a mistake occurs, it would be difficult to find out which rule caused the mistake. Thus a sensitivity test of each rule would be beneficial.

References

Antonarakis, A. S., Richards, K. S., and Brasington, J. 2008. Object-based land cover classification using airborne LIDAR. *Remote Sensing of Environment* **112,** 2988–2998.

Aplin, P., and Atkinson, P. M. 2001. Sub-pixel land cover mapping for per-field classification. *International Journal of Remote Sensing* **22,** 2853–2858.

Aplin, P., Atkinson, P. M., and Curran, P. J. 1999a. Per-field classification of land use using the forthcoming very fine spatial resolution satellite sensors: Problems and potential solutions. In *Advances in Remote Sensing and GIS Analysis*, edited by P. M. Atkinson and N. J. Tate, pp. 219–239. New York: Wiley.

Aplin, P., Atkinson, P. M., and Curran, P. J. 1999b. Fine spatial resolution simulated satellite sensor imagery for land cover mapping in the United Kingdom. *Remote Sensing of Environment* **68,** 206–216.

Baatz, M., and Schäpe, A. 2000. Multiresolution segmentation: An optimization approach for high quality multi-scale image segmentation. In *Angewandte Geographiche Informationsverarbeitung*, Vol. XII, edited by J. Strobl and T. Blaschke, pp. 12–23. Heidelberg: Wichmann-Verlag.

Barnsley, M. J., and Barr, S. L. 1997. Distinguishing urban land-use categories in fine spatial resolution land-cover data using a graph-based, structural pattern recognition system. *Computers, Environments and Urban Systems* **21,** 209–225.

Benz, U. C., Hofmann, P., Willhauck, G., et al. 2004. Multi-resolution, object-oriented fuzzy analysis of remote sensing data for GIS-ready information. *ISPRS Journal of Photogrammetry & Remote Sensing* **58,** 239–258.

Blaschke, T., Lang, S., Lorup, E., et al. 2000. Object-oriented image processing in an integrated GIS/remote sensing environment and perspectives for environmental applications. *Environmental Information for Planning, Politics and the Public, Metropolis, Marburg* **2,** 555–570.

Cao, G., and Jin, Y. Q. 2007. Automatic detection of main road network in dense urban area using microwave SAR images. *Imaging Science Journal* **55,** 215–222.

Carlotto, M. J. 1998. Spectral shape classification of Landsat Thematic Mapper imagery. *Photogrammetric Engineering and Remote Sensing* **64,** 905–913.

Chalifoux, S., Cavayas, F., and Gray, J. T. 1998. Mapping-guided approach for the automatic detection on Landsat TM images of forest stands damaged by the spruce budworm. *Photogrammetric Engineering and Remote Sensing* **64,** 629–635.

Cushnie, J. L. 1987. The interactive effect of spatial resolution and degree of internal variability within land-cover types on classification accuracies. *International Journal of Remote Sensing* **8,** 15–29.

Dean, A. M., and Smith, G. M. 2003. An evaluation of per-parcel land cover mapping using maximum likelihood class probabilities. *International Journal of Remote Sensing* **24,** 2905–2920.

Geneletti, D., and Gorte, B. G. H. 2003. A method for object-oriented land cover classification combining Landsat TM data and aerial photographs. *International Journal of Remote Sensing,* **24,** 1273–1286.

Gitas, I. Z., Mitri, G. H., and Ventura, G. 2004. Object-based image classification for burned area mapping of Creus Cape Spain using NOAA-AVHRR imagery. *Remote Sensing of Environment* **92,** 409–413.

Goetz, S. J., Wright, R. K., Smith, A. J., et al. 2003. IKONOS imagery for resource management: tree cover, impervious surfaces, and riparian buffer analyses in the mid-Atlantic region. *Remote Sensing of Environment* **88,** 195–208.

Guo, D., Weeks, A., and Klee, H. 2007. Robust approach for suburban road segmentation in high-resolution aerial images. *International Journal of Remote Sensing* **28,** 307–318.

Harris, P. M., and Ventura, S. J. 1995. The integration of geographic data with remotely sensed imagery to improve classification in an urban area. *Photogrammetric Engineering and Remote Sensing* **61,** 993–998.

Herold, M., Liu, X., and Clarke, K. C. 2003. Spatial metrics and image texture for mapping urban land use. *Photogrammetric Engineering and Remote Sensing* **69,** 991–1001.

Hsieh, P.-F., Lee, L. C., and Chen, N.-Y. 2001. Effect of spatial resolution on classification errors of pure and mixed pixels in remote sensing. *IEEE Transactions on Geoscience and Remote Sensing* **39,** 2657–2663.

Hurtt, G., Xiao, X., Keller, M., et al. 2003. IKONOS imagery for the large scale biosphere: Atmosphere experiment in Amazonia (LBA). *Remote Sensing of Environment* **88,** 111–127.

Irons, J. R., Markham, B. L., Nelson, R. F., et al. 1985. The effects of spatial resolution on the classification of Thematic Mapper data. *International Journal of Remote Sensing* **6,** 1385–1403.

Janssen, L., and Molenaar, M. 1995. Terrain objects, their dynamics and their monitoring by integration of GIS and remote sensing. *IEEE Transactions on Geoscience and Remote Sensing* **33,** 749–758.

Jin, X., and Paswaters, S. 2007. A fuzzy rule base system for object-based feature extraction and classification. In: K. Ivan (ed.). *SPIE,* p. 65671H.

Karimi, H. A., and Liu, S. 2004. Developing an automated procedure for extraction of road data from high-resolution satellite images for geospatial information systems. *Journal of Transportation Engineering ASCE* **130,** 621–631.

Laliberte, A. S., et al. 2004. Object-oriented image analysis for mapping shrub encroachment from 1937 to 2003 in southern New Mexico. *Remote Sensing of Environment* **93,** 198–210.

Liu, Z. J., Wang, J., and Liu, W. P. 2005. Building extraction from high resolution imagery based on multi-scale object oriented classification and probabilistic Hough transform. In *Proceedings of the Geoscience and Remote Sensing Symposium, IGARSS'05,* pp. 2250–2253. New York: IEEE International.

Lloyd, C. D., Berberoglu, S., Curran, P. J., and Atkinson, P. M. 2004. A comparison of texture measures for the per-field classification of Mediterranean land cover. *International Journal of Remote Sensing* **25,** 3943–3965.

Lobo, A., Chic, O., and Casterad, A. 1996. Classification of Mediterranean crops with multisensor data: Per-pixel versus per-object statistics and image segmentation. *International Journal of Remote Sensing* **17,** 2385–2400.

Lu, D., and Weng, Q. 2007. A survey of image classification methods and techniques for improving classification performance. *International Journal of Remote Sensing* **28,** 823–870.

Mallinis, G., Koutsias, N., Tsakiri-Strati, M., and Karteris, M. 2008. Object-based classification using Quickbird imagery for delineating forest vegetation polygons in a Mediterranean test site. *ISPRS Journal of Photogrammetry and Remote Sensing* **63,** 237–250.

Mayer, H., Laptev, I., and Baumgartner, A. 1998. Multi-scale and snakes for automatic road extraction. In *Proceedings of the Computer Vision Symposium, ECCV'98,* pp. 720–733.

Mayer, H., Laptev, I., Baumgartner, A., and Steger, C. 1997. Automatic road extraction based on multi-scale modeling, context and snakes. *International Archives of Photogrammetry and Remote Sensing* **32,** 106–113.

Miliaresis, G., and Kokkas, N. 2007. Segmentation and object-based classification for the extraction of the building class from LIDAR DEMs. *Computers & Geosciences* **33,** 1076–1087.

Mitri, G. H., and Gitas, I. Z. 2008. Mapping the severity of fire using object-based classification of IKONOS imagery. *International Journal of Wildland Fire* **17,** 431–442.

Pal, N. R., and Pal, S. K. 1993. A review on image segmentation techniques. *Pattern Recognition* **26,** 1277–1294.

Pedley, M. I., and Curran, P. J. 1991. Per-field classification: An example using SPOT HRV imagery. *International Journal of Remote Sensing* **12,** 2181–2192.

Shackelford, A. K., and Davis, C. H. 2003. A combined fuzzy pixel-based and object-based approach for classification of high-resolution multispectral data over urban areas. *IEEE Transactions on Geoscience and Remote Sensing* **41,** 2354–2363.

Sharma, K. M. S., and Sarkar, A. 1998. A modified contextual classification technique for remote sensing data *Photogrammetric Engineering & Remote Sensing* **64,** 273–280.

Smith, G. M., and Fuller, R. M. 2001. An integrated approach to land cover classification: An example in the Island of Jersey. *International Journal of Remote Sensing* **22,** 3123–3142.

Sugumaran, R., Zerr, D., and Prato, T. 2002. Improved urban land cover mapping using multitemporal IKONOS images for local government planning. *Canadian Journal of Remote Sensing* **28,** 90–95.

Thomas, N., Hendrix, C., and Congalton, R. G. 2003. A comparison of urban mapping methods using high-resolution digital imagery. *Photogrammetric Engineering and Remote Sensing* **69,** 963–972.

van Coillie, F. M. B., Verbeke, L. P. C., and De Wulf, R. R. 2007. Feature selection by genetic algorithms in object-based classification of IKONOS imagery for forest mapping in Flanders, Belgium. *Remote Sensing of Environment* **110,** 476–487.

van der Sande, C. J., de Jong, S. M., and de Roo, A. P. J. 2003. A segmentation and classification approach of IKONOS-2 imagery for land cover mapping to assist flood risk and flood damage assessment. *International Journal of Applied Earth Observation and Geoinformation* **4,** 217–229.

Voode, T. V. D., et al. 2004. Extraction of land use/land cover-related information from very high resolution data in urban and suburban areas. In *Remote Sensing in Transition: Proceedings of the 23rd Symposium of the European Association of Remote Sensing Laboratories,* edited by R. Goossens, pp. 237–244. Rotterdam: Millpress.

Walker, J. S., and Briggs, J. M. 2007. An object-oriented approach to urban forest mapping in Phoenix. *Photogrammetric Engineering and Remote Sensing* **73,** 577–583.

Walter, V. 2004. Object-based classification of remote sensing data for change detection. *ISPRS Journal of Photogrammetry & Remote Sensing* **58,** 225–238.

Wang, L., Sousa, W. P., Gong, P., and Biging, G. S. 2004. Comparison of IKONOS and QuickBird images for mapping mangrove species on the Caribbean coast of panama. *Remote Sensing of Environment* **91,** 432–440.

Wei, Y., Zhao, Z., and Song, J. 2004. Urban building extraction from high-resolution satellite panchromatic image using clustering and edge detection. *Proceedings of 2004 IEEE International Geoscience and Remote Sensing Symposium,* **3**, 20–24, Anchorage: IEEE GRSS.

Xu, B., Gong, P., Seto, E., and Spear, R. 2003. Comparison of gray-level reduction and different texture spectrum encoding methods for land-use classification using a panchromatic ikonos image. *Photogrammetric Engineering and Remote Sensing* **69,** 529.

Yun, L., and Uchimura, K. 2007. Using self-organizing map for road network extraction from Ikonos imagery. *International Journal of Innovative Computing Information and Control* **3,** 641–656.

Zhang, Q., and Wang, J. 2003. A rule-based urban land use inferring method for fine-resolution multispectral imagery. *Canadian Journal of Remote Sensing* **29,** 1–13.

Zhou, W., and Troy, A. 2008. An object-oriented approach for analyzing and characterizing urban landscape at the parcel level. *International Journal of Remote Sensing* **29,** 3119–3135.

Zhou, Y., and Wang, Y. Q. 2007. An assessment of impervious surface areas in Rhode Island. *Northeastern Naturalist* **14,** 643–650.

Zhou, Y., and Wang, Y. Q. 2008. Extraction of impervious surface areas from high spatial resolution imagery by multiple agent segmentation and classification. *Photogrammetric Engineering and Remote Sensing* **74,** 857–868.

CHAPTER 6

Building Extraction from LiDAR Data

Accurate and timely spatial information about buildings (and associated attributes) in urban areas is needed as the basis to assist decision making in understanding, managing, and planning the continuously changing environment. Construction of new neighborhoods, transportation networks, and drainage systems are a few situations in which information about buildings is essential to the planning process. However, frequently, useful quantitative spatial information cannot be obtained directly from existing data sources and must rely on extraction techniques. Conventional field survey methods are time-consuming and costly. Planning maps are used widely in various urban applications because of their high accuracy, although producing and updating them are time-consuming and hard to automate.

Recently available high-resolution remotely sensed data have provided the potential to achieve effective and efficient building information through automated extraction methods. Aerial photographs are a popular data source for building extraction owing to their high resolution. However, they often cover selected areas only and lack multiple times of coverage for the same area, making it difficult to update the building database once created. High-resolution commercial satellite image data have been used increasingly in building extraction, but the issues of shadowing and distortion can affect the resulting accuracy to a certain degree. A new data source, light detection and ranging (LiDAR) data, is used in this research, which provides land-surface elevation information by emitting a laser pulse and providing high vertical and horizontal resolutions of less than 1 m.

Object-oriented classifiers have been employed to classify and extract terrain features in recent studies. When combined with a LiDAR data source, an object-oriented method shows great potential for extracting buildings, which are identified by height information. However, it is a challenging task to define a proper rule to identify meaningful image objects and belonging classes owing to the fact that the object identification is influenced significantly by the researcher's knowledge and the quality of the data. Moreover, urban trees tend to be confused with buildings because of similar heights of trees and low-rise buildings. Therefore, how to extract urban buildings accurately and efficiently from LiDAR data based on object-oriented classifiers is one of the main issues in current research.

The objective of this research is to extract buildings from LiDAR data based on an object-oriented classification technique. The downtown area of Indianapolis, Indiana, was selected for this research; it is located in the center of Marion County between 39°36′ to 39°56′N and 85°56′ to 86°19′W, covering an area of 4 mi² (Fig. 6.1). This area is also

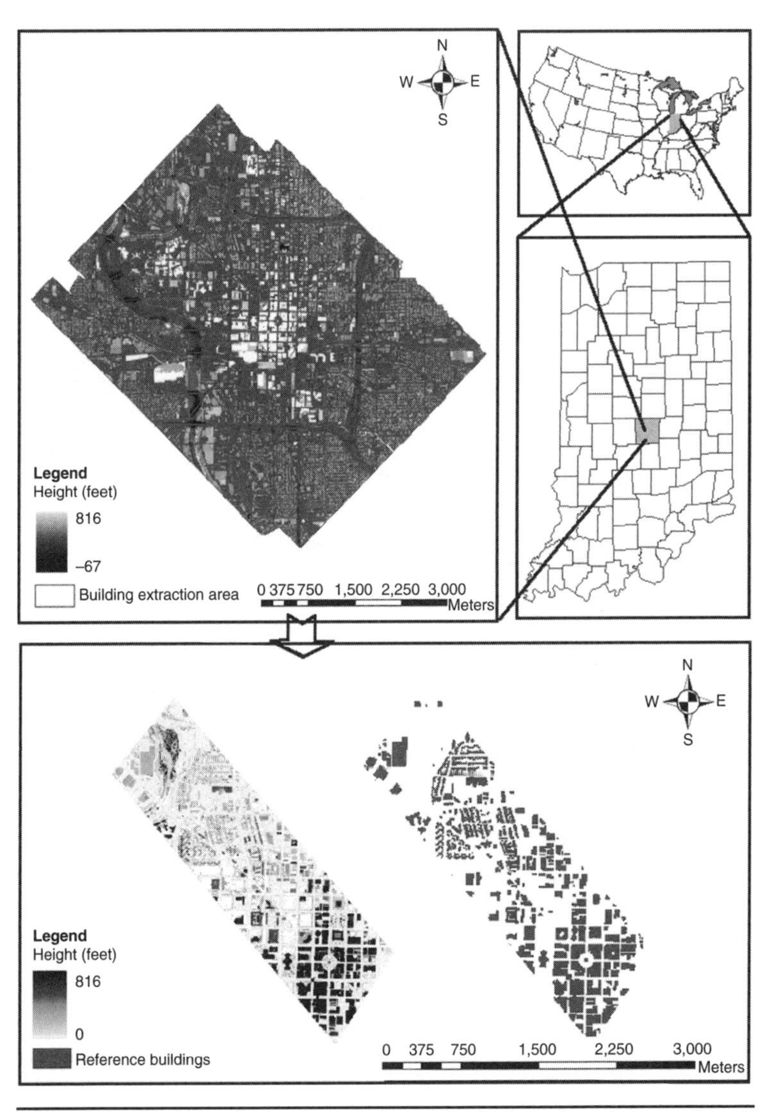

FIGURE 6.1 Study area in downtown Indianapolis, Indiana. See also color insert.

regarded as the central business district (CBD) Indianapolis, with the Monument Circle as its center. It is expected that the methodology for building extraction can be applied to other urban features and can be extrapolated to other parts of Indianapolis and other cities with similar surface characteristics.

6.1 The LiDAR Technology

LiDAR technology dates back to the 1970s, when airborne laser scanning system was developed. The first system was developed by NASA, and it operated by emitting a laser pulse and precisely measuring its return time to calculate the range (height) by using the speed of light. Later, with advent of the Global Positioning Satellite (GPS) system in the late 1980s and rapidly pulsing laser scanners, the necessary positioning accuracy and orientation parameters were achieved. The LiDAR system now had become comprehensive, typically with four primary components, including a rapidly pulsing laser scanner, precise kinematic GPS positioning, inertial measurement units (IMUs) for capturing the orientation parameters (i.e., tip, tilt, and roll angles), and a precise timing clock (Renslow et al., 2000). With all the components, along with software and computer support, assembled in a plane, and by flying along a well-defined flight plan, LiDAR data are collected in the form of records of the returned-pulse range values and other parameters (e.g., scanner position, orientation parameters, etc.). After filtering noise and correcting, laser point coordinates are calculated, and a basic ASCII file of x, y, and z values for LiDAR data is formed, which may be transformed into a local coordinate system. LiDAR data can provide very good vertical and horizontal resolutions as high as 0.3 m (Webster et al., 2004).

Since the inception of LiDAR systems, LiDAR technology has been used widely in many geospatial applications owing to its high data resolution, short acquisition time, and low cost compared with traditional methods. Unlike other remotely sensed data, LiDAR data focus solely on geometry rather than radiometry. Some typical products derived from it contain digital elevation model (DEM), surface elevation model (SEM), triangulated irregular network (TIN), and intermediate-return information. Moreover, researchers working on monitoring and predicting forest biomass characteristics also regard LiDAR data as a critical source in urban studies. As a result, feature classification and extraction based on LiDAR data are extensive (Clode et al., 2007; Filin, 2004; Forlani et al., 2006; Lee et al., 2008). Moreover, LiDAR data show great potential in urban building extraction because elevation data can be derived quickly and at high resolution in comparison with photogrammetric techniques (Miliaresis and Kokkas, 2007).

6.2 Building Extraction

Building extraction is a feature classification and detection technique, as well as a pattern recognition technique. The quality of building-extraction results depends both on the *content* of target areas and the *observability* of different objects from imagery. To get better observability of building objects, some prerequisites need to be considered. One of the most important prerequisites is spatial resolution (Mayer, 1999). Aerial imagery is a popular data source for building extraction owing to its high spatial resolution of 1 m or less. Another important prerequisite is the number of images in which a scene can be found. Mono-, stereo-, and multispectral images can be distinguished and used in reconstructing objects (Mayer, 1999). Multiple views enable a better geometric reconstruction of building objects and also reduce problems with occlusion. Recent commercial multispectral high resolution satellite imagery such as Quickbird and IKONOS is regarded as a replaceable data source for building-extraction techniques with an acceptable resolution. Quickbird panchromatic imagery has a resolution of 0.6 to 0.7 m and multispectral imagery of 2.4 and 2.8 m with four channels (i.e., blue, green, red, and near-infrared), whereas the respective resolutions of IKONOS are 1 and 4 m. Furthermore, this type of satellite imagery provides a renewable data source that can be updated every few days (1 to 3.5 days for Quickbird, 3 to 5 days for IKONOS) and covers almost every region of the earth. Extremely interesting for an automatic interpretation of buildings are airborne LiDAR data. As a unique data collection and structure, LiDAR data have potential over satellite imagery and aerial photographs for accurate capture of urban features with absolute height information, especially for building extraction. LiDAR data have been used in urban planning, telecommunications network planning, and vehicle navigation, all of which are of increasing importance in urban areas (Kokkas, 2005).

A wide range of models and methods has been developed for building extraction. The following discussion will focus on the state-of-the-art automatic building-extraction techniques. Most of the extraction algorithms use edge-based techniques that consist of linear feature detection for perpendicular outline and flat roofs, grouping for parallelogram structure hypothesis extraction, and building polygon verification using knowledge such as geometric structure, shadows and walls, illuminating angles, substructures such as doors or windows, and so forth (Jin and Davis, 2005). These edge-based techniques mainly use the geometric properties of buildings, whereas other techniques are combined with consideration of image attributes and multiple scales. Apart from geometry, detailed image models with rich radiometric attributes can exploit much better information contained in the image (Braun et al., 1995; Henricsson, 1998; Lang and Forstner, 1996; Moons et al., 1998; Steger et al., 1995). With data fusion algorithms, both panchromatic and multispectral images from different sensors may be used (Garnesson et al., 1990; Henricsson, 1998; Moons et al., 1998; Weidner, 1997). In addition, by using multiple scales, one can start with reliable structures on a coarse scale and use them to focus extraction on specific areas and object types on fine scales (Mayer, 1999).

Object oriented image classification algorithms have been employed increasingly in building extraction. Object oriented image classification at its simplest form is the classification of homogeneous image primitives, or objects, rather than individual pixels (Platt and Rapoza, 2008). If carefully derived, image objects can be closely related to real-world objects, whose topological relationships with other objects, statistical summaries of spectral and textural values, and shape characteristics all can be employed in the classification procedures (Benz et al., 2004). The idea of object-based image analysis came into existence around the 1970s. By the 1980s, specialized object-oriented software packages became available to extract roads and other linear features on a limited basis. Commencing in the 1990s, increasing computing power and the availability of high spatial-resolution imagery promoted a new era of object-oriented techniques (Platt and Rapoza, 2008). Recently, object oriented image classification has been used successfully to identify logging and other forest-management activities with Landsat Enhanced Thematic Mapper Plus (ETM+) imagery (Flanders et al., 2003). It also has been employed in mapping vegetation communities in urban areas with IKONOS multispectral imagery (Mathieu and Aryal, 2005). Furthermore, it is used to map fuel types with Landsat Thematic Mapper and IKONOS imagery (Giakoumakis et al., 2002) and to detect changes in land use from imagery (Walter, 2004). Compared with the traditional per-pixel methods, object-oriented classification has the advantages of eliminating misclassification of pixels, which is easily caused by the complexity of ground features (e.g., edges, different materials, small-scale features, and shadows), and noise.

In addition, since elevations and intensities can be derived from LiDAR data (Miliaresis and Kokkas, 2007), LiDAR data have been applied to quite a few studies of object-oriented classification in urban land use and land cover (LULC). Zhou and Troy (2008) combined aerial imagery and LiDAR data for an object-oriented classification of urban LULC and achieved an overall accuracy of 92.3 percent. Brennan and Webster (2006) proposed an objected-oriented LULC classification method using solely LiDAR data derived surfaces, including DSM, DEM, intensity, multiple echoes, and normalized height. In the classification, the average accuracy of 10 classes reached 94 percent, and it can be improved to 98 percent if the data are aggregated into 7 classes. Owing to LiDAR data's characteristics, object-oriented classification techniques have a great potential in extracting buildings and houses in urban areas. The similarity of urban feature classes (e.g., buildings) in spectral, spatial, and textural information, such as height, shape, and texture, allows one to merge discrete points or pixels into objects with similar properties close to real spatial objects, which then can be interpreted into a high-qualified images based on knowledge (Gitas et al., 2004). Miliaresis and Kokkas (2007) developed an object-oriented method to extract buildings using the LiDAR DEM, which was based on seed cells and region-growing criteria. However, this method required a certain level of user interface for crucial parameters and was time-consuming.

6.3 Case Study

6.3.1 Datasets

Figure 6.2 shows procedures for the rule-based object-oriented building extraction. LiDAR data of the first and last returns, dated in March and April 2003, were used to derive the normalized height model (NHM).

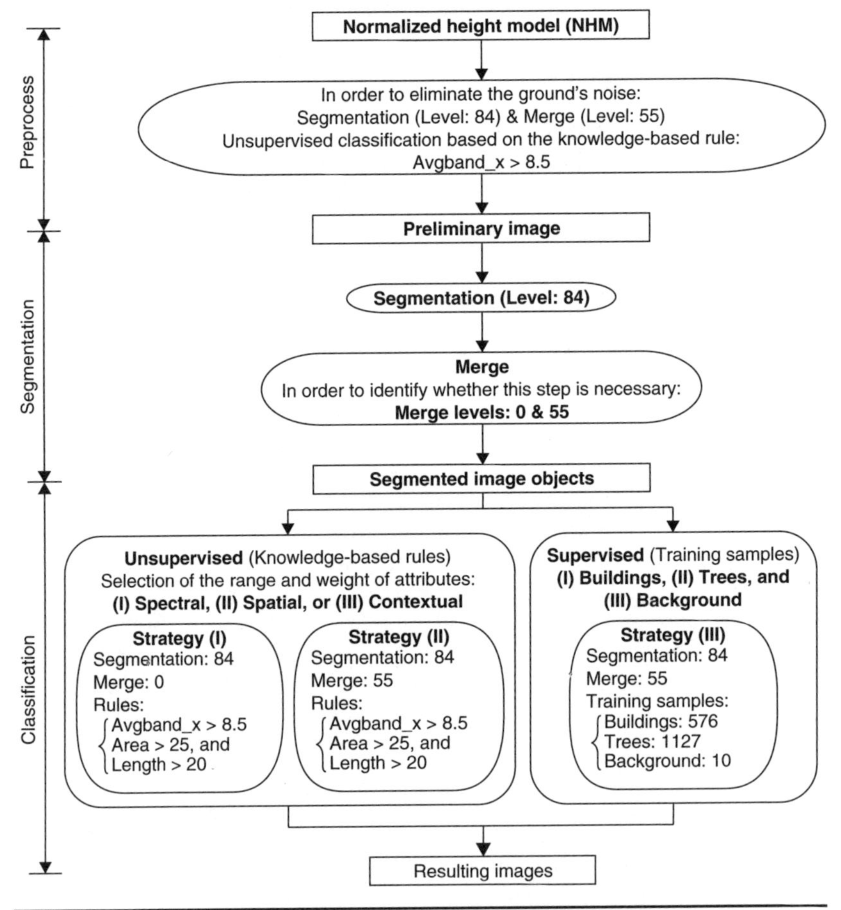

FIGURE 6.2 Procedure for rule-based object-oriented building extraction with LiDAR data.

The data were acquired by Laser Mapping Specialists, Inc. (Raymond, MS), for the area of Marion County, Indiana, covering the city proper of Indianapolis. The LiDAR sensor is known as the Optech ALTM 2033 and, for this survey, was operated at 2700 ft above ground level. A subset of the data covering 4 mi^2 of downtown Indianapolis was used as the main data source in this study. The data had a resolution of one cloud point per square meter on a 45-degree flight line. The Indiana State Plane East (NAD83, NAVD88) system was the coordinate system.

The 2005 statewide digital elevation model (DEM) of Marion County with a resolution of 5 ft was used to eliminate the elevation influence from the digital surface model (DSM) derived from the LiDAR data so that a normalized height model could be computed. The DEM was created from the 2005 Orthophotography of Indiana (6-in resolution) imagery collected during March and April. The DEM was coordinated with the state plane coordinate system (1983).

The Indianapolis planning map of 2000 was used as the reference for accuracy assessment (in a shape-file format), and it contained information of each building and house in the downtown area. A comparison between this map and the raw LiDAR data indicated that the buildings and houses in the area had changed little from 2000 to 2003. Thus, in this research it was assumed that the planning map could be used as the "ground truth" in assessing the accuracy of building-extraction results.

6.3.2 Generation of the Normalized Height Model

The digital surface model (DSM) derived from LiDAR data provides height, shape, and context information that confers has a great advantages in distinguishing building objects from other features on the ground owing to their height differences. Moreover, with LiDAR data, the complexity of separately extracting buildings of different colors in aerial photographs and satellite imagery does not need to be considered because it is the height information rather than spectral information that is used as pixel values. The building extraction in this research was performed based on the normalized height model (NHM), which was calculated by subtracting the statewide DEM from the DSM (derived from LiDAR data). In the NHM image, only objects above the terrain were considered in object-oriented classifications. The following steps detail derivation of the NHM.

The LiDAR system has the ability to capture more than one return for each height point (Alharthy and Bethel, 2002). Among all the returns, the first- and last-return points are the most important. Their individual characteristics also determine their respective applications. The first-return points reflect the surfaces of all ground objects, including solid objects and transparent objects. This property is extremely useful for detecting penetrable objects, especially trees

(Popescue et al., 2003; Secord and Zakhor, 2007; Voss and Sugumaran, 2008). In contrast, the last-return points hit through the leakage among tree branches and leaves to nonpenetrable objects such as the ground and buildings. Figure 6.3 shows that by comparing the first-return DSM with the last-return DSM, we can see a smaller number of urban trees and clearer branch layers. Urban trees were not completely invisible in the last-return DSM image because thick branches and leaves stopped the laser pulse from hitting the ground. The DSM image derived from the last-return points was selected to produce an

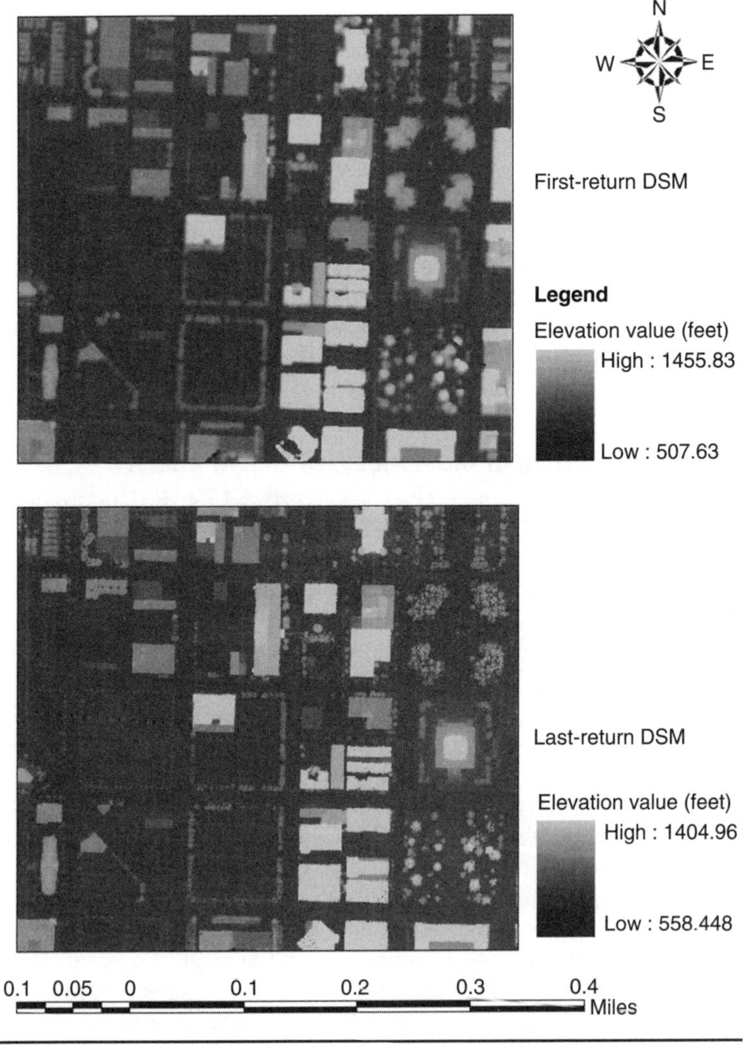

FIGURE 6.3 A comparison of the first return DSM and last return DSM. See also color insert.

NHM image owing to its minimized height information on urban trees. Before obtaining the DSM image, a process of rasterizing the height information from the raw LiDAR data was performed. A 3-ft^2 (approximately 0.91×0.91 m^2) neighborhood function was selected for converting vector-points data into a DSM grid image so that the DSM had the same resolution as the DEM.

The NHM then was calculated using the following formula:

$$\text{NHM} = \text{DSM} - \text{DEM} \tag{6.1}$$

The NHM image contained absolute height information on the objects that were above the ground surface. Some pixel values of the NHM were less than 0. According to Fig. 6.4, it is clear that most of them were located near water bodies and fly-over highways (where the pixels are light yellow). Water pixels usually had negative values because laser points were absorbed and did not reflect back to the scanner, causing the problem of point lost. Moreover, owing to a variable scanning angle of 1 to 75 degrees, points emitting to fly-over highways can easily hit underbridge areas, where the reflected-range values were usually lower than the surface-elevation values. Pixel values in the range of –2 to 0 ft were regarded as common errors and thus considered acceptable. A correction process was needed to filter out abnormal points based on the following criteria:

$$\text{If IV of NHM} < 0\text{, then IV of NHM} = 0;$$
$$\text{Else IV of NHM} = \text{IV of NHM}$$

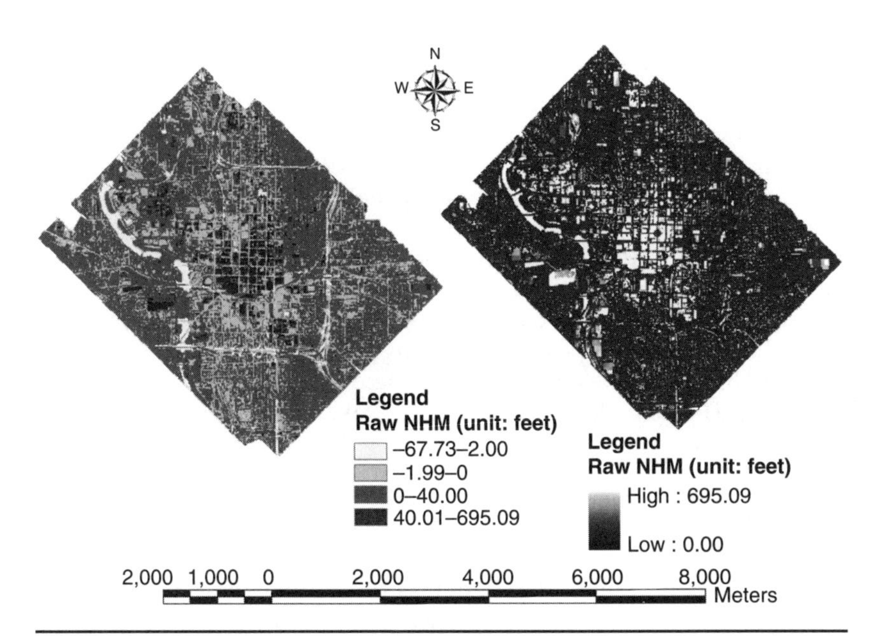

Figure 6.4 The original and corrected NHM image. See also color insert.

where IV is image value. After correction, the resulting NHM image was used in the object-oriented classification.

6.3.3 Object-Oriented Building Extraction

Segmentation and Merging of Ground Objects

In this research, the function of feature extraction in the ENVI Feature Extraction module was used to perform object-oriented classification. It includes a series of procedures for object-oriented classification:

1. *Segmentation.* This process divides the image into segments by setting various scale levels (0 to 100).
2. *Merge.* This process is used for solving the oversegmentation problem and grouping small segments together.
3. *Object-oriented classification.*

The selection of segmentation and merge scales was based on the perception of segmenting as many meaningful ground objects as possible. In other words, segmented ground objects should match as closely as possible to objects in the real world. For example, buildings are expected to be segmented into one or only a few ground objects depending on the disparity level of their roofs, while urban trees are expected to be segmented into individual objects or larger objects that represent assembled tree groups. Figure 6.5 presents examples of the resulting objects at segmentation level 10 and merge level 86 for the CBD and a residential area.

Analysis of Ground Objects

After segmentation and merging of the corrected NHM image, the spectral, spatial, and textural attributes of each object were computed and summarized. These data served as the basis for later image classification. Table 6.1 lists all the spectral, spatial, and textural attributes used and defines each attribute's meaning. In the process of determining the attributes and their thresholds, the characteristics of buildings and other objects that may be confused with buildings in the extraction need to be examined carefully. Urban trees were considered to be the main possible confusing objects. Technically, fragmented objects that belonged to the same object also can result in confusion. Therefore, these factors must be taken into account in setting the decision rules.

Spectral attributes Since pixel values of the NHM image were absolute heights, spectral Avgband_x and Minband_x can be used directly to remove objects lower than one-floor buildings. Stdband_x can be used to distinguish trees from buildings because buildings usually have much smoother surfaces, whereas tree crowns usually are rough and have large discrepancies in height.

Residential area | Central business district

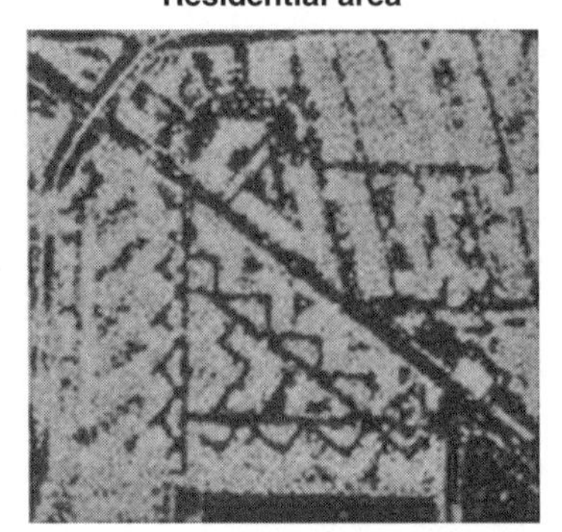

Segmentation (scale 10)

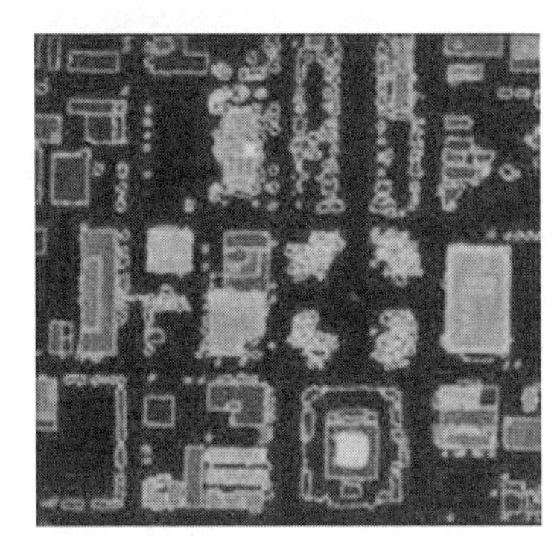

Merge (scale 86)

Resulting ground objects

FIGURE 6.5 An example of segmentation and merge of the corrected NHM image. See also color insert.

Attribute	Description
Spectral	
Minband_x	Minimum value of the pixels comprising the region in band *x*.
Maxband_x	Maximum value of the pixels comprising the region in band *x*.
Avgband_x	Average value of the pixels comprising the region in band *x*.
Stdband_x	Standard deviation value of the pixels comprising the region in band *x*.
Spatial	
Area	Total area of the polygon minus the area of the holes; map units.
Length	The combined length of all boundaries of the polygon, including the boundaries of the holes; map units.
Compact	A shape measure that indicates the compactness of the polygon. A circle is the most compact shape, with a value of $1/\pi$. The compactness value of a square is $\frac{1}{2}\sqrt{\pi}$. $Compact = \sqrt{4 \times \text{area}/\pi}/\text{outer contour length}$
Convexity	Polygons are either convex or concave. This attribute measures the convexity of the polygon. The convexity value for a convex polygon with no holes is 1.0, whereas the value for a concave polygon is less than 1.0. Convexity = length of convex hull/length
Solidity	A shape measure that compares the area of the polygon with the area of a convex hull surrounding the polygon. The solidity value for a convex polygon with no holes is 1.0, whereas the value for a concave polygon is less than 1.0. Solidity = area/area of convex hull
Roundness	A shape measure that compares the area of the polygon with the square of the maximum diameter of the polygon. The maximum diameter is the length of the major axis of an oriented bounding box enclosing the polygon. The roundness value for a circle is 1, and the value for a square is $4/\pi$. Roundness = $(4 \times \text{area})/(\pi \times \text{MAXAXISLEN2})$

TABLE 6.1 Object Oriented Attributes and Their Description

Attribute	Description
Formfactor	A shape measure that compares the area of the polygon with the square of the total perimeter. The form-factor value of a circle is 1, and the value of a square is $\pi/4$. Form factor = $(4 \times \pi \times \text{area})/(\text{total perimeter})^2$
Elongation	A shape measure that indicates the ratio of the major axis of the polygon to the minor axis of the polygon. The major and minor axes are derived from an oriented bounding box containing the polygon. The elongation value for a square is 1.0, whereas the value for a rectangle is greater than 1.0. Elongation = Maxaxislen/Minaxislen
Rect_fit	A shape measure that indicates how well the shape is described by a rectangle. This attribute compares the area of the polygon with the area of the oriented bounding box enclosing the polygon. The rectangular fit value for a rectangle is 1.0, whereas the value for a nonrectangular shape is less than 1.0. Rect_fit = area/(Maxaxislen × Minaxislen)
Maindir	The angle subtended by the major axis of the polygon and the *x* axis in degrees. The main direction value ranges from 0 to 180 degrees; 90 degrees is North-South, and 0 to 180 degrees is East-West.
Majaxislen	The length of the major axis of an oriented bounding box enclosing the polygon. Values are map units of the pixel size. If the image is not georeferenced, then pixel units are reported.
Minaxislen	The length of the minor axis of an oriented bounding box enclosing the polygon. Values are map units of the pixel size. If the image is not georeferenced, then pixel units are reported.
Numholes	The number of holes in the polygon; integer value.
Holesolrat	The ratio of the total area of the polygon to the area of the outer contour of the polygon. The hole solid ratio value for a polygon with no holes is 1.0. Holesolrat = area/outer contour area
Texture	
Tx_range	Average data range of the pixels comprising the region inside the kernel. A kernel is an array of pixels used to constrain an operation to a subset of pixels. Refer to the texture kernel size preference.

Table 6.1 (*Continued*)

Attribute	Description
Tx_mean	Average value of the pixels comprising the region inside the kernel.
Tx-variance	Average variance of the pixels comprising the region inside the kernel.
Tx_entropy	Average entropy value of the pixels comprising the region inside the kernel. ENVI Zoom computes entropy, in part, from the max bins in histogram preference

From ENVI, 2007.

TABLE 6.1 Object Oriented Attributes and Their Description (*Continued*)

Spatial attributes Compact, Roundness, and Rect_fit can be helpful parameters for detecting round and rectangular shapes and thus can be used to differentiate individual trees from well-shaped buildings. Area can be used to eliminate objects with small areas, but special attention needs to be paid to fragmented small objects associated with buildings. The Convexity parameter can detect convex and concave, which may be useful in identifying tree crowns and house roofs.

Texture attributes These attributes can be used to examine the texture quality within an object.

Knowledge-Based Extraction Rules

Through examination of a series of individual attributes, proper ranges of attribute thresholds were determined and integrated into the decision rules. Different weights can be set for each attribute. Besides, factors such as fragment level, knowledge-based integration, and edge detection all affected extraction accuracy.

6.3.4 Accuracy Assessment

The accuracy of the extraction result was assessed using the planning map as the ground truth. The resulting building image was compared with a rasterized planning map. The evaluation measurements used were widely accepted for building extraction (Lee et al., 2003; Shufelt, 1999), which categorized all pixels into four types as a result of comparing two images pixel by pixel:

1. *True positive (TP).* Both the extraction and the reference indicated that a pixel belonged to a building.
2. *True negative (TN).* Both the extraction and the reference indicated that a pixel belonged to the background.
3. *False positive (FP).* The extraction incorrectly identified a pixel as belonging to a building.

4. *False negative (FN).* The extraction did not correctly identify a pixel that truly belonged to a building.

To evaluate the extraction accuracy, the numbers of pixels that fell into each of the four categories (TP, TN, FP, and FN) were computed, and the following measures were calculated:

$$\text{Branching factor} = \text{FP}/\text{TP} \tag{6.2}$$

$$\text{Miss factor} = \text{FN}/\text{TP} \tag{6.3}$$

$$\text{Detection percentage} = 100 \times \text{TP}/(\text{TP} + \text{FN}) \tag{6.4}$$

$$\text{Quality percentage} = 100 \times \text{TP}/(\text{TP} + \text{FP} + \text{FN}) \tag{6.5}$$

Interpretation of these measures follows below. The detection percentage denoted the percentage of building pixels correctly labeled by the extraction. The branching factor was a measure of the commission error, where the system incorrectly labeled the background pixels as buildings. The miss factor measured the omission error, where the system incorrectly labeled building pixels as background. The miss factor can be derived from the detection percentage. Among these statistics, the quality percentage measured the absolute quality of the extraction and was the most stringent measure. To obtain 100 percent quality, the extraction algorithm must correctly label every object pixel (FN = 0) without mislabeling any background pixel (FP = 0) (Jin and Davis, 2005).

6.3.5 Strategies for Object-Oriented Building Extraction

In this section, different strategies for object-oriented extraction of buildings are analyzed and compared. Object-oriented classification involves three steps: *segmentation, merge,* and *classification of objects* by creating rules (rule-based) or by selecting training samples (supervised). In order to examine the effectiveness and efficiency of object-oriented extraction, three strategies for building extraction were compared: (1) rule-based extraction with segmentation (strategy I), (2) rule-based extraction with segmentation and merge (strategy II), and (3) supervised object-oriented extraction (strategy III).

A focus area was selected along a narrow strip from the northwest part of the study area to the center of the CBD. This area contained all typical landscapes within the whole study area in the NHM image: (1) buildings, including both residential and commercial and (2) urban trees and shrubs, including discrete individual trees (usually distributed within the residential areas or along the roads) and patterned trees (usually distributed in parks).

Rule-Based Building Extraction with Segmentation

The first step for object-oriented classification is to segment pixel images into objects that are as meaningful as possible in the real world.

With the ENVI feature-extraction module, the generation of objects is performed through two steps: segmentation and merge. Ideally, the scale level of segmentation should be selected to delineate the boundaries of features as well as possible. A larger value for the segmentation scale is preferred, which can minimize the number of segmented fragments. Smaller scale levels of segmentation should be avoided because of the oversegmentation problem, but the scale should not be too large to segment out the smallest buildings. Merge is an optional step to aggregate small segments into larger, textured ones and to overcome the oversegmentation problem.

Based on our experiment, a scale level of 84 in segmentation was selected. Figure 6.6 shows the result of image classification using the segmentation scale of 84 and the merge level of 0 (no application of merge). It was found that (1) trees were basically segmented into small pieces that may represent one or a few trees, (2) residential buildings and regular commercial buildings basically were segmented into individual and complete objects, and (3) high-rises usually consisted of many segments, sometimes containing large numbers of tiny segments or segments with irregular shapes. Among them, high-rises and trees tended to cause errors in the final result.

The segmentation step is important in the design of rules for object-oriented extraction because objects vary substantially even with

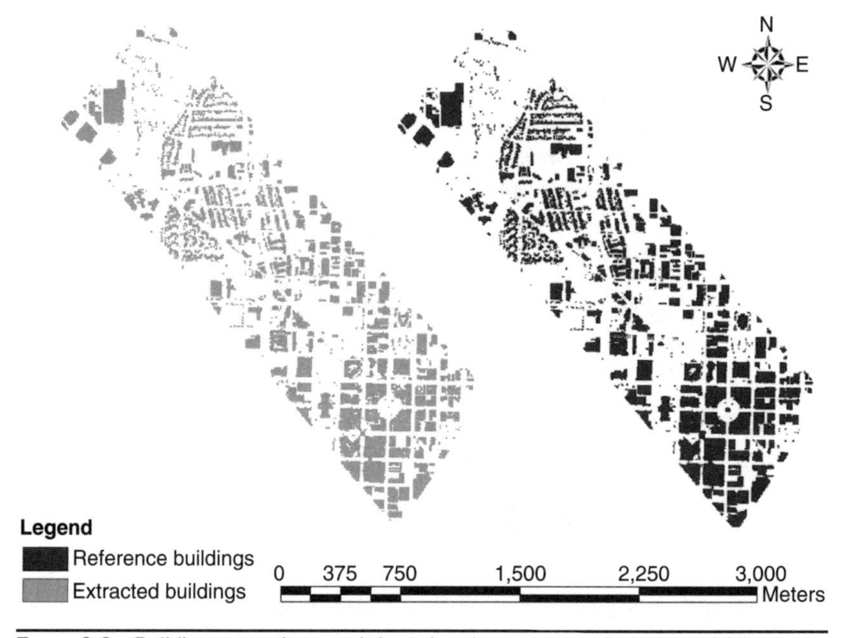

Figure 6.6 Building extraction result by using the segmentation scale of 84 and the merge level of 0 (no application of merge). See also color insert.

a tiny change at the scale level, thus directly affecting the attributes for each object. In order to develop proper rules for extracting buildings, the spectral, spatial, and textural attributes of segmented objects must be scrutinized. A pre-run was performed in the training area to improve the final result. The resulting image mainly covered trees and buildings and contained 95.4 percent of the buildings. This pre-run helped to select the attributes for rule development. Table 6.2 shows the 21 selected attributes that were grouped into seven categories. The rule development was based on the assumptions that (1) through segmentation, the image can be fragmented into individual building objects and individual tree objects, and (2) trees can be eliminated by delimitation of the height and size of objects. The rule was finally set as segmentation level 84, merge level 0, Avgband_x > 8.5, area > 25, and length > 20. Accuracy assessment for the strategy I indicated that branching factor (BF), miss factor (MF), detection percentage (DP), and quality percentage (QP) were 10.34, 12.95, 88.5, and 81.1 percent, respectively (Table 6.3). The accuracy was high enough, so this extraction method may be applied to the whole study area.

Rule-Based Building Extraction with Segmentation and Merge

Although merge can be used to aggregate the segments into larger objects to solve THE oversegmentation problem to a certain degree, its application needs to be carried out carefully. In other words, the step merge should be applied only when it can improve the analysis of attributes of the segmented objects. Otherwise, it would increase the complexity of the segmented image. In this study, we tested a

Attribute Groups	Attributes
Image values	Minband_x, Maxband_x, Avgband_x, Tx_mean
Variance attributes of image values	Stdband_x, Tx_variance
Basic geometric attributes	Area, length, Majaxislen, Minaxislen
Shape attributes	Compact, roundness, form factor, Rect_fit, elongation
Hole-related attributes	Numholes, Holesolrat
Convexity attributes	Convexity, solidity
Others	Tx_range, Tx_entropy

Table 6.2 Selected Attributes for Developing Rules in Objected-Oriented Feature Extraction

Assessment	Branching Factor (BF)	Miss Factor (MF)	Detection Percentage (QP)	Quality Percentage (QP)
Strategy I (unsupervised/ no merge operation)	10.34%	12.95%	88.5%	81.1%
Strategy II (unsupervised/ with merge operation)	12.08%	20.02%	80.02%	71.36%
Strategy III (supervised)	9.46%	7.05%	93.42%	85.83%

TABLE 6.3 Accuracy Assessment Results of Three Building-Extraction Strategies

series of merge levels at the same segmentation scale of 84 and found that at the merge level of 55 there was the fewest number of building segments. Figure 6.7 shows the result of image classification based on the segmentation scale of 84 and the merge level of 55. Some high-rises were merged together, but residential and regular commercial buildings did not change much. Almost all the trees were aggregated into larger objects regardless of their locations and densities.

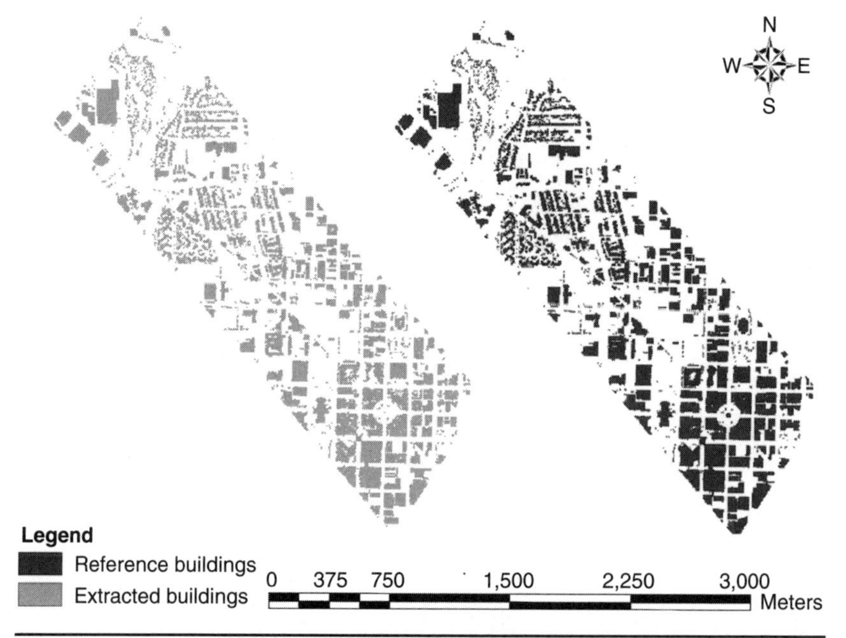

FIGURE 6.7 Building extraction result by using the segmentation scale of 84 and the merge level of 55. See also color insert.

Rules When the Merge Level = 55 After merging, some building segments were merged into larger segments. Almost all trees were aggregated into larger tree-group objects. Therefore, at this merge level, objects that need to be considered contained (1) individual trees, (2) tree groups, (3) residential buildings, and (4) commercial buildings, including regular commercial buildings (usually individual building objects) and high-rises and/or irregular-surface commercial buildings (usually building fragments). Since the merge created new objects, such as tree-group objects, rules for building extraction should be set up for various types of objects. Thus the statistics of attributes for each type of object were analyzed to check their feasibility for building extraction. One attribute from each attribute group (except for convexity attribute group) was selected and examined to check the separation between the four types of objects. Figure 6.8 shows coincident plots illustrating the ranges of the attributes for each type of object. From the figure, we can conclude that the attributes other than spectral and geometric ones did not have obvious advantages in object-oriented extraction. This is so because high-rises and irregular-surface buildings usually had a large variance of spectral values. With the segmentation step, they were segmented into building fragments with various shapes and variance attributes, which had more similarity to tree-group objects in shape, size, and variance attributes. Accuracy assessment for the strategy II indicated that branching factor (BF), miss factor (MF), detection percentage (DP), and quality percentage (QP) were 12.08, 20.02, 80.02, and 71.36 percent, respectively. There was an overall lower accuracy than with strategy I.

Supervised Object-Oriented Extraction

Supervised object-oriented classification also was performed so that the effectiveness and efficiency of both methods could be compared. The resulting image is shown in Fig. 6.9. Accuracy assessment yielded a BF of 9.46 percent, MF of 7.05 percent, detection percentage of 93.42 percent, and quality percentage of 85.83 percent. The disadvantages of supervised object-oriented classification are similar to those of traditional supervised classification. Training objects may have various attributes. In order to make training objects representative and complete, a large number of objects need to be selected, which can be time-consuming.

6.3.6 Error Analysis

Table 6.3 provides a comparison of quantitative accuracy among the three strategies. Errors in building extraction occurred in two fronts: (1) nonbuilding pixels (mainly urban trees) mislabeled as buildings and (2) building pixels mislabeled as background. In order to examine the error sources, urban trees were categorized as street trees,

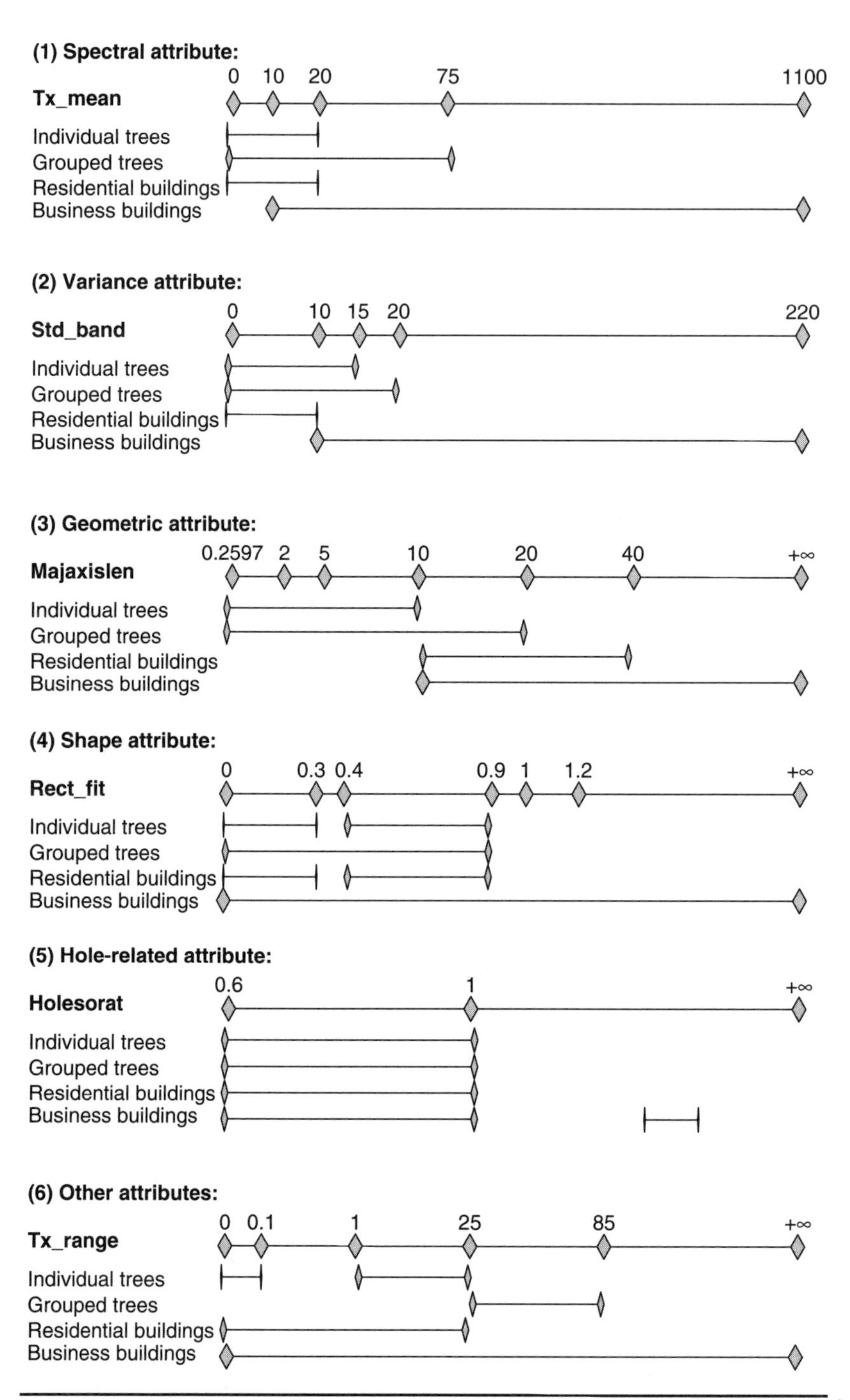

Figure 6.8 Coincident plots in the selected attributes for four types of objects.

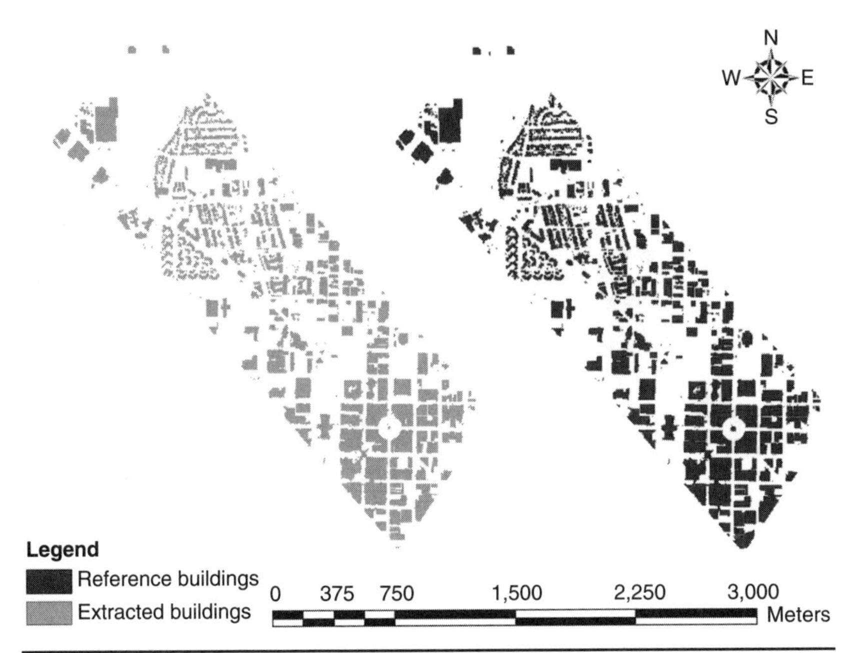

Figure 6.9 Building extraction result by supervised object-oriented classification. See also color insert.

yard trees, and urban forest based on their pattern and distribution. Buildings were identified as residential buildings (with ridged roofs), regular business buildings (basically with flat roofs), and irregular-roof business buildings.

Figure 6.10 illustrates nonbuilding pixels mislabeled as buildings. It can be found that strategy III eliminated the largest number of urban trees, leading to higher extraction accuracy. Strategy I can remove more urban trees than strategy II, especially in the categories of urban forest and yard trees. Owing to the merge operation with the strategy II, neighboring trees, especially those of urban forest, were aggregated into bigger objects that cannot be eliminated by the original thresholds of object size. Errors from street trees were similar in both strategies I and II because street trees often are planted sparsely, and thus the merge operation did not influence the objects' sizes and shapes much.

Figure 6.11 shows the errors from mislabeled building pixels. Strategy III had the least number of building pixels lost during the extraction and thus possessed the highest accuracy. The number of extracted residential buildings and the number of regular business buildings were similar with strategies I and II. Missed pixels of irregular-roof business buildings accounted for most of the errors in both strategies I and II. Large variations in height of the irregular roofs can easily lead to oversegmentation and produced a large

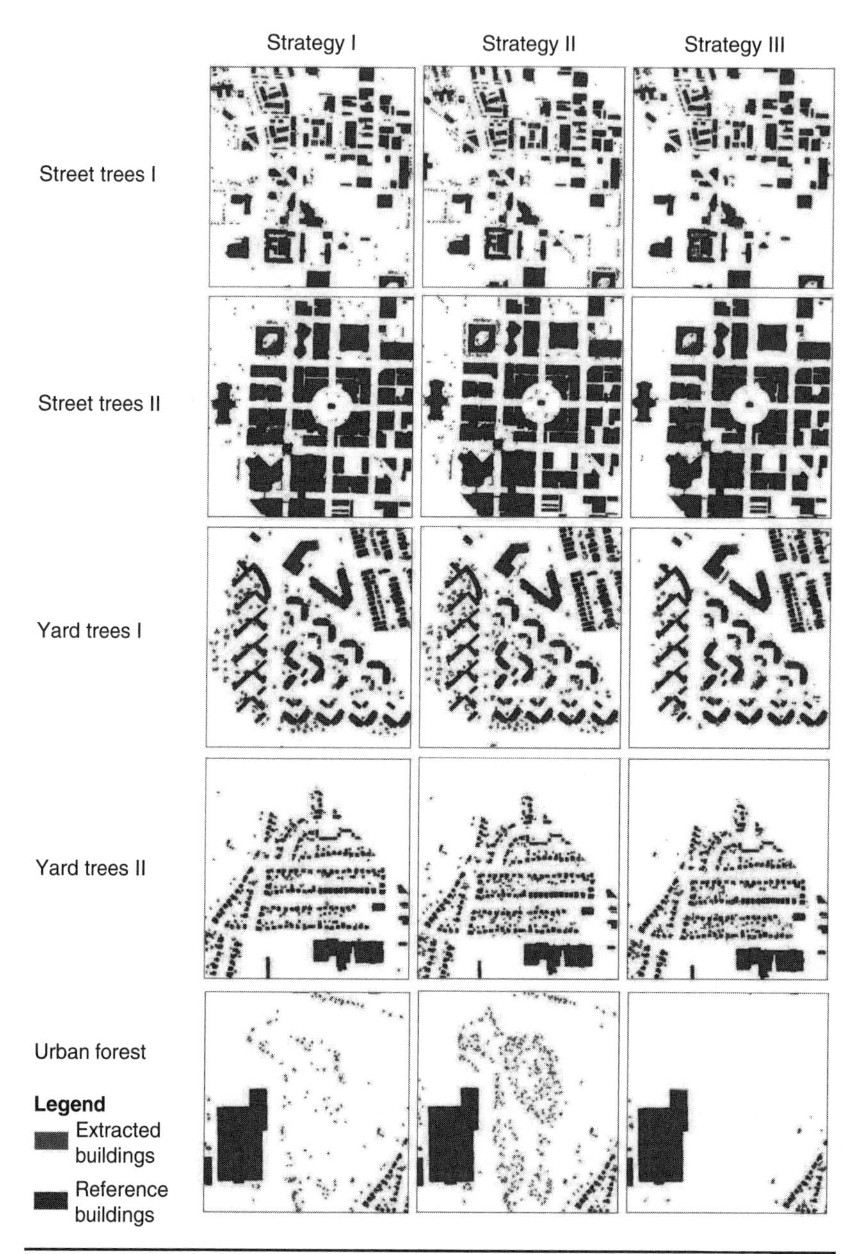

Figure 6.10 Nonbuilding pixels mislabeled as buildings in the three strategies of building extraction (blue pixels not covered by orange ones). See also color insert.

number of fragmented small objects that were eliminated in the process of building extraction. In addition, strategy II had the largest number of missed pixels of irregular-roof business buildings because the merge operation combined fragmented small objects into larger

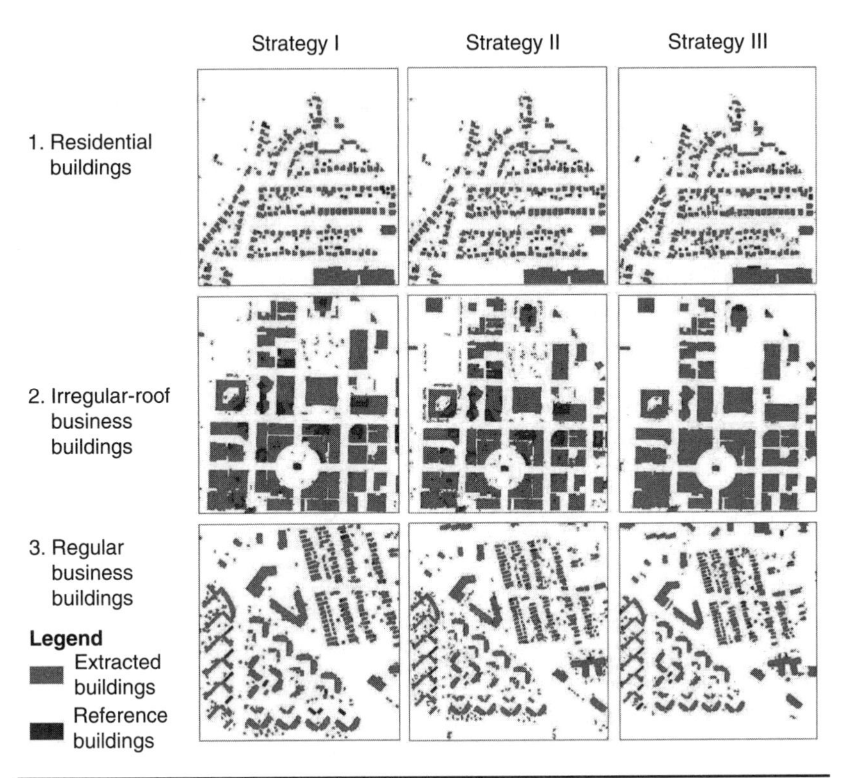

FIGURE 6.11 Building pixels mislabeled as the background in the three strategies of building extraction (orange pixels not covered by blue ones). See also color insert.

objects, which frequently were not large enough to be kept, leading to the missing of more pixels.

6.4 Discussion and Conclusions

Accuracy of extraction reached 88.5 percent in detection percentage and 81.7 percent in quality percentage by using proper segmentation methods and object-oriented rules related to three-dimensional (3D) geometric properties of buildings. Supervised object-oriented classification improved the detection percentage and quality percentage to 93.42 and 85.83 percent for the same dataset.

More research is needed to improve the methodology applied in this study for object-oriented building extraction/classification. Using object-oriented feature-extraction/classification techniques to extract buildings is still a challenging task owing to the characteristics of remotely sensed data, the complexity of objects, the characteristics of object attributes, and the surface characteristics of the study area. The design of segmentation strategy and the selection of rules for building extraction are directly related to all the factors.

One of the ways for more effective extraction is to combine different types of remotely sensed data, such as the combined use of LiDAR and IKONOS data. The incorporation of LiDAR data into IKONOS may provides complementary information needed for object-oriented feature extraction/classification. LiDAR data have the advantages of providing height information at each point on the earth's surface, whereas IKONOS data offer the spectral information about ground features, which can effectively deal with the spectral similarity between buildings and other impervious features.

The segmentation process has an effect on the design of the rules for target features. The ideal segmentation result should be that individual buildings are segmented into corresponding individual objects without fragments. However, the oversegmentation problem may arise. In this study, residential buildings were found to be easily merged with surrounded trees with a high segmentation level, whereas commercial buildings with irregular surfaces tended to be fragmented into smaller objects with a low segmentation level. In addition, although merge is regarded as a solution to oversegmentation, it needs to be considered carefully. It tended to merge trees more than buildings fragments, which, in turn, increases the complexity of segmented objects. Proper segmentation-merge strategies must be selected to provide good results of segmentation for final object-oriented classification.

The attributes of segmented objects can vary greatly based on the setup of the segmentation parameters. After the segmentation result is obtained, the meaningfulness of segmented "objects" needs to be examined. In this study, four kinds of objects were identified to represent individual trees, tree groups, and residential and commercial buildings. Errors were detected mainly in tree-group objects and commercial buildings with irregular surfaces. These buildings were oversegmented and formed various building fragments, with large variations in their attributes.

References

Alharthy, A. and Bethel, J., 2002. Heuristic filtering and 3d feature extraction from LIDAR data. In: *International Archives of Photogrammetry and Remote Sensing (IAPRS)*, Graz, Austria, Vol. XXXIV, Part 3A, ISSN 1682–1750, pp. 29–34.

Benz, U. C., Hofmann, P., Willhauck, G., et al. 2004. Multi-resolution, object-oriented fuzzy analysis of remote sensing data for GIS-ready information., *ISPRS Journal of Photogrammetry and Remote Sensing* **58,** 239–258.

Braun, C., Kolbe, T. H., Lang, F., et al. 1995. Models for photogrammetric building reconstruction. *Computers & Graphics* **19,** 109–118.

Brennan, R., and Webster, T. L. 2006. Object-oriented land cover classification of LiDAR-derived surfaces. *Canadian Journal of Remote Sensing* **32,** 162–172.

Clode, S., Rottensteinerb, F., Kootsookosc, P., and Zelniker, E. 2007. Detection and vectorization of roads from LiDAR data. *Photogrammetric Engineering and Remote Sensing* **73,** 517–535.

FIGURE 3.2 Landsat ETM+ image (bands 4, 5, and 3 as red, green, and blue, respectively) of Indianapolis acquired on June 22, 2000.

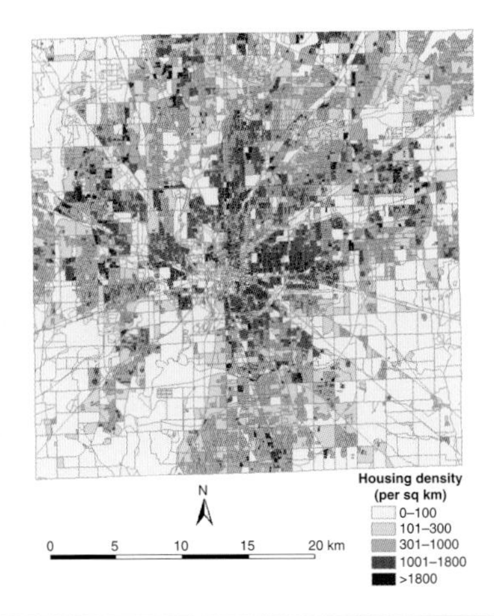

FIGURE 3.4 Housing density distribution at Census block, Marion County, Indiana, in 2000.

FIGURE 3.5 Centroids of Census block.

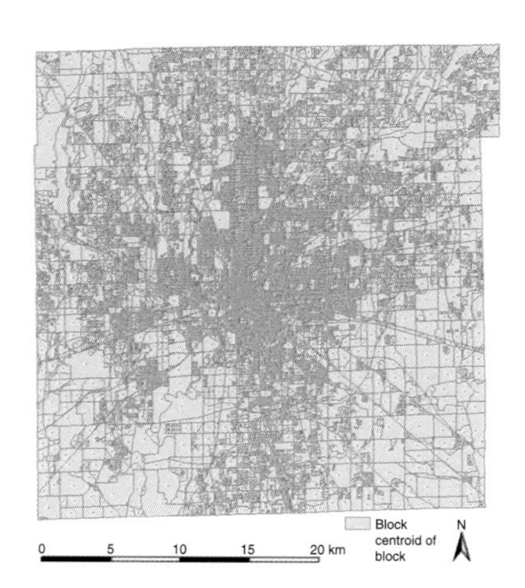

FIGURE 3.6 Housing surface generated by IDW interpolation.

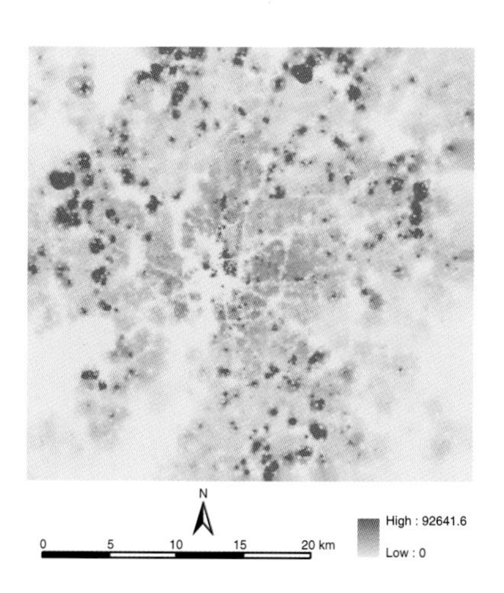

FIGURE 3.7 Rasterized housing density.

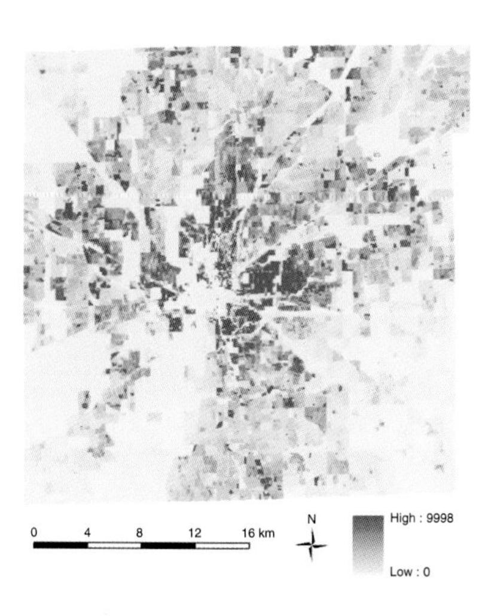

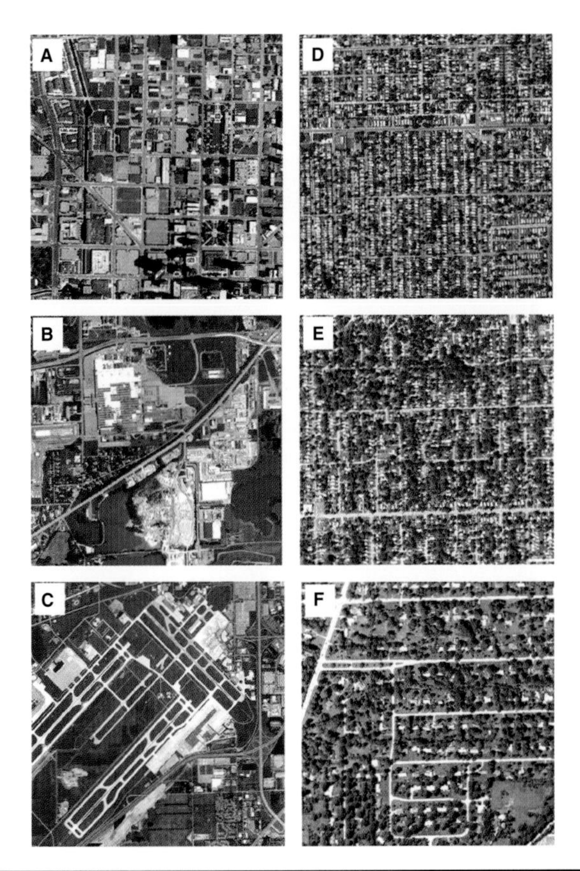

Figure 3.8 Examples of typical land use: (*a*) commercial, (*b*) industrial, (*c*) transportation, (*d*) high-density residential, (*e*) medium-density residential, (*f*) low-density residential.

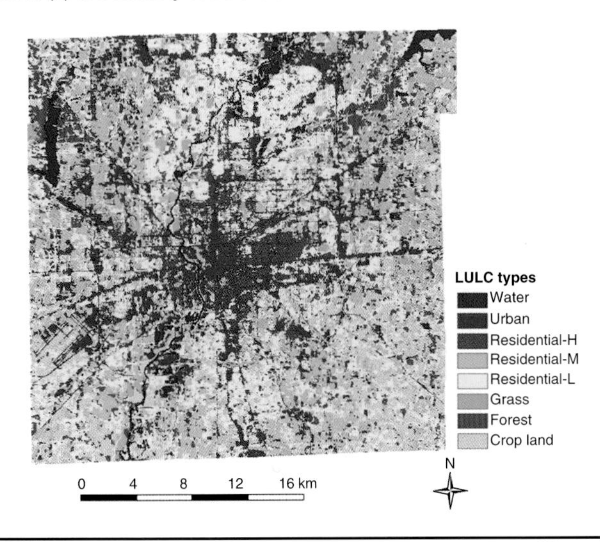

Figure 3.9 LULC classification image based on intergration of housing and the ETM+ image at the pre-classification stage.

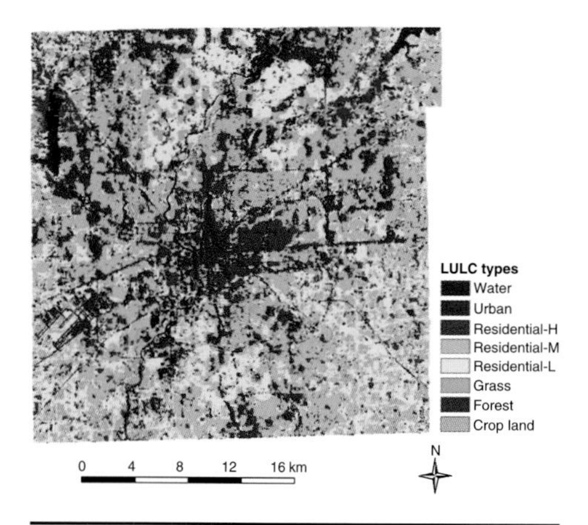

FIGURE 3.10 LULC map based on integration of housing surface into the ETM+ image as an additional layer (during the classification).

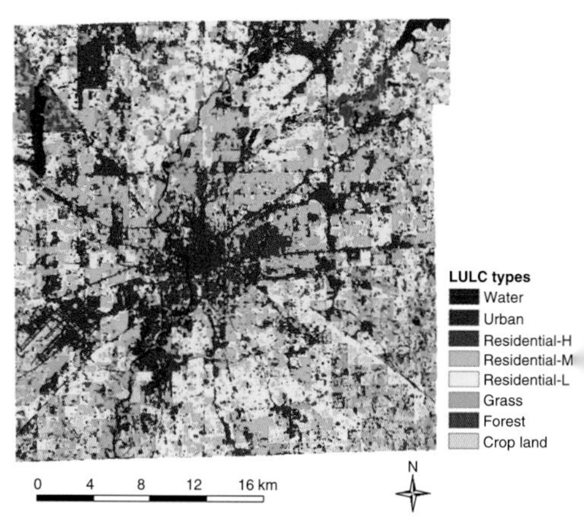

FIGURE 3.11 LULC map based on integration of housing data at the post-classification stage.

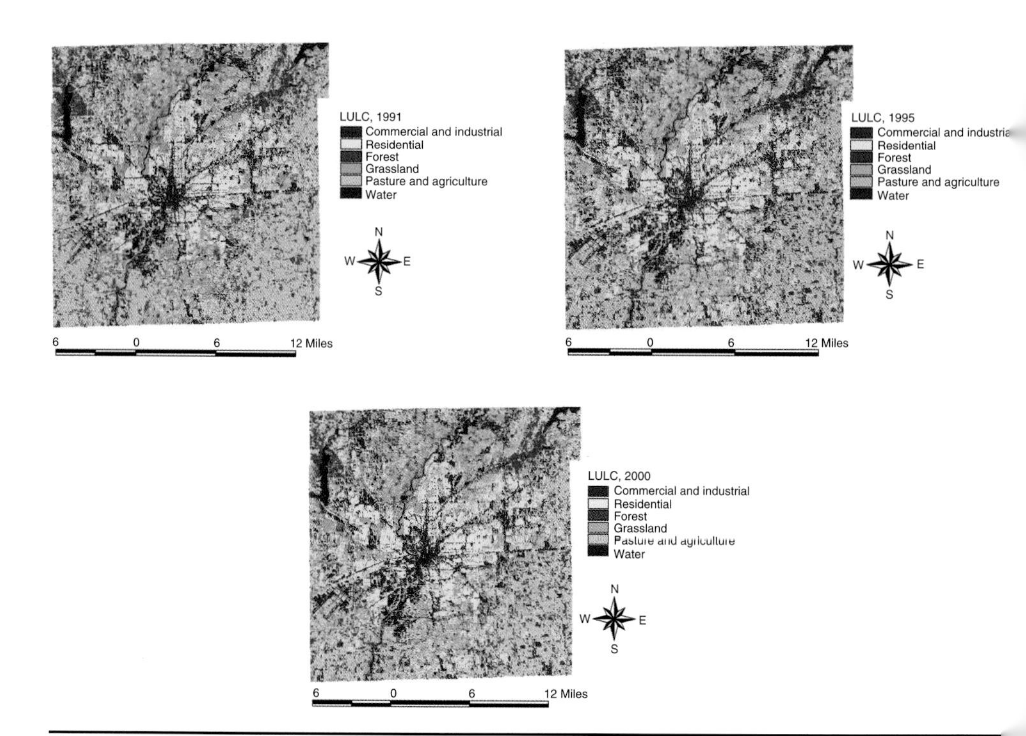

FIGURE 4.6 LULC maps of 1991, 1995, and 2000. (*Adapted from Weng and Lu, 2006.*)

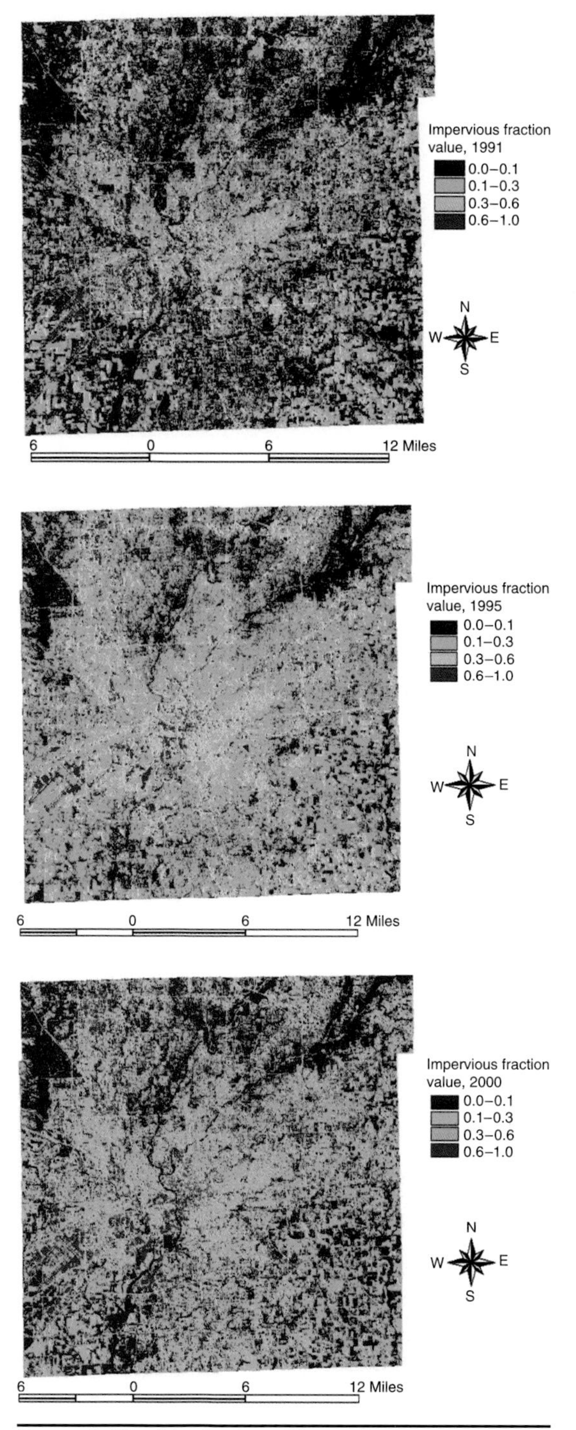

FIGURE 4.10 Distribution of impervious coverage by year. (*Adapted from Weng and Lu, 2009.*)

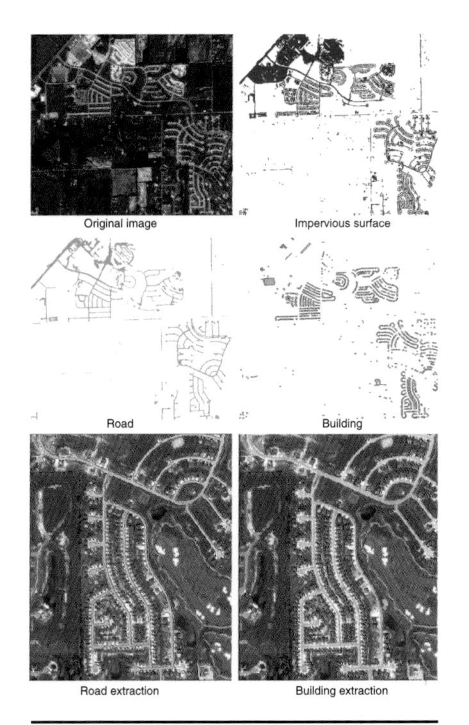

FIGURE 5.1 Feature extraction from residential area.

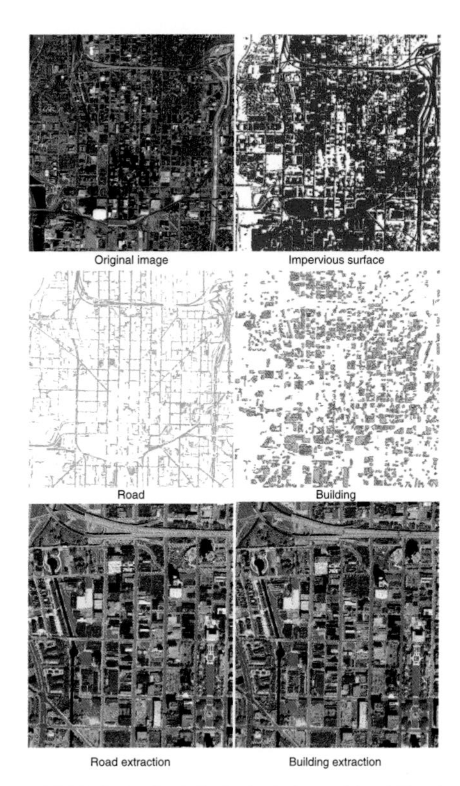

FIGURE 5.2 Feature extraction from CBD area.

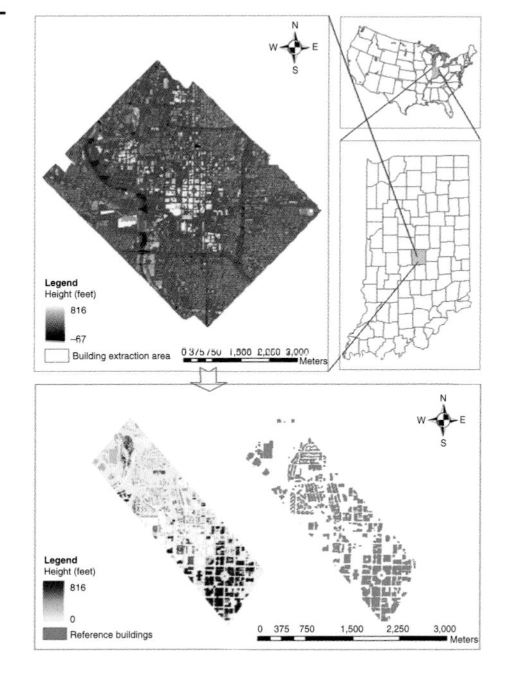

FIGURE 6.1 Study area in downtown Indianapolis, Indiana.

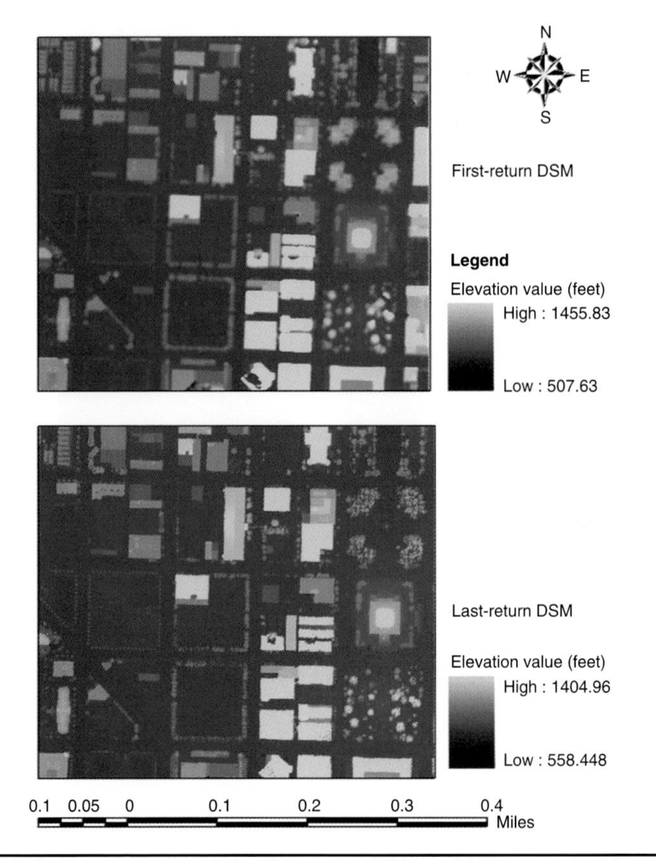

Figure 6.3 A comparison of the first return DSM and last return DSM.

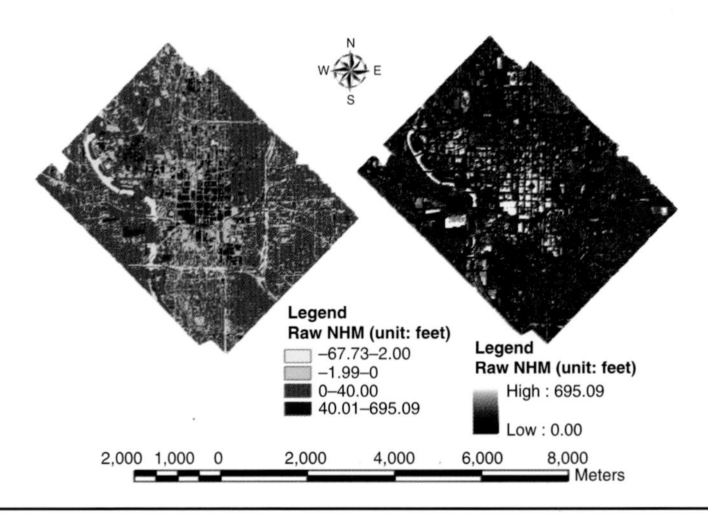

Figure 6.4 The original and corrected NHM image.

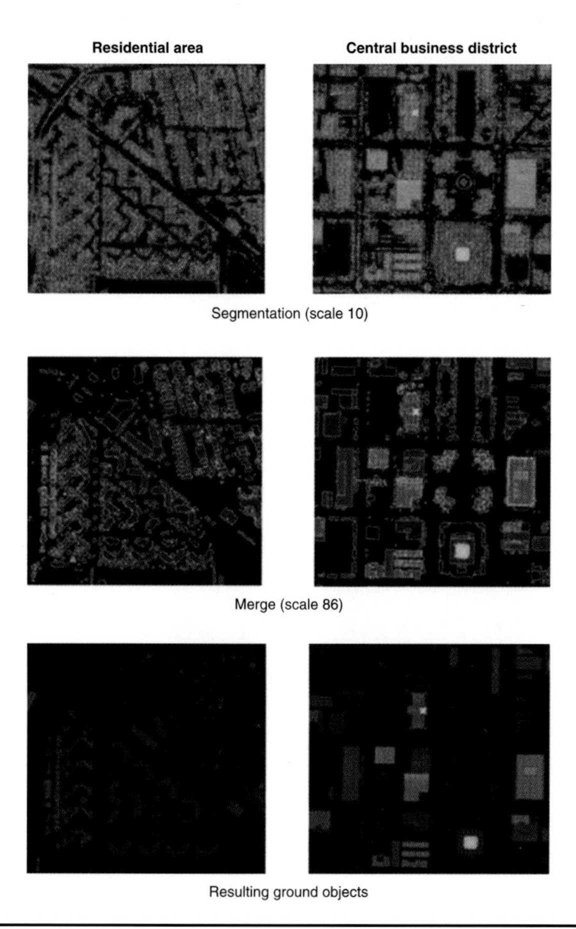

Figure 6.5 An example of segmentation and merge of the corrected NHM image.

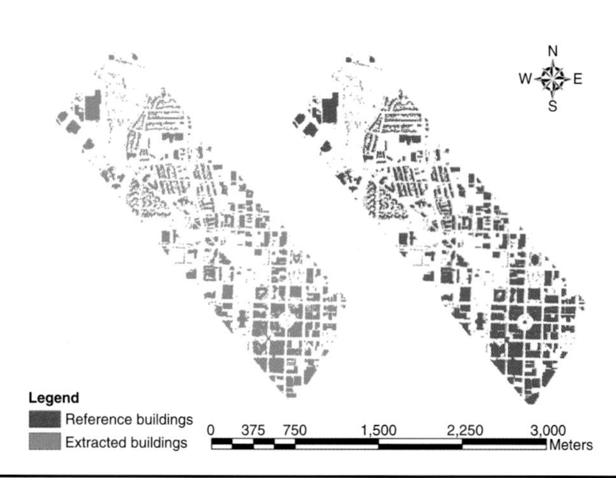

Figure 6.6 Building extraction result by using the segmentation scale of 84 and the merge level of 0 (no application of merge).

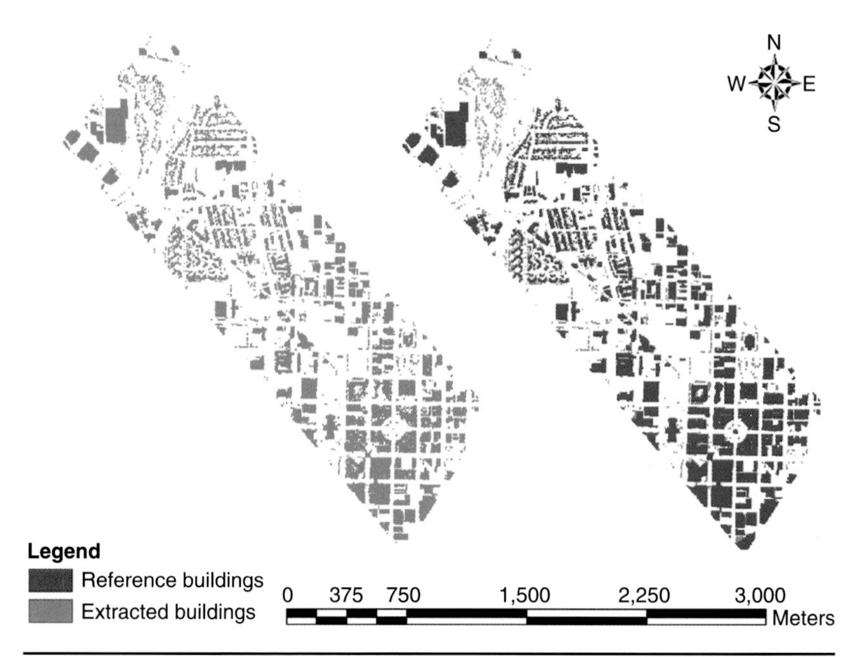

FIGURE 6.7 Building extraction result by using the segmentation scale of 84 and the merge level of 55.

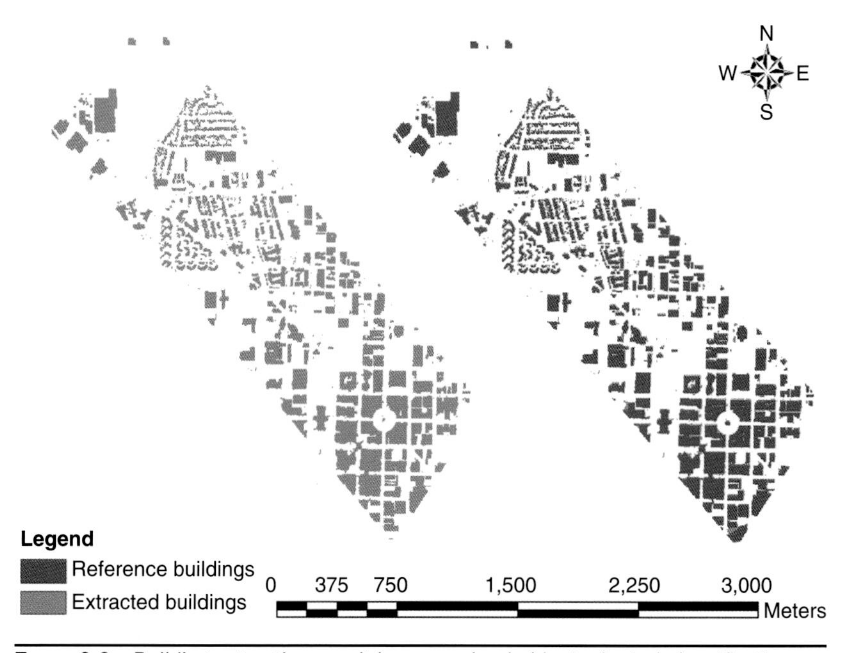

FIGURE 6.9 Building extraction result by supervised object-oriented classification.

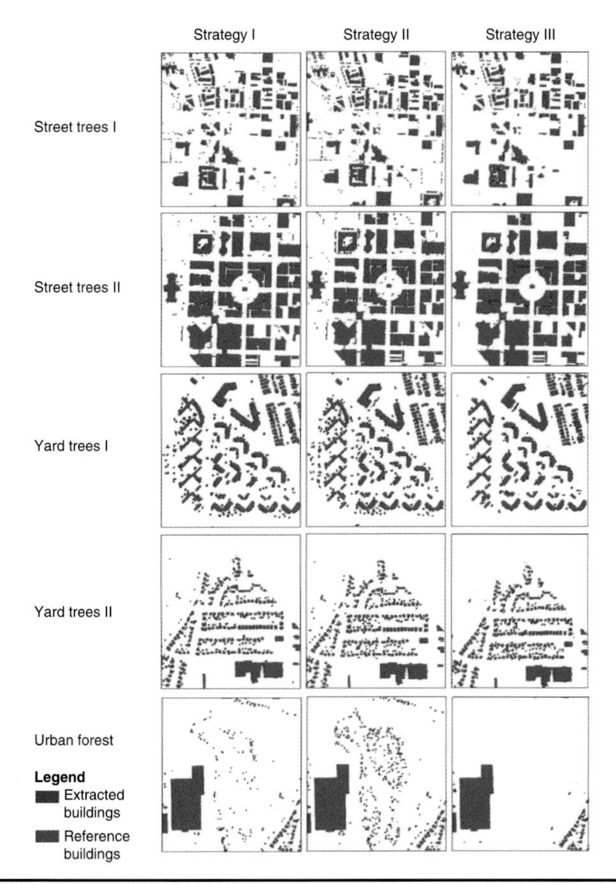

FIGURE 6.10 Nonbuilding pixels mislabeled as buildings in the three strategies of building extraction (blue pixels not covered by orange ones).

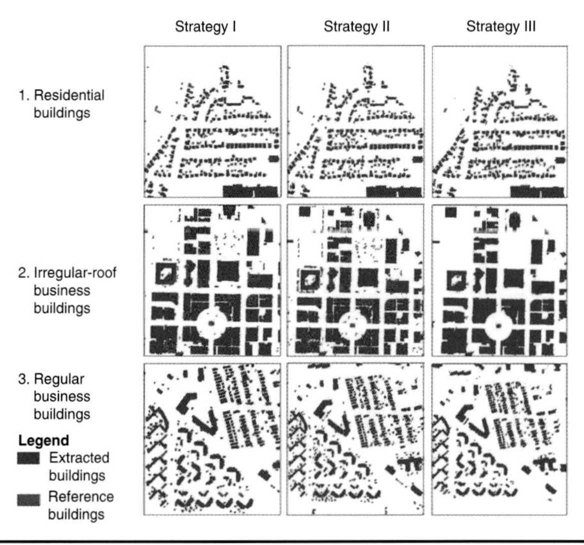

FIGURE 6.11 Building pixels mislabeled as the background in the three strategies of building extraction (orange pixels not covered by blue ones).

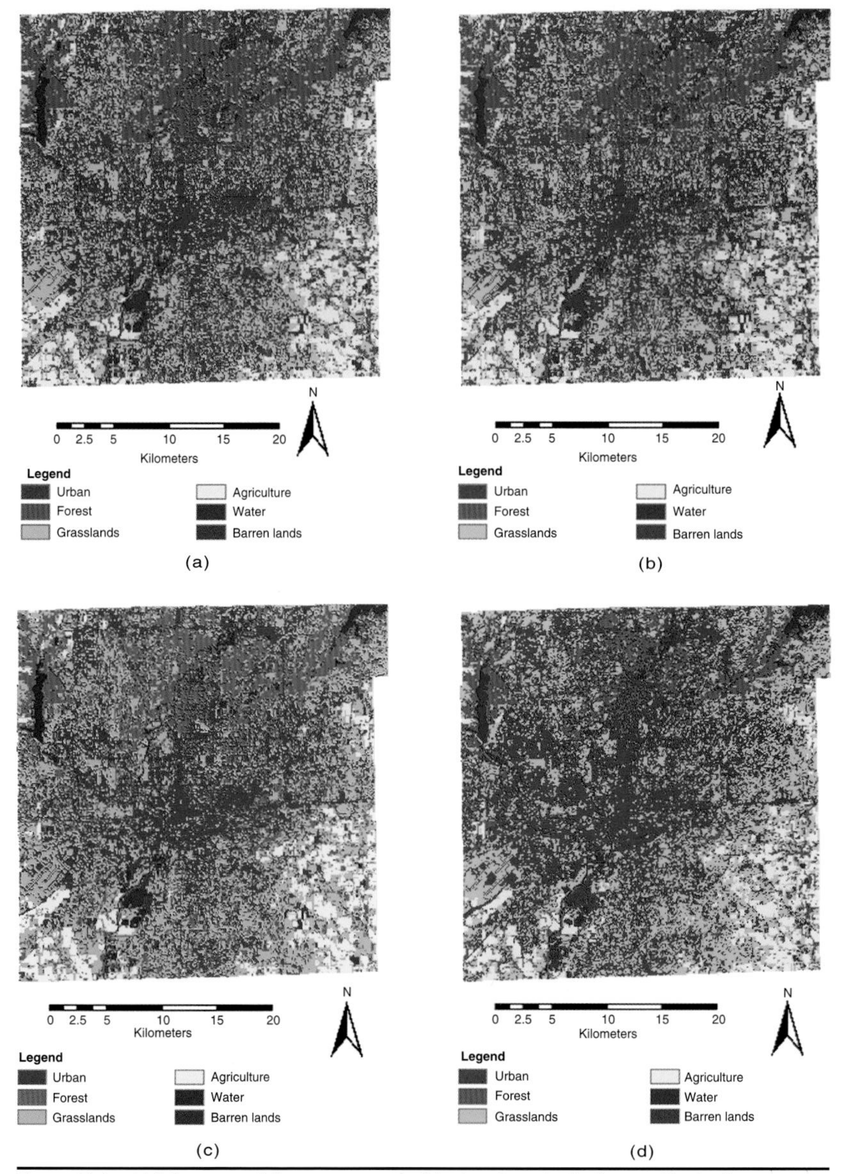

Figure 7.1 LULC maps of four seasons in Marion County, Indianapolis, Indiana, derived from ASTER images: (*a*) October 3, 2000; (*b*) June 16, 2001; (*c*) April 5, 2004; and (*d*) February 6, 2006. (*Adapted from Weng et al., 2008.*)

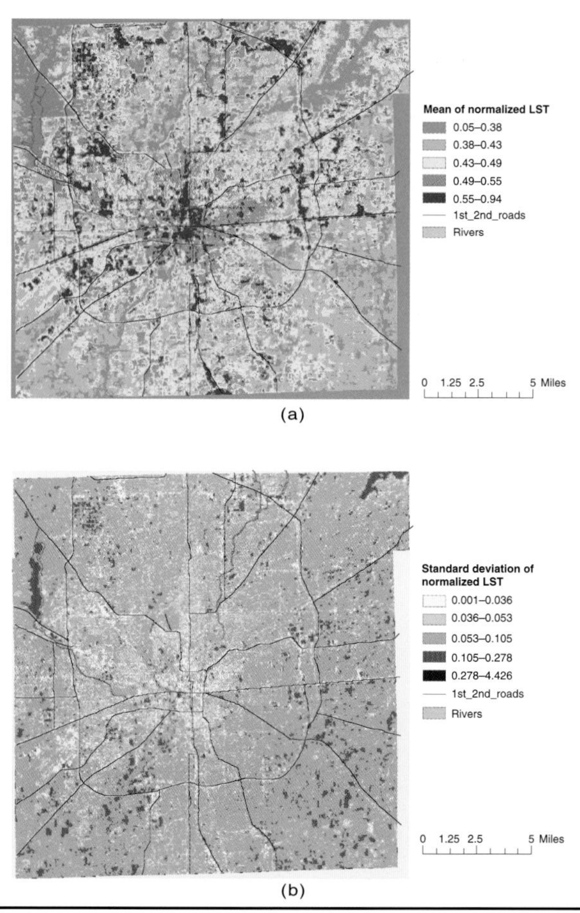

Figure 7.2 Normalized land surface temperatures computed based on the four dates of images: (*a*) mean value of T_N and (*b*) standard deviation values of T_N. (*Adapted from Weng et al., 2008.*)

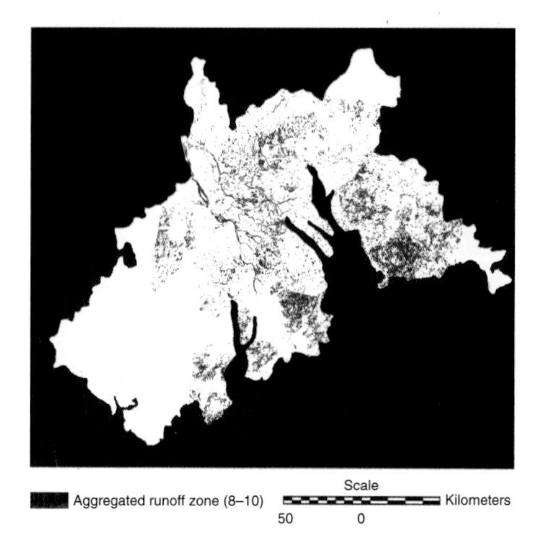

Figure 8.5 Surface runoff changes in the Zhujiang Delta, 1989–1997.

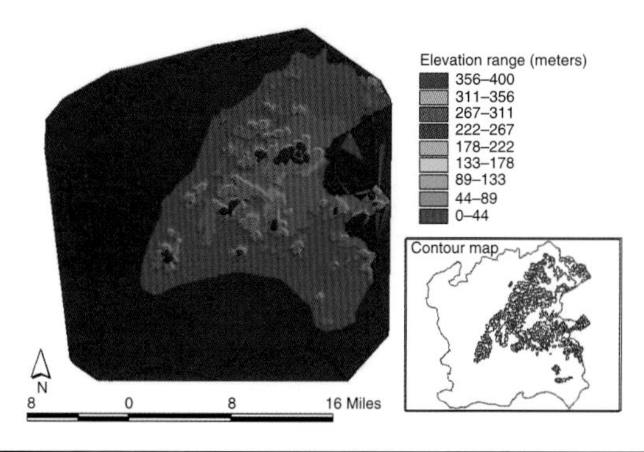

FIGURE 9.2 Digital elevation model of Guangzhou. (Inset: Contour map.)

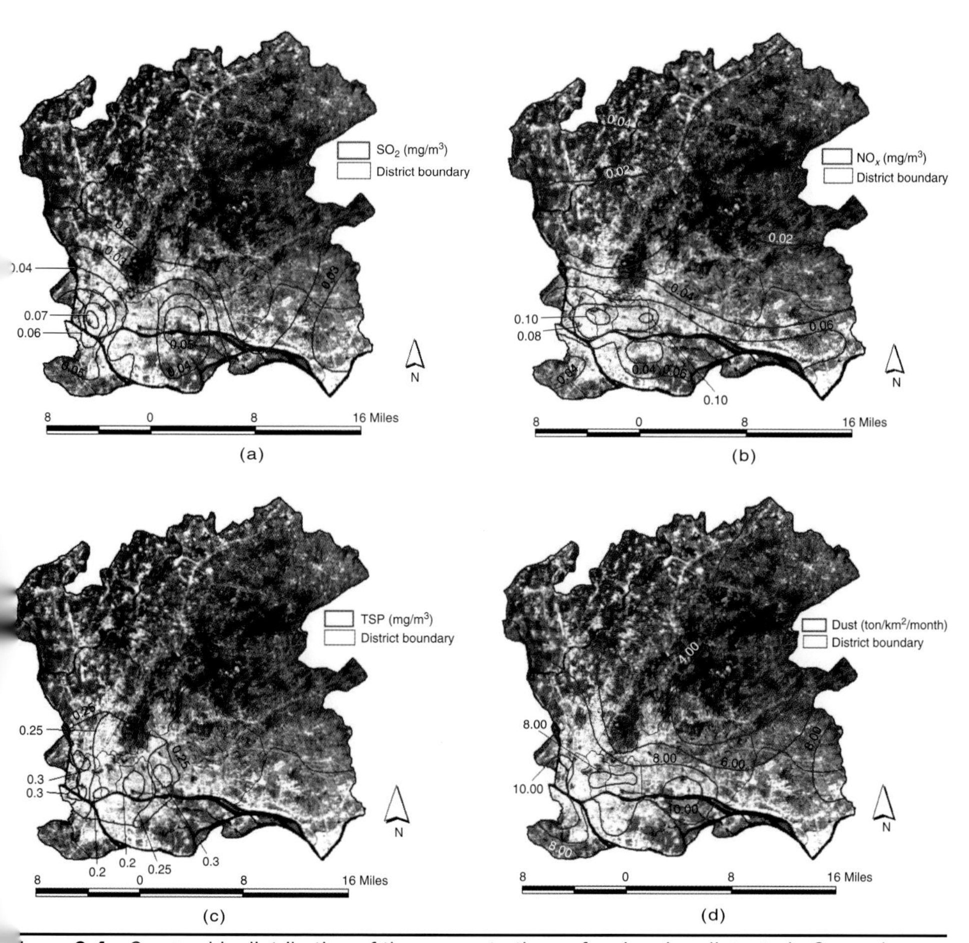

IGURE 9.4 Geographic distribution of the concentrations of major air pollutants in Guangzhou. ;ontours are interpolated from point data observed in the monitoring stations. Data represent nnual averages of pollutant measurements. The base map is the LST map of August 29, 1997, erived from Landsat TM thermal infrared data.

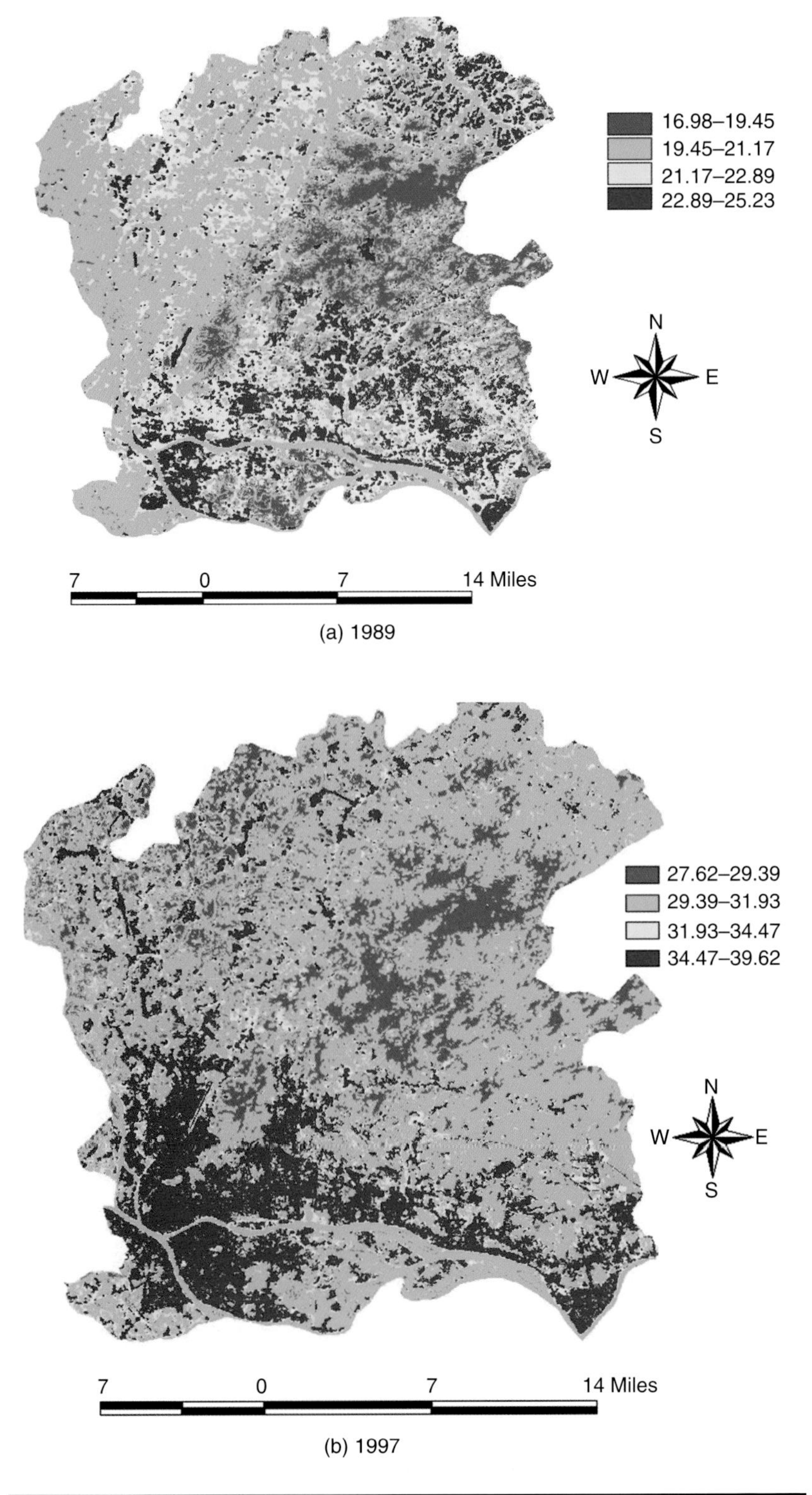

FIGURE 9.7 A choropleth map showing the geographic distribution of LSTs in 1989 and 1997.

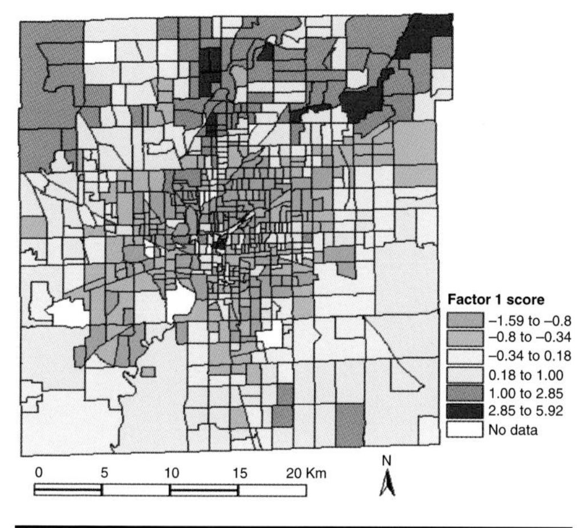

FIGURE 11.2 The first factor score—economic index. (*Adapted from Li and Weng, 2007.*)

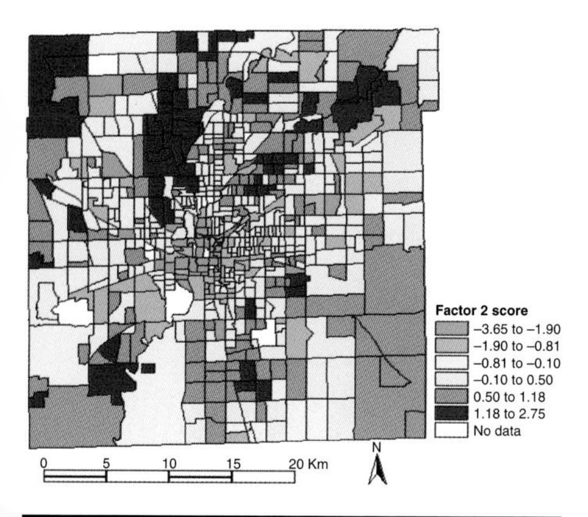

FIGURE 11.3 The second factor score—environmental index. (*Adapted from Li and Weng, 2007.*)

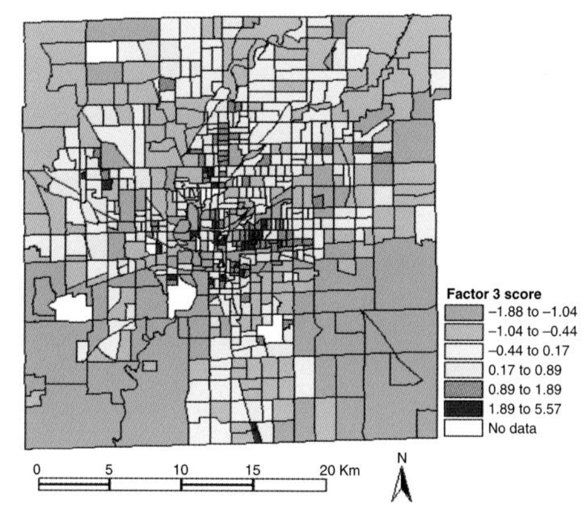

FIGURE 11.4 The third factor score—crowdedness. (*Adapted from Li and Weng, 2007.*)

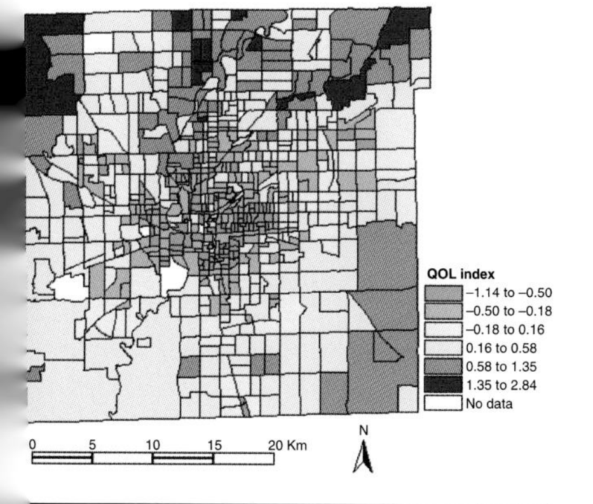

FIGURE 11.5 Synthetic quality of life index. (*Adapted from Li and Weng, 2007.*)

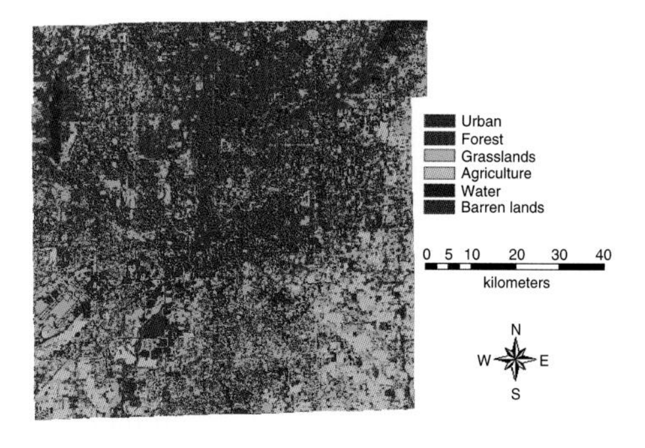

FIGURE 13.1 LULC map of Indianapolis, Indiana, on October 13, 2006.

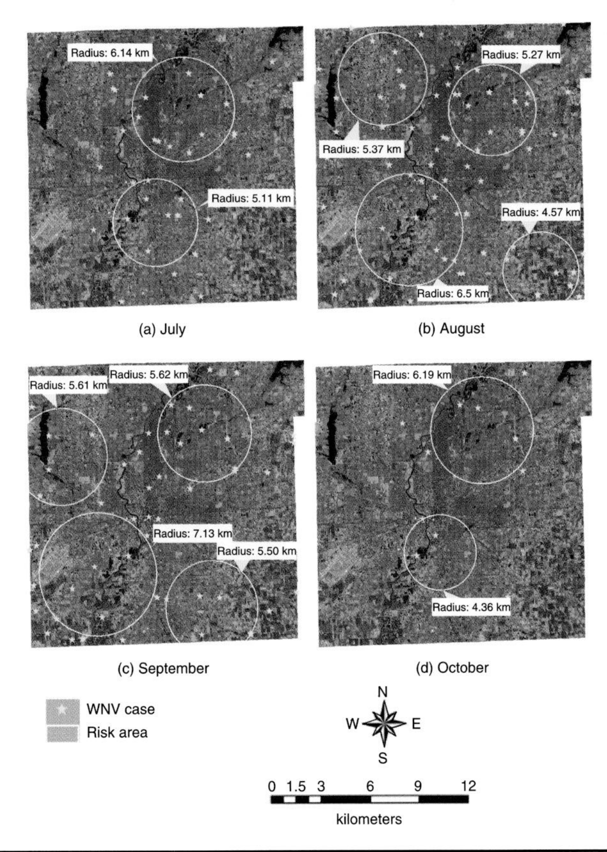

FIGURE 13.4 Risk areas in July, August, September, and October in the years 2002 through 2007.

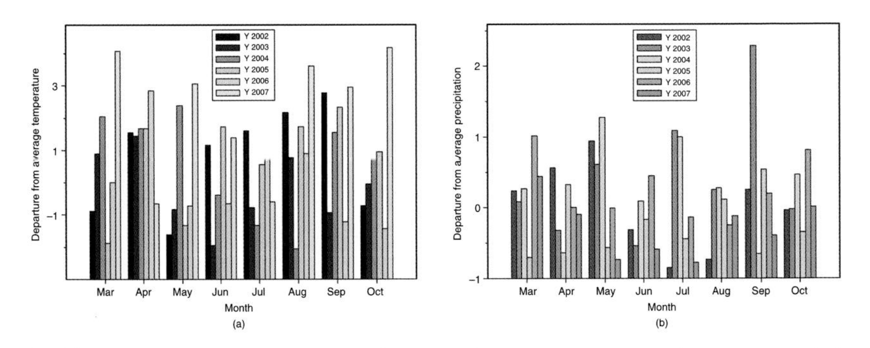

FIGURE 13.5 Departure from average temperature and precipitation in Indianapolis from March through October in 2002 through 2007.

ENVI. 2007. *ENVI Feature Extraction Module User's Guide.* Denver, ITT Visual Information Solutions.

Filin, S. 2004. Surface classification from airborne laser scanning data. *Computers & Geosciences* **30,** 1033–1041.

Flanders, D., Hall-Beyer, M., and Pereverzoff, J. 2003. Preliminary evaluation of e-cognition object-based software for cut block delineation and feature extraction. *Canadian Journal of Remote Sensing* **29,** 441–452.

Forlani, G., Nardinocchi, C., Scaioni, M., and Zingaretti, P. 2006. Complete classification of raw LiDAR data and 3D reconstruction of buildings. *Pattern Analysis and Applications* **8,** 357–374.

Garnesson, P., Giraudon, G., and Montesinos, P. 1990. An image analysis system application for aerial imagery interpretation. In *Proceedings of the 10th International Conference on Pattern Recognition,* pp. 210–212, Atlantic City, June 1990.

Giakoumakis, M. N., Gitas, I. Z., and San-Miguel, J. 2002. Forest fire research and wildland fire safety. In *Proceedings of IV International Conference on Forest Fire Research, 2002 Wildland Fire Safety Summit,* edited by D. X. Viegas. Luso, Coimbra, Portugal, November 18–23.

Gitas, I. Z., Mitri, G. H., and Ventura, G. 2004. Object-based image classification for burned area mapping of Creus Cape, Spain, using NOAA-AVHRR imagery. *Remote Sensing of Environment* **92,** 409–413.

Henricsson, O. 1998. The role of color attributes and similarity grouping in 3D building reconstruction. *Computer Vision and Image Understanding* **72,** 163–184.

Jin, X., and Davis, C. H. 2005. Automated building extraction from high-resolution satellite imagery in urban areas using structural, contextual, and spectral information. *EURASIP Journal on Applied Signal Processing* **14,** 2196–2206.

Kokkas, N. 2005. City modeling and building reconstruction with Socet Set v.5.2 BAE Systems. Customer presentation at the 2005 GXP Regional User Conference, Cambridge, England, September 19–21.

Lang, F., and Forstner, W. 1996. Surface reconstruction of man-made objects using polymorphic mid-level features and generic scene knowledge. *International Archives of Photogrammetry and Remote Sensing* **31,** 415–420.

Lee, D. H., Lee, K. M., and Lee, S. U. 2008. Fusion of LiDAR and imagery for reliable building extraction. *Photogrammetric Engineering and Remote Sensing* **74,** 215–225.

Lee, D. S., Shan, J., and Bethel, J. S. 2003. Class-guided building extraction from IKONOS imagery. *Photogrammetric Engineering and Remote Sensing* **69,** 143–150.

Mathieu, R. and Jagannath, A. 2005. *Object-oriented classification and Ikonos multispectral imagery for mapping vegetation communities in urban areas.* In: 17th Annual Colloquium of the Spatial Information Research Centre (SIRC 2005: A Spatio-temporal Workshop), 24-25 November 2005, Dunedin, New Zealand, pp. 181–188.

Mayer, H. 1999. Automatic object extraction from aerial imagery: A survey focusing on buildings. *Computer Vision and Image Understanding* **74,** 138–139.

Miliaresis, G., and Kokkas, N. 2007. Segmentation and object-based classification for the extraction of the building class from LiDAR DEMs. *Computers & Geosciences* **33,** 1076–1087.

Moons, T., Frere, D., Vandekerckhove, J., and Van Gool, L. 1998. Automatic modeling and 3D reconstruction of urban house roofs from high resolution aerial imagery. In Computer Vision—ECCV'98, 5th European Conference on Computer Vision, Freiburg, Germany, June, 2–6, 1998 Proceedings, Vol. I, pp. 410–425.

Platt, R. V., and Rapoza, L. 2008. An evaluation of an object-oriented paradigm for land use/land cover classification. *Professional Geographer* **60,** 87–100.

Popescue, S. C., Wynne, R. H., and Nelson, R. F. 2003. Measuring individual tree crown diameter with LiDAR and assessing its influence on estimating forest volume and biomass. *Canadian Journal of Remote Sensing* **29,** 564–577.

Renslow, M., Greenfield, P., amd Guay, T. 2000. Evaluation of multi-return LiDAR for forestry applications. Project Report for the Inventory and Monitoring Steering Committee, RSAC-2060/4810-LSP-0001-RPT1.

Secord, J., and Zakhor, A. 2007. Tree detection in urban regions using aerial lidar and image data. *IEEE Geoscience and Remote Sensing Letters* **4,** 196–200.

Shufelt, J. A. 1999. Performance evaluation and analysis of monocular building extraction from aerial imagery. *IEEE Transactions on Pattern Analysis and Machine Intelligence* **21,** 311–326.

Steger, C., Glock, C., Eckstein, W., et al. 1995. Model-based road extraction from images. In *Automatic Extraction of Man-Made Objects from Aerial and Space Images,* Vol. III, pp. 275–284.

Steger, C., Glock, C., Eckstein, et al. 1995. Model-based road extraction from images. In: A. Gruen, O. Kuebler and P. Agouris (eds), Automatic Extraction of Man-Made Objects from Aerial and Space Images, Birkhäuser Verlag, pp. 275–284.

Voss, M, and Sugumaran, R. 2008. Seasonal effect on tree species classification in an urban environment using hyperspectral data, LiDAR, and an object-oriented approach. *Sensors* **8,** 3020–3036.

Walter, V. 2004. Object-based classification of remote sensing data for change detection. *ISPRS Journal of Photogrammetry and Remote Sensing* **58,** 225–238.

Webster, T. L., Forbes, D. L., Dickie, S., and Shreenan, R. 2004. Using topographic LiDAR to map flood risk from storm-surge events for Charlottetown, Prince Edward Island, Canada. *Canadian Journal of Remote Sensing* **30,** 64–76.

Weidner, U. 1997. Digital surface models for building extraction. In *Automatic Extraction of Man-Made Objects from Aerial and Space Images,* Vol. II, pp. 193–202.

Weidner, U. 1997. Digital Surface Models for Building Extraction, in Proceedings, *Automatic Extraction of Man-Made Objects from Aerial and Space Images (II).* Birkhäuser Verlag, Basel, Switzerland, pp. 193–202.

Zhou, W., and Troy, A. 2008. An object-oriented approach for analysing and characterizing urban landscape at the parcel level. *International Journal of Remote Sensing* **29,** 3119–3135.

CHAPTER 7

Urban Land Surface Temperature Analysis

Urban climate studies have long been concerned about the magnitude of the difference in observed ambient air temperature between cities and their surrounding rural regions, which collectively describes the urban heat island (UHI) effect (Landsberg, 1981). Pertinent to the methods of temperature measurement, two types of UHIs can be distinguished: the urban canopy-layer (UCL) heat island and the urban boundary layer (UBL) heat island (Oke, 1979). The former consists of air between the roughness elements (e.g., buildings and tree canopies) with an upper boundary just below roof level. The latter situates above the former, with a lower boundary subject to the influence of urban surface. UHI studies traditionally have been conducted for isolated locations and with in situ measurements of air temperatures. The advent of satellite remote-sensing technology has made it possible to study UHIs both remotely and on continental and global scales (Streutker, 2002). Remotely sensed thermal imagery has the advantage of providing a time-synchronized dense grid of temperature data over a whole city.

Land surface temperature (LST) is an important parameter in understanding the urban thermal environment and its dynamics. LST modulates the air temperature of the lower layer of urban atmosphere and is a primary factor in determining surface radiation and energy exchange, the internal climate of buildings, and human comfort in cities (Voogt and Oke, 1998). LSTs are regarded as a function of four surface and subsurface properties: albedo, emissivity, the thermal properties of urban construction materials (including moisture), and the composition and structure of the urban canopy (Goward, 1981). Because the receipt and loss of radiation by urban surfaces correspond closely to the distribution of land use and land cover (LULC) characteristics (Lo et al., 1997; Weng, 2001), there has been a tendency to use thematic LULC data, not quantitative surface descriptors, to describe urban thermal landscapes (Voogt and Oke, 2003). Previous research has focused more on the relationship between LST variation and the thermal properties and composition of urban construction materials than on the effect of urban morphology and its component surfaces. Remote sensing-derived biophysical attributes bring great potential to parameterize urban construction materials and the composition and structure of urban canopies, as well as to linking with pixel-based LST measurements in understanding of the surface-energy budget and the UHI phenomenon. Recent advancements in landscape ecology facilitate the characterization of urban surface components and their quantitative links to environmental processes (e.g., the UHI process).

This chapter analyzes the spatial patterns of LSTs and explores factors that have contributed to LST variations based on a case study in Indianapolis, Indiana. For a detailed description of the study area, please refer to Sec. 3.2.1 of Chap. 3. Four advanced spaceborne thermal emission and reflection radiometer (ASTER) images, one for each season, are used in conjunction with other types of spatial data for the analysis. The potential factors are grouped into four categories: LULC composition, biophysical conditions, intensity of human activity, and landscape pattern. Statistical analysis is conducted to determine the relative importance of each group of variables. Moreover, by employing satellite images of different seasons, the dynamics of the LST and its interaction with the explanatory factors are investigated. Finally, the spatial variations in LST are examined at the levels of residential and general zoning to assess the environmental implications of urban planning.

7.1 Remote Sensing Analysis of Urban Land Surface Temperatures

Previous studies of urban LSTs and UHI have been conducted with National Oceanographic and Atmospheric Administration (NOAA) Advanced Very High Resolution Radiometer (AVHRR) data (Balling and Brazell, 1988; Gallo and Owen, 1998; Gallo et al., 1993; Kidder and Wu, 1987; Roth et al., 1989; Streutker, 2002). However, for all these studies, the 1.1-km spatial resolution AVHRR data is found to be suitable only for large-area urban temperature mapping, not for establishing accurate and meaningful relationships between image-derived values and those measured on the ground. The 120-m resolution Landsat Thematic Mapper (TM) thermal infrared (TIR) data and later, 60-m resolution, Enhanced Thematic Mapper Plus (ETM+) TIR data have been used extensively to derive LSTs and to study UHIs (Carnahan and Larson, 1990; Kim, 1992; Nichol, 1994, 1998; Weng, 2001, 2003; Weng et al., 2004). More recently, Lu and Weng (2006) used spectral mixture analysis to derive hot-object and cold-object fractions from the TIR bands of ASTER images and biophysical variables from nonthermal bands in Indianapolis. Statistical analyses were conducted to examine the relationship between LST and the five derived fraction variables across resolution from 15 to 90 m.

Studies using satellite-derived LSTs have been termed *surface-temperature UHIs* (Streutker, 2002). The concept of a satellite-derived heat island is largely an artifact of the low-spatial-resolution imagery used, and the term *surface-temperature patterns* is more meaningful than surface heat island (Nichol, 1996). Moreover, satellite-derived LSTs are believed to correspond more closely with the canopy-layer heat islands, although a precise transfer function between LST and the near-ground air temperature is not yet available (Nichol, 1994).

To study urban LST, some sophisticated numerical and physical models have been developed. These include energy balance models, by far the main methods (Oke et al., 1999; Tong et al., 2005); laboratory models (Cendese and Monti, 2003); three-dimensional simulations (Saitoh et al., 1996); Gaussian models (Streutker, 2003); and other numerical simulations. Statistical analysis may play an important role in linking LSTs to related factors, especially on larger scales. Previous studies have focused primarily on biophysical and meteorologic factors, such as built-up area and height (Bottyán and Unger, 2003), urban and street geometry (Eliasson, 1996), LULC (Dousset and Gourmelon, 2003; Weng, 2001); and vegetation. The relationship between LST and vegetation cover has been studied extensively by using vegetation indices such as normalized difference vegetation index (NDVI) (Carlson et al., 1994; Gallo and Owen, 1998; Gillies et al., 1997; Lo et al., 1997; Weng, 2001). Most recent advances include development and use of quantitative surface descriptors for assessing the interplay between urban material fabric and urban thermal behavior (Lu and Weng, 2006; Weng and Lu, 2008; Weng et al., 2004, 2006). Moreover, the landscape ecology approach was employed to assess this interplay across various spatial resolutions and to identify the operational scale where both LST and LULC processes interacted to generate the urban thermal landscape patterns (Liu and Weng, 2008; Weng et al., 2007). Because ASTER sensors collect both daytime and nighttime TIR images, analysis of LST spatial patterns also has been conducted for a diurnal contrast (Nichol, 2005).

7.2 Case Study: Land-Use Zoning and LST Variations

7.2.1 Satellite Image Preprocessing

ASTER images, acquired on October 3, 2000 (fall), June 25, 2001 (summer), April 5, 2004 (spring), and February 6, 2006 (winter), respectively, were used in this research. ASTER data have 14 bands with different spatial resolutions, that is, two visible bands and one near-infrared (VNIR) band with 15-m spatial resolution, six shortwave infrared (SWIR) bands with 30-m spatial resolution, and five thermal infrared (TIR) bands with 90-m spatial resolution. The level 1B ASTER data were purchased, and they consisted of the image data, radiance conversion coefficients, and ancillary data (Fujisada, 1998). Although these ASTER images have been corrected radiometrically and geometrically, the root-mean-square error (RMSE) was so large that the images could not meet our research needs. We conducted a geometric correction for each image using 1:24,000 U.S. Geological Survey (USGS) digital raster graphs (DRG) as reference maps. At least 30 ground control points were selected for each georectification, and the nearest-neighbor resampling method was applied with a pixel size of 15 m for all VNIR, SWIR, and TIR bands. An RMSE of less than 0.5 pixel was achieved for all georectifications.

7.2.2 LULC Classification

LULC maps were produced with an unsupervised classification algorithm called iterative self-organizing data analysis (ISODATA). The images (VNIR and SWIR bands) were classified into six LULC types, including agricultural land, aquatic systems, barren land, developed land, forest land, and grassland. Agricultural land is characterized by herbaceous vegetation that has been planted or is intensively managed for the production of food. This category includes croplands such as corn, wheat, soybeans, and cotton and also includes fallow land. Aquatic systems refer to all areas of open water, including lakes, rivers, streams, ponds, and outdoor swimming pools. Barren land is characterized by bare rock, gravel, sand, silt, clay, or other earthen material and no green vegetation present, including quarries, bare dunes, construction sites, and mines. Developed lands are defined as the areas that have a high percentage (30 percent or greater) of constructed materials, such as asphalt, concrete, and buildings. This category includes commercial, industrial, transportation, and low and high residential uses. Forest lands are the areas characterized by tree cover (natural or seminatural woody vegetation), including natural deciduous forest, evergreen forest, mixed forest, urban and suburban forest, and shrubs. Grasslands refer to the areas covered by herbaceous vegetation, including pasture/hay planted for livestock grazing or the production of hay. It also includes the urban/recreational grasses planted in developed settings for recreation, erosion control, or aesthetic purposes, such as parks, lawns, golf courses, airport grasses, and industrial-site grasses.

Remotely sensed data are often highly correlated between the adjacent spectral wavebands, and redundant bands would slow down the image processing if all bands were used. Principal components analysis (PCA) was applied to identify which bands of the nine VNIR and SWIR bands would contain more information. Based on the correlation coefficients between the bands and the first component, five to six bands normally would be selected for classification. Because the visible (green and red) and near-infrared bands were found useful in all cases, these three bands were always used in the image classifications.

Owing to the complexity of LULC types in the study area, each image scene was first stratified into two subscenes, one for the urban area and the other for the surrounding rural area. Image classification was performed separately for each subscene, and the results then were merged. Fifty spectral clusters were generated with ISODATA. Next, spectral classes were labeled after referencing to high-resolution aerial photographs and other geospatial data. After the first unsupervised classification, if we could not label all spectral classes, we then would mask out those confused classes and run a second unsupervised classification. The same procedure would be repeated for a third classification if confusion still existed after the second classification.

Accuracy assessment of classification images was conducted by using an error matrix. Some important measures, such as overall accuracy, producer's accuracy, and user's accuracy, can be calculated from the error matrix (Foody, 2002). In this study, a total of 350 points were checked on each classified image using a stratified random-sampling method. Digital orthophotographs from 2003 were used as the reference data. The color orthophotographs were provided by the Indianapolis Mapping and Geographic Infrastructure System, which was acquired in April 2003 for the entire county. The orthophotographs had a spatial resolution of 0.14 m. The coordinate system belonged to Indiana State Plane East, Zone 1301, with North American Datum of 1983. The orthophotographs were reprojected and resampled to 1-m pixel size for the sake of quicker display and shorter computing time. Figure 7.1 shows the resulting classified LULC maps. Overall classification accuracy of 87 percent (October 3, 2002, image), 88.33 percent (June 16, 2001, image), 92 percent (April 5, 2004, image), and 87.33 percent (February 6, 2006, image) was achieved, respectively. We made a further comparison among the four classified maps to ensure consistency within the classes and found that the magnitude and spatial pattern of each class corresponded well with each other but also reflected the seasonal and temporal differences.

7.2.3 Spectral Mixture Analysis

Spectral mixture analysis (SMA) is regarded as a physically based image processing technique that supports repeatable and accurate extraction of quantitative subpixel information (Mustard and Sunshine, 1999; Roberts et al., 1998; Smith et al., 1990). It assumes that the spectrum measured by a sensor is a linear combination of the spectra of all

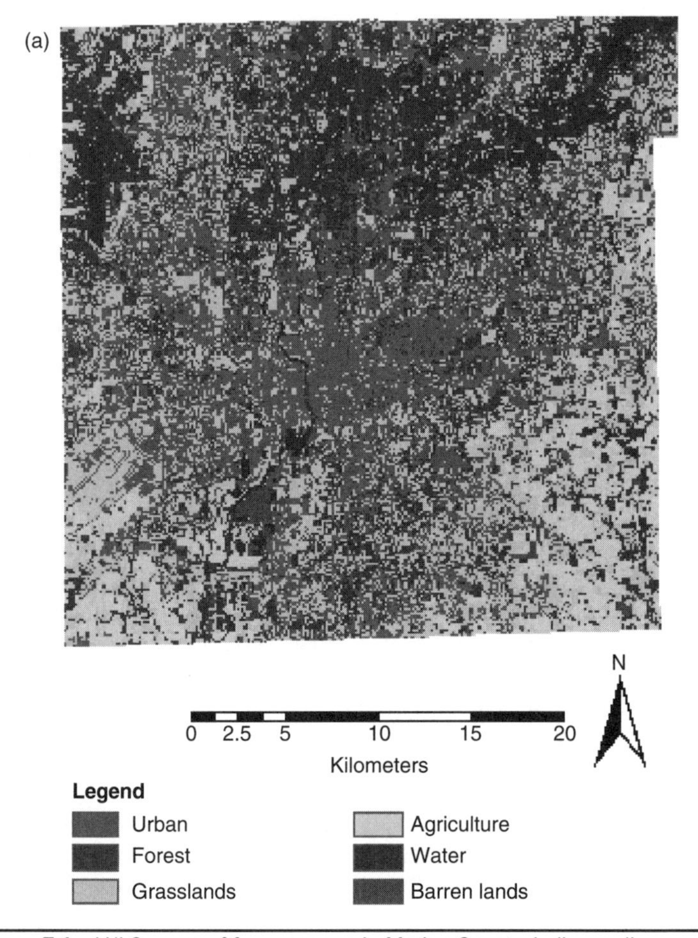

Figure 7.1 LULC maps of four seasons in Marion County, Indianapolis, Indiana, derived from ASTER images: (*a*) October 3, 2000; (*b*) June 16, 2001; (*c*) April 5, 2004; and (*d*) February 6, 2006. (*Adapted from Weng et al., 2008.*) See also color insert.

components within the pixel (Adams et al., 1995; Roberts et al., 1998). Because of its effectiveness in handling spectral mixture problems, SMA has been used widely in estimation of vegetation cover (Asner and Lobell, 2000; McGwire et al., 2000; Small, 2001), in vegetation or land-cover classification and change detection (Aguiar et al., 1999; Cochrane and Souza, 1998), and in urban studies (Lu and Weng, 2004, 2006a, 2000b; Phinn et al., 2002; Rashed et al., 2001; Wu and Murray, 2003). In this study, SMA was used to develop green vegetation, soil, and impervious surface fraction images. Endmembers were identified initially from high-resolution aerial photographs. An improved image-based dark-object subtraction model has proved effective and

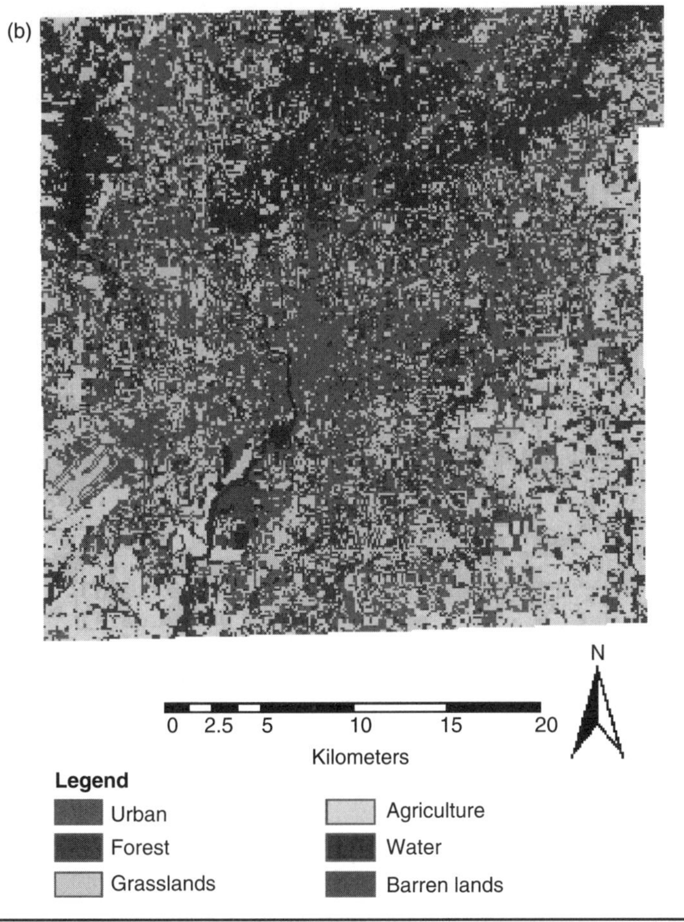

FIGURE 7.1 (*Continued*)

was applied to reduce the atmospheric effects (Chavez, 1996; Lu et al., 2002). After implementation of the atmospheric correction and geometric rectification of the ASTER images, a constrained least-squares solution was applied to unmix the nine VNIR and SWIR bands of the ASTER imagery into fraction images, including high-albedo, low-albedo, vegetation, and soil fractions. An impervious surface then was estimated based on the relationship between high- and low-albedo fractions and impervious surfaces. For more details about the derivation of fraction images, please refer to the article by Lu and Weng (2006b).

7.2.4 Estimation of LSTs

Various algorithms have been developed for converting ASTER TIR measurements into surface kinetic temperatures (i.e., LSTs), as

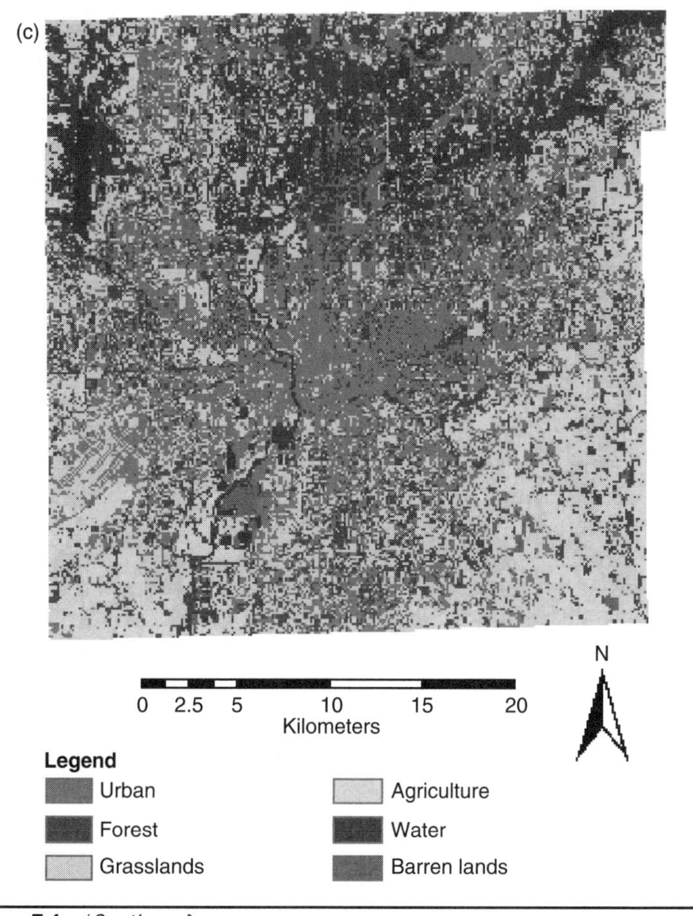

Figure 7.1 (*Continued*)

reported by the ASTER Temperature/Emissivity Working Group (1999) and Gillespie and colleagues (1998). However, a universally accepted method is not available currently for computing LSTs from multiple bands of TIR data such as those found on ASTER. In this study, we selected ASTER band 13 (10.25 to 10.95 μm) to compute LSTs because the spectral width of this band is close to the peak radiation of the blackbody spectrum given off by the urban surface of the study area. Two steps were taken to compute LSTs: (1) converting spectral radiance to at-sensor brightness temperature (i.e., blackbody temperature) and (2) correcting for spectral emissivity. We adopted the most straightforward approximation to replace the sensor-response function with a delta function at the sensor's central wavelength to invert LSTs with the assumption of uniform emissivity

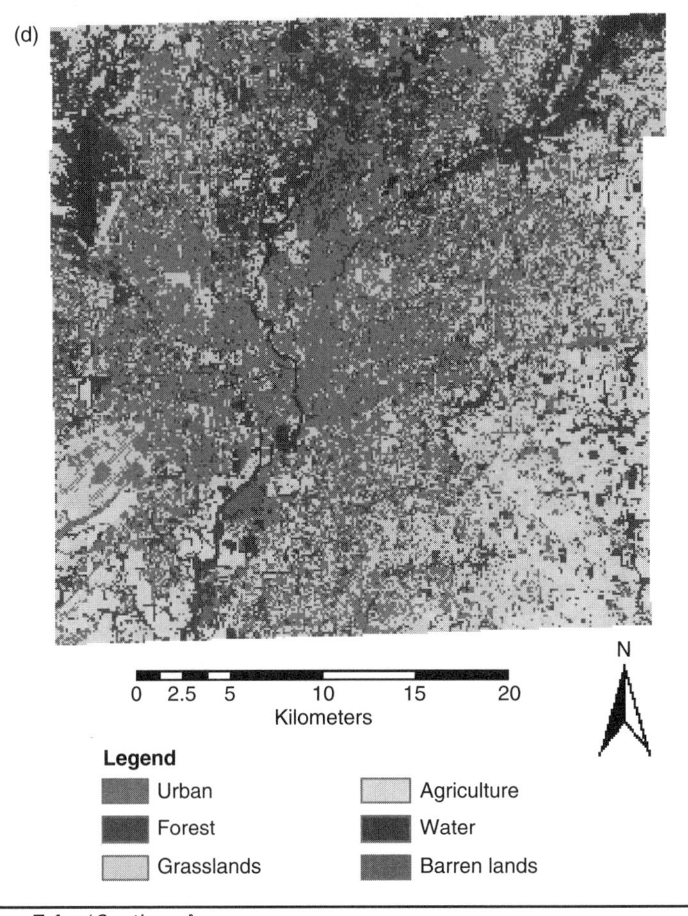

Figure 7.1 (*Continued*)

(Dash et al., 2002; Li et al., 2004; Schmugge et al., 2002). The conversion formula is

$$T_c = \frac{C_2}{\lambda_c \ln\left(\dfrac{C_1}{\lambda_c^5 \pi L_\lambda} + 1\right)} \tag{7.1}$$

where T_c = brightness temperature in kelvins (K) from a central wavelength

L_λ = spectral radiance in W m^{-3} sr^{-1} μm^{-1}, λ_c is the sensor's central wavelength

C_1 = the first radiation constant (3.74151×10^{-16} W m^{-2} sr^{-1} μm^{-1})

C_2 = the second radiation constant (0.0143879 m · K)

The temperature values thus obtained are referenced to a blackbody. Therefore, corrections for spectral emissivity ε became necessary according to the nature of the land cover. Each of the land-cover categories was assigned an emissivity value according to the emissivity classification scheme by Snyder and colleagues (1998). The emissivity-corrected LST was computed as follows (Artis and Carnhan, 1982):

$$\text{LST} = \frac{T_c}{1 + (\lambda * T_c / \rho)\ln\varepsilon} \tag{7.2}$$

where λ = the wavelength of emitted radiance (for which the peak response and the average of the limiting wavelengths λ = 10.6 μm) (Markham and Barker, 1985)
ρ = $h \times c/\sigma$ (1.438×10^{-2} mK)
σ = Boltzmann's constant (1.38×10^{-23} J/K)
h = Planck's constant (6.626×10^{-34} J · s)
c = the velocity of light (2.998×10^{8} m/s)

Relative LST is sufficient for mapping of the spatial variations of urban land surface temperatures. Therefore, the effects of atmosphere and surface roughness on LST were not taken into account in this study. Lack of atmospheric correction may introduce a temperature error of 4 to 7°C for the midlatitude summer atmosphere (Voogt and Oke, 1998). The magnitude of atmospheric correction depends on the image bands used, as well as atmospheric conditions and the height of observation. Errors owing to urban effective anisotropy depend on surface structure and relative sensor position and can yield a temperature difference of up to 6 K or higher in downtown areas (Voogt and Oke, 1998).

7.2.5 Statistical Analysis

Based on previous research results (Oke, 1982, 1988), we hypothesized that the spatial variations of LST were related to four groups of factors that described LULC composition (six variables), biophysical conditions (four variables), intensity of human activity (four variables), and landscape pattern (four variables) (Table 7.1). We computed the mean (and standard deviation) values of LST and each potential factor per general zoning polygon and per residential zoning polygon. Tables 7.2 and 7.3 show the definitions for each zoning category and relevant attributes. Multiple stepwise regressions then were applied to obtain independent variables with statistical significance ($P < 0.001$). Variables that were removed from the stepwise regressions were not considered to be explanatory factors of LST variation and therefore were excluded in the subsequent statistical analyses. Next, factor analysis (specifically PCA) was conducted to transform the identified independent variables into a set of uncorrelated principal components. Factors with eigenvalues greater than 1 were extracted (Kaiser, 1960), and the factor loadings of each original variable were examined.

Categories of Variables	Variable	Meaning of Variable
LULC composition	Per_Ur	Percentage of built-up land
	Per_Ba	Percentage of barren land
	Per_Gr	Percentage of grassland
	Per_Ag	Percentage of agricultural land
	Per_Fo	Percentage of forest land
	Per_Wa	Percentage of water bodies
Biophysical conditions	NDVI	Mean value of normalized difference vegetation index
	GV	Mean value of green-vegetation fraction derived from SMA
	IMP	Mean value of impervious surface fraction derived from SMA
	SOIL	Mean value of soil fraction derived from SMA
Intensity of human activity	PAVEMENT	Percentage of pavement area per zoning polygon
	BLDG_ARE	Percentage of building area per zoning polygon
	MEAN_POP	Mean population in a zoning polygon
	POP_DEN	Population density in a zoning polygon
Landscape patterns	SHAPE_IN	Shape index of a zoning polygon
	FRACTAL	Fractal dimension of a zoning polygon
	LANDSIM	Landscape similarity index (percent)
	DIVERS	Shannon's diversity index of LULC composition within a zoning polygon

Table 7.1 Variables Applied to Examine LST Variations

7.2.6 Landscape Metrics Computation

Landscape-pattern metrics have been employed frequently in landscape ecology to characterize the arrangement of species, communities, and habitat patches within landscapes (Read and Lam, 2002). Their potential for monitoring ecosystem changes and linking with ecologic and environmental processes has been recognized (Li and Reynolds, 1994). These metrics can be applied to create quantitative measures of spatial patterns found on a map or remotely sensed imagery. When applying landscape metrics to remotely sensed data,

Zoning Category	Code	Description	Number of Polygons	Percent of Landscape	Mean Polygon Size (ha)	Mean Shape Index	Mean Fractal Dimension
Historical preservation	HP	Historic preservation district including a variety of land uses (mostly residential)	1	0.01	8.64	1.44	1.29
Special uses	SU	Wide variety of uses, such as schools, utility infrastructure, cemeteries, libraries, community centers, charitable organizations, golf courses, and penal institutions	1232	17.84	14.09	1.38	1.32
University	UQ	Variety of land uses typical of higher-education institutions, including classroom, office, dormitory, facility maintenance, and parking structures	13	0.19	19.24	1.28	1.35
Agriculture	DA	Agriculture and single family, very low density	781	11.31	23.76	1.47	1.32

Residential	D	Variety of residential categories summarized in Table 7.3	1926	27.88	18.72	1.41	1.30
CBD	CBD	Central business district: core activities of all types with a variety of related land uses	44	0.64	24.60	1.38	1.29
Commercial	C	Includes office buffer, high-intensity office/apartment, neighborhood commercial, thoroughfare service, and corridor commercial districts	2185	31.63	4.58	1.38	1.34
Hospital	HD	Major hospital complexes and campuses	33	0.48	17.45	1.44	1.30
Industrial	I	Variety of industrial uses, including urban and suburban and light, medium, and heavy industry	498	7.21	22.12	1.51	1.32
Park	PK	Permits all sizes and ranges of public park land and facilities, including park peripheral areas ensuring compatibility of adjacent land use	166	2.40	42.92	1.53	1.31

Table 7.2 General Zoning (Polygon) Attributes

Zoning Category	Code	Description	Number of Polygons	Percent of Landscape	Mean Polygon Size (ha)	Mean Shape Index	Mean Fractal Dimension
Airport	A	Public airports, municipally owned or operated, including all necessary navigation and flight-operation facilities and accessory uses	28	0.41	97.67	1.53	1.33

Note: A zoning geographic information system (GIS) data layer was provided by the Indianapolis Mapping and Geographic Infrastructure System, City of Indianapolis. Information on zoning category, code, and description was provided by the Metropolitan Planning Department, City of Indianapolis. Other attributes in the table are computed and compiled by the authors. We also referred to Wilson and colleagues (2003) in compiling this table.

TABLE 7.2 General Zoning (Polygon) Attributes (*Continued*)

Zoning Category	Typical Density (acre)	Primary Use	Minimum Open Space	Typical Lot Size (ft^2)	Comprehensive Planning Classification	Number of Polygons	Percent of Landscape	Mean Polygon Size (ha)	Mean Shape Index	Mean Fractal Dimension
DP	Varies	Planned unit development	Varies	Varies	Varies	89	13.46	38.73	1.44	1.28
DS	0.5	Suburban single family	85 percent	1 acre	Very low density	67	4.71	29.56	1.44	1.30
D1	0.9	Suburban single family	80 percent	24,000	Very low density	104	4.61	20.07	1.37	1.29
D2	1.9	Suburban single family	75 percent	15,000	Very low density	226	16.22	35.14	1.46	1.30
D3	2.6	Low- or medium-intensity single family	70 percent	10,000	Low density	296	19.14	30.73	1.53	1.30
D4	4.2	Low- or medium-intensity single family	65 percent	7200	Low density	195	10.95	26.99	1.48	1.30
D5	4.5	Medium-intensity single family	65 percent	5000	Low and medium density	231	19.74	33.35	1.56	1.32
D5II	5.0	Medium-intensity single or two family	65 percent	3200	Low and medium density	17	D5II was merged into D5 for computing landscape metrics.			

TABLE 7.3 Residential Zoning (Polygon) Attributes

Zoning Category	Typical Density (acre)	Primary Use	Minimum Open Space	Typical Lot Size (ft^2)	Comprehensive Planning Classification	Number of Polygons	Percent of Landscape	Mean Polygon Size (ha)	Mean Shape Index	Mean Fractal Dimension
D6	6–9	Low-intensity multifamily	3.85	Varies	Medium density	81	5.10	12.44	1.36	1.28
D6II	9–12	Medium-intensity multifamily	2.65	Varies	Medium density	118	D6II was merged into D6 for computing landscape metrics.			
D7	12–15	Medium-intensity multifamily	2.10	Varies	Medium density	153	2.93	9.64	1.47	1.32
D8	5–26	Urban multiuse residential	2.65	Varies	High density	112	1.76	6.00	1.48	1.36
D9	12–120	Suburban high-rise apartment	0.29–1.45	Varies	High density	29	0.34	5.68	1.35	1.33
D10	20–140	Central and inner-city high-rise apartment	0.27–1.18	Varies	High density	9	0.04	1.43	1.23	1.33
D11	6.0	Mobile dwellings	Varies	Varies	Medium density	23	0.93	22.40	1.34	1.26
D12	5.0	Low-density two family	65 percent	9000	Low density	13	0.08	3.28	1.36	1.31

Note: See the Note to Table 7.2.

TABLE 7.3 Residential Zoning (Polygon) Attributes (*Continued*)

each unique pixel value represents a patch type (Read and Lam, 2002). In this study, four landscape metrics were computed for each zoning polygon, including a shape index of a zoning polygon, fractal dimension of a zoning polygon, landscape similarity index, and Shannon's diversity index of LULC composition within a zoning polygon. FRAGSTATS, a software program designed to compute a wide variety of landscape metrics for categorical map patterns, was selected for use (McGarigal and Marks, 1995). Selection of these metrics was based on a research result by Riitters and colleagues (1995). Shape index, a simple measure of shape complexity, is computed by dividing patch perimeter by the minimum perimeter possible for a maximally compact patch of the corresponding patch area. In our study, patches refer to zoning polygons. Fractal dimension of a zoning polygon equals two times the logarithm of patch perimeter divided by the logarithm of the patch area. A fractal dimension greater than 1 for a two-dimensional (2D) patch indicates an increase in shape complexity. Fractal dimension approaches 1 for shapes with very simple perimeters and approaches 2 for shapes with highly convoluted, plane-filling perimeters. Fractal dimension index is appealing because it reflects shape complexity across a range of spatial scales. Similarity index equals 0 if all the patches within the specified neighborhood have a zero similarity coefficient and increases as the neighborhood is increasingly occupied by patches with greater similarity coefficients and as those similar patches become closer and more contiguous and less fragmented in distribution. Shannon's diversity index equals 0 when the landscape contains only one patch (i.e., no diversity) and increases as the number of different patch types increases and/or the proportional distribution of area among patch types becomes more equitable.

7.2.7 Factors Contributing to LST Variations

As a rule in interpreting factor analysis results, the suitability of data for factor analysis was first checked based on Kaiser-Meyer-Olkin (KMO) and Bartlett's test values (Tabachnick and Fidell, 1996). Only when the KMO value was greater than 0.5 and the significant level of Bartlett's test was less than 0.1 was the data acceptable for factor analysis. The second step was to validate the variables based on the communality of variables. Among the eight PCAs conducted (four for residential zoning and another four for general zoning), 11 to 14 variables finally entered into the factor analysis. Based on the rule that the minimum eigenvalue should not be less than 1, five factors were extracted from each factor analysis, with the exception of the general zoning of October 3, 2000, which had only four factors. Tables 7.4 through 7.7 show rotated factor loading matrices of residential zoning for the four images, whereas Tables 7.8 through 7.11 show loading matrices for general zoning. From the residential zoning matrices it

Variable	Component 1	Component 2	Component 3	Component 4	Component 5
SHAPE_IN	0.035	0.022	0.883	–0.024	0.013
FRACTAL	–0.129	0.045	0.862	0.061	–0.115
LANDSIM	0.165	–0.202	0.021	0.052	–0.545
NDVI	0.963	–0.085	–0.038	0.063	–0.022
GV	0.875	–0.135	–0.066	0.175	–0.045
IMP	–0.613	0.654	–0.080	0.060	0.054
SOIL	–0.095	–0.877	–0.080	–0.118	–0.033
Per_Wa	–0.124	0.168	–0.027	–0.792	–0.023
Per_Ur	–0.876	0.213	0.058	0.247	0.057
Per_Fo	0.772	0.496	–0.017	–0.179	0.017
Per_Gr	0.193	–0.836	–0.042	0.157	–0.083
PAVEMENT	–0.327	0.327	0.016	0.530	–0.200
BLDG_ARE	–0.469	0.440	–0.011	0.460	0.230
POP_DEN	0.087	–0.086	–0.073	0.031	0.872
Initial eigenvalues	4.41	2.17	1.61	1.23	1.07
Percent of variance	31.521	15.521	11.489	8.764	7.668
Cumulative percent	31.521	47.043	58.531	67.295	74.964

TABLE 7.4 Rotated Factor Loading Matrix for Residential Zoning, October 3, 2000

Variable	Component 1	Component 2	Component 3	Component 4	Component 5
SHAPE_IN	–0.012	–0.009	–0.036	0.912	–0.011
FRACTAL	0.202	–0.121	–0.665	0.329	0.238
LANDSIM	–0.169	0.539	–0.089	0.409	–0.071
BLDG_ARE	0.597	–0.451	0.142	0.135	0.284
NDVI	–0.963	–0.094	0.055	0.047	0.079
GV	–0.932	–0.106	0.072	0.048	0.115
IMP	0.828	–0.354	–0.089	0.024	–0.086
SOIL	0.120	0.853	0.052	–0.103	0.234
Per_Ur	0.924	0.008	–0.009	–0.039	0.141
Per_Wa	0.070	–0.131	0.051	0.027	–0.940
MEAN_POP	0.033	–0.097	0.881	0.095	0.087
Initial eigenvalues	3.858	1.496	1.330	1.074	0.988
Percent of variance	35.072	13.603	12.090	9.761	8.978
Cumulative percent	35.072	48.675	60.765	70.526	79.504

TABLE 7.5 Rotated Factor Loading Matrix for Residential Zoning, June 16, 2001

Variable	Component 1	Component 2	Component 3	Component 4	Component 5
FRACTAL	–0.164	0.862	0.015	–0.043	0.087
LANDSIM	0.224	–0.008	0.186	–0.020	0.688
GV	0.857	–0.040	–0.114	–0.110	0.014
IMP	–0.876	–0.067	–0.140	–0.191	–0.076
SOIL	0.034	–0.145	0.663	0.446	0.175
SHAPE_IN	0.060	0.869	–0.017	0.018	0.034
NDVI	0.790	–0.077	0.261	–0.183	0.023
Per_Ag	0.056	0.070	0.073	0.762	–0.230
Per_Wa	–0.006	–0.076	–0.811	0.166	0.052
Per_Ba	–0.026	–0.062	–0.087	0.532	0.233
POP_DEN	0.128	–0.131	0.111	–0.051	–0.725
Per_Ur	–0.850	0.094	0.156	–0.151	0.023
Initial eigenvalues	3.018	1.594	1.387	1.105	1.055
Percent of variance	25.148	13.279	11.560	9.211	8.791
Cumulative percent	25.148	38.427	49.987	59.198	67.988

TABLE 7.6 Rotated Factor Loading Matrix for Residential Zoning, April 5, 2004

Variable	Component 1	Component 2	Component 3	Component 4	Component 5
FRACTAL	–0.295	–0.158	–0.589	0.159	0.377
LANDSIM	0.290	0.190	–0.236	0.167	0.336
BLDG_ARE	–0.664	–0.350	0.276	0.202	0.123
GV	0.879	–0.201	0.074	–0.031	0.073
IMP	–0.876	–0.202	–0.011	–0.106	–0.006
SOIL	0.073	0.625	–0.057	0.552	–0.086
Per_Wa	–0.072	0.098	0.032	–0.874	–0.019
MEAN_POP	–0.021	0.025	0.864	0.024	0.057
Per_Ag	0.044	0.623	–0.014	–0.045	0.017
Per_Ba	–0.044	0.523	0.086	–0.044	0.065
NDVI	0.736	–0.176	0.257	0.260	0.033
SHAPE_IN	0.001	0.037	0.011	–0.068	0.933
Initial eigenvalues	2.803	1.511	1.306	1.163	1.007
Percent of variance	23.361	12.591	10.883	9.695	8.393
Cumulative percent	23.361	35.951	46.834	56.529	64.922

TABLE 7.7 Rotated Factor Loading Matrix for Residential Zoning, February 6, 2006

Variable	Component 1	Component 2	Component 3	Component 4
SHAPE_IN	0.064	0.065	0.846	0.195
FRACTAL	−0.051	−0.100	0.836	−0.212
LANDSIM	0.550	0.047	0.081	0.037
PAVEMENT	−0.258	−0.395	0.059	−0.397
MEAN_POP	0.263	0.013	−0.196	0.261
DIVERS	0.413	0.272	−0.027	0.539
Per_Ag	−0.086	0.635	0.063	0.009
Per_Wa	−0.142	−0.136	0.067	0.818
Per_Fo	0.720	−0.323	−0.003	0.137
NDVI	0.944	0.060	−0.063	0.077
GV	0.894	0.043	−0.106	0.032
IMP	−0.733	−0.455	−0.040	−0.008
SOIL	0.105	0.865	−0.111	−0.002
Initial eigenvalues	3.693	1.624	1.458	1.158
Percent of variance	28.408	12.493	11.217	8.911
Cumulative percent	28.408	40.900	52.118	61.028

Table 7.8 Rotated Factor Loading Matrix for General Zoning, October 3, 2000

became known that the first factor (factor 1) explained about 23 to 35 percent of the total variance, the second factor (factor 2) accounted for 13 to 15 percent, the third factor (factor 3) explained 11 to 12 percent, the fourth factor (factor 4) explained 9 to 10 percent, and the fifth factor (factor 5) explained 8 to 9 percent. Together, the first five factors explained approximately 65 to 80 percent of the variance. From the general zoning matrices, the first factor (factor 1) explained about 28 to 36 percent of the total variance, the second factor (factor 2) accounted approximately for 12 percent, the third factor (factor 3) explained 9 to 11 percent, the fourth factor (factor 4) explained 8 to 9 percent, and the fifth factor (factor 5) explained about 7 percent. Overall, the first four/five factors explained between 61 and 73 percent of the variance.

Interpreting factor loadings is the key in factor analysis. Factor loadings involve the measurement of the relationships between variables and factors. Generally speaking, only variables with loadings greater than 0.32 should be considered (Tabachnick and Fidell, 1996).

Variable	Component 1	Component 2	Component 3	Component 4	Component 5
SHAPE_IN	0.014	0.035	0.040	0.941	0.084
FRACTAL	–0.109	–0.677	0.064	0.451	–0.208
PAVEMENT	–0.259	–0.447	0.133	–0.042	–0.312
MEAN_POP	0.033	0.707	0.082	0.013	–0.172
BLDG_ARE	–0.664	0.010	0.107	0.032	–0.194
DIVERS	0.355	0.581	–0.057	0.274	–0.015
Per_Ag	0.068	–0.029	–0.081	0.048	0.877
Per_Ur	–0.912	–0.222	–0.130	–0.033	–0.079
Per_Fo	0.519	0.157	0.752	–0.012	0.107
Per_Gr	0.692	0.116	–0.562	0.032	–0.243
NDVI	0.903	0.214	0.249	0.020	0.008
GV	0.867	0.193	0.241	–0.007	–0.032
IMP	–0.864	–0.119	0.106	–0.023	–0.074
SOIL	0.027	0.107	–0.752	–0.069	0.266
Initial eigenvalues	5.066	1.711	1.286	1.100	1.041
Percent of variance	36.187	12.221	9.186	7.854	7.437
Cumulative percent	36.187	48.409	57.594	65.449	72.886

TABLE 7.9 Rotated Factor Loading Matrix for General Zoning, June 16, 2001

Variable	Component 1	Component 2	Component 3	Component 4	Component 5
FRACTAL	–0.073	–0.758	0.084	–0.120	–0.106
PAVEMENT	–0.272	–0.446	0.110	–0.179	–0.196
MEAN_POP	0.033	0.673	0.101	–0.310	–0.156
BLDG_ARE	–0.683	–0.006	0.009	–0.290	–0.196
GV	0.805	0.199	0.355	–0.085	0.031
IMP	–0.868	–0.134	–0.018	0.037	–0.135
SOIL	0.103	0.170	–0.758	–0.015	0.218
DIVERS	0.367	0.583	0.083	0.256	0.009
Per_Ag	0.029	0.066	–0.078	–0.040	0.918
Per_Wa	–0.091	0.171	0.185	0.759	–0.099
Per_Ba	0.021	–0.015	–0.111	0.507	0.032
Per_Fo	0.348	0.205	0.797	–0.047	0.076
Per_Gr	0.720	0.139	–0.534	–0.037	–0.186
NDVI	0.812	0.232	0.029	–0.188	–0.183
Initial eigenvalues	3.946	1.739	1.398	1.097	1.016
Percent of variance	28.187	12.421	9.985	7.833	7.259
Cumulative percent	28.187	40.608	50.593	58.426	65.685

TABLE 7.10 Rotated Factor Loading Matrix for General Zoning, April 5, 2004

Variable	Component 1	Component 2	Component 3	Component 4	Component 5
FRACTAL	–0.109	0.099	–0.724	–0.151	–0.101
PAVEMENT	–0.277	0.123	–0.482	–0.281	–0.186
MEAN_POP	0.059	0.136	0.732	–0.182	–0.145
BLDG_ARE	–0.671	0.082	0.026	–0.309	–0.138
GV	0.841	0.268	0.188	–0.066	–0.017
IMP	–0.851	0.035	–0.110	0.004	–0.124
SOIL	0.070	–0.736	0.187	–0.022	0.244
LANDSIM	0.540	0.239	–0.001	–0.036	0.132
Per_Ag	0.056	–0.085	0.036	–0.032	0.917
Per_Wa	–0.064	0.230	0.075	0.776	–0.056
Per_Ba	0.006	–0.127	0.003	0.530	0.007
Per_Fo	0.382	0.744	0.230	–0.030	0.090
Per_Gr	0.707	–0.499	0.124	–0.017	–0.241
NDVI	0.778	–0.004	0.299	–0.166	–0.169
Initial eigenvalues	3.890	1.637	1.267	1.120	1.021
Percent of variance	27.788	11.690	9.052	8.002	7.293
Cumulative percent	27.788	39.479	48.530	56.532	63.825

Table 7.11 Rotated Factor Loading Matrix for General Zoning, February 6, 2004

Comrey and Lee (1992) suggested a range of values to interpret the strength of the relationships between variables and factors. Loadings of 0.71 and higher are considered excellent, 0.63 is considered very good, 0.55 is considered good, 0.45 is considered fair, and 0.32 is considered poor. Tables 7.4 through 7.7 present factor loadings on each variable for residential zoning. No matter which date of image is considered, factor 1 always had strong loadings on the variables of biophysical conditions, including vegetation and impervious surface fractions, and NDVI. In most cases, factor 1 also had a strong loading on the percentage of built-up land. Apparently, factor 1 was associated with biophysical conditions. Factor 2 had a high loading on soil fraction for three images, except for the image of April 5, 2004, where factor 2 had a strong positive loading on two landscape-pattern variables (i.e., shape index and fractal dimension). Factor 3 showed high factor loadings either on population, landscape-pattern variables, or percentage of water bodies. Similarly, factor 4 displayed high loadings on different variables, such as percentage of water bodies, shape index, and percentage of agricultural land. Factor 5 had high factor loadings on population density, percentage of water bodies, and landscape-pattern variables. Apparently, among factors 3 through 5, one of the factors was associated with the percentage of water, another with the landscape-pattern variable(s), and still another with population variable(s). Despite the slight differences, each factor explained approximate 10 percent of the total variance.

Tables 7.8 through 7.11 display factor loadings for general zoning. Factor 1 was found to have a strong positive loading on vegetation fraction and NDVI for all four images (for the February and April images, there were strong positive loadings on percentage of grassland too), whereas factor 1 was found to have a strong negative loading on impervious surface fraction (also on percentage of built-up land for the June image). It may be concluded that regardless residential or general zoning, factor 1 can be described as a biophysical conditions factor and that, on average, it explained 30 percent of the total variation in LST pattern. Factor 2 showed strong loadings on soil fraction, fractal dimension, or mean population. Factor 3 had a strong loading on landscape-pattern variables, soil fraction, or percentage of forest land. Factor 4 had strong loadings on percentage of water for the three images but had the strongest loading on shape index for the June image. Factor 5 had the strongest loading on percentage of agricultural land for all images, with the exception of the October image, where this variable did not enter into the PCA.

7.2.8 General Zoning, Residential Zoning, and LST Variations

To assess the zoning effect on LST, the variance explained and the factor loadings of residential and general zoning of each image were examined. Results indicate that the cumulative variance accounted for by all factors always was larger in residential zoning than that in general zoning. This finding suggests that factors contributing to the spatial variation of LST in general zoning were more complex or that the variables used for this study were more appropriate for explaining residential LST patterns. Regardless of residential or general zoning, factor 1 always had strong loadings (on average, 29 percent of the variance in residential zoning and 30 percent of the variance in general zoning) on the variables that described biophysical conditions. These variables included vegetation fraction, impervious surface fraction, and NDVI. With a few exceptions, factor 2 was found to be closely associated with soil fraction in both zoning schemes, another biophysical variable. These findings are consistent with most of the previous research, pinpointing the correlation between the spatial pattern of LST and biophysical conditions in a specific region. Apparently, zoning as a tool of urban planners has a profound impact on the biophysical characteristics of urban landscapes by imposing such restrictions as maximum building height and density and the extent of impervious surfaces and open space, land-use types, and activities. These variables control surface energy exchange, surface and subsurface hydrology, micro- to meso-scale weather, and climate systems in general and LST in particular (Wilson et al., 2003).

Previous studies have proved that the spatial arrangement and areal extent of different LULC types largely regulate the variations of spectral radiance and texture in LST (Lo et al., 1997, Weng, 2001, Weng et al., 2004). This study finds that the LULC composition variables generally played a less significant role in the spatial patterns of LST. However, the percentage of water bodies consistently had a strong impact on LST variation in either residential or general zoning. While the percentage of built-ups was found to be highly correlated with LST variation in residential zoning, both the percentage of forest land and the percentage of agricultural land had an important influence on LST variation in general zoning. Therefore, statistical models may be built based on the composition of LULC types within a zoning unit because the spatial arrangement and areal extent of different LULC types largely regulate the variations in spectral radiance and texture in LST (Weng et al., 2004). Using a surface energy balance model, the spatial variations in LST may be modeled based on such factors as imperviousness, green-vegetation coverage and abundance, and other biophysical variables that have been identified to relate directly or indirectly to the radiative, thermal, moisture, and aerodynamic properties in the urban surface and subsurface (Oke, 1982). This is so because thermal spectral response for each pixel is largely controlled by the dynamic relationship among these biophysical variables. This dynamic would produce an aggregated effect on the surface energy balance for each zoning unit.

Population-related variables (e.g., mean population and population density within a zoning polygon) had more explanatory power in the spatial variation of residential zoning than in general zoning. There was always a population-related factor in the factor matrices of residential zoning. These factors accounted for between 8 and 12 percent of the variance when a population-related variable was the single most important variable.

Tables 7.4 through 7.11 further indicate that the landscape metrics variables may play an important role in LST spatial patterns. All the tables of factor loading matrices contain a factor that is mainly associated with landscape metrics variable(s). The landscape-pattern factors explained from 8 to 13 percent of the variance, with strong loadings on shape index and the fractal dimension of zoning polygons. In most cases, shape index contributed more to the landscape-pattern factors than did fractal dimension. To further examine the relationship between LST and the landscape metrics of zoning polygons, the zoning polygons were grouped into categories (see Tables 7.2 and 7.3). Mean and standard deviation values of LSTs were computed for each zoning category. Six metrics were calculated for the zoning categories, which included mean patch size, mean shape index, area-weighted mean shape index, mean fractal dimension, double log fractal dimension, and area-weighted mean patch fractal dimension. Tables 7.12 and 7.13 show the results of correlation analysis between LST and landscape metrics in the residential and non-residential

	MEAN_SIZ	SHA_INDE	AWMSI	FRACTAL	DLFD	AWMPFD
TS_M_00	–0.53	–0.04	–0.05	0.36	0.41	0.44
TS_S_00	–0.55*	–0.62*	–0.56*	0.16	0.39	–0.02
TS_M_01	–0.59*	–0.09	–0.04	0.46	0.48	0.50
TS_S_01	–0.56*	–0.65*	–0.55*	0.17	0.44	0.02
TS_M_04	–0.44	0.02	0.02	0.33	0.36	0.46
TS_S_04	–0.51	–0.62*	–0.53	0.17	0.37	–0.02
TS_M_06	–0.62*	–0.20	–0.07	0.57*	0.43	0.41
TS_S_06	–0.40	–0.35	–0.30	0.25	0.24	0.11

Note: Correlation is significant at the 0.01 level (two-tailed).
*Correlation is significant at the 0.05 level (two-tailed).
MEAN_SIZ = mean patch size; SHA_INDE = mean shape index; AWMSI = area-weighted mean shape index; FRACTAL = fractal dimension; DLFD = double log fractal dimension; AWMPFD = area-weighted mean patch fractal dimension; TS_M_00 = mean LST, October 3, 2000; TS_S_00 = standard deviation of LST, October 3, 2000.

TABLE 7.12 Pearson Correlation Coefficients between LST and Landscape Metrics in Residential Zoning Categories

	MEAN_SIZ	SHA_INDE	FRACTAL	AWMSI	DLFD	AWMPFD
TS_M_00	−0.24	−0.11	0.02	−0.14	−0.29*	0.27
TS_S_00	0.24	0.09	−0.25	0.10	−0.03	−0.32*
TS_M_01	−0.21	−0.15	−0.02	−0.10	−0.43*	0.28*
TS_S_01	0.12	−0.04	−0.15	−0.03	0.04	−0.29*
TS_M_04	−0.24	−0.15	0.00	−0.17	−0.28*	0.27*
TS_S_04	0.28*	0.11	−0.26	0.14	0.04	−0.34*
TS_M_06	−0.06	−0.02	0.06	0.13	−0.31*	0.35*
TS_S_06	0.40*	−0.02	−0.23	0.04	−0.02	−0.26

Note: See the note to Table 7.12.

Table 7.13 Pearson Correlation Coefficients between LST and Landscape Metrics in Non-residential Zoning Categories

zoning categories, respectively. For residential zoning, mean patch size was found to be negatively correlated with the standard deviation values of LST, implying that smaller zoning polygons tended to associate with larger temperature variations. Moreover, shape index and area-weighted shape index were both discovered to have a negative correlation with the standard deviation values of LST. That is to say, the more complex in shape a residential zoning category is, the more intrapolygon variation in LST there tended to be. Besides, fractal dimension and area-weighted mean patch fractal dimension values had a weaker positive correlation with the mean values of LST. In contrast, no significant correlation was found in the non-residential zoning categories. These observations should be viewed as experimental. Further research is warranted in investigation of the relationship between the shape complexity of zoning polygons and spatial variation in LST.

7.2.9 Seasonal Dynamics of LST Patterns

The ASTER images were taken in different seasons in different years. To compare and analyze these images together, normalized LST was computed according to following formula:

$$T_N = \frac{\mathrm{LST} - \mathrm{LST}_o}{\mathrm{LST}_s - \mathrm{LST}_o} \tag{7.3}$$

where LST_o is the minimum value and LST_s is the maximum value of LST in an image. As a result, the spatial pattern of LST and T_N would be similar, but all pixel values of T_N would be in the range of 0 to 1.

Figure 7.2*a* shows the distribution of mean T_N values calculated from the images of the four seasons. To show the spatial pattern better, major highways and water bodies were added to the map. This map has values ranging from 0.045 to 0.938, with a mean of 0.443 and standard deviation of 0.066. This choropleth map was produced based on Jenk's natural-breaks classification scheme, in which classes were established among the largest breaks in the data array (Smith, 1986). It is evident from the map that there was a thermal gradient that progressed from the CBD out into the countryside. Some hot spots can be easily identified. The most extensive hot area was distributed in the central part of the city. Hot spots extended along Highway 465 (the bypass of the city) which runs on north, west, and east sides of the city and became disconnected in the southern part of the city. Overall, a hot ring was evident in Fig. 7.2*a*. Since urban development in Indianapolis took place along major radial transportation routes, hot corridors also existed along many highways that radiated from the city center. In contrast, vegetated areas, rivers, streams, and reservoirs showed low values of mean normalized LST.

Figure 7.2*b* displays the standard deviation values calculated from the four images. This map shows the pixel-based spatial variability of T_N among the four seasons. It can be observed that LST fluctuated less in the central part of the city, especially in the CBD area, and in the areas with extensive development, such as those along major highways. As we moved into the countryside, the values of the standard deviation became larger. Numerous clusters of high standard deviations were clearly seen in the surrounding agricultural areas. In order to better understand the spatial variations in LST and to assess the environmental consequences of planning decisions, the standard deviation of T_N was computed for each general and residential zoning polygon. At the general zoning level, LST fluctuated most in dwellings (several types of residential categories-D), agriculture and single-family areas (DA), and parks (PK). In contrast, the low values of standard deviation (i.e., steadier LSTs throughout the year) were largely discovered in commercial areas (C), some of the low- to medium-density single- or multifamily residential areas (largely D5), and a few industrial zones (I). At the residential zoning level, the most pronounced LST fluctuation occurred in the planned-unit development area (DP), with some in newly developed suburban single-family areas (D2) and low- or medium-density single-family housing areas (D3 and D4). Residential categories with the least LST variability mainly included the medium-density single-family area (D5) but also could be found in low- or medium-density multifamily areas (D6 and D7), low- or medium-density single-family areas (D4), urban multiuse residential areas (D8), and suburban high-rise apartment areas (D9). The vast majority of these residential polygons were located in the central part of the city, with few exceptions in the northern part of the city toward the township of Carmel (a bedroom township

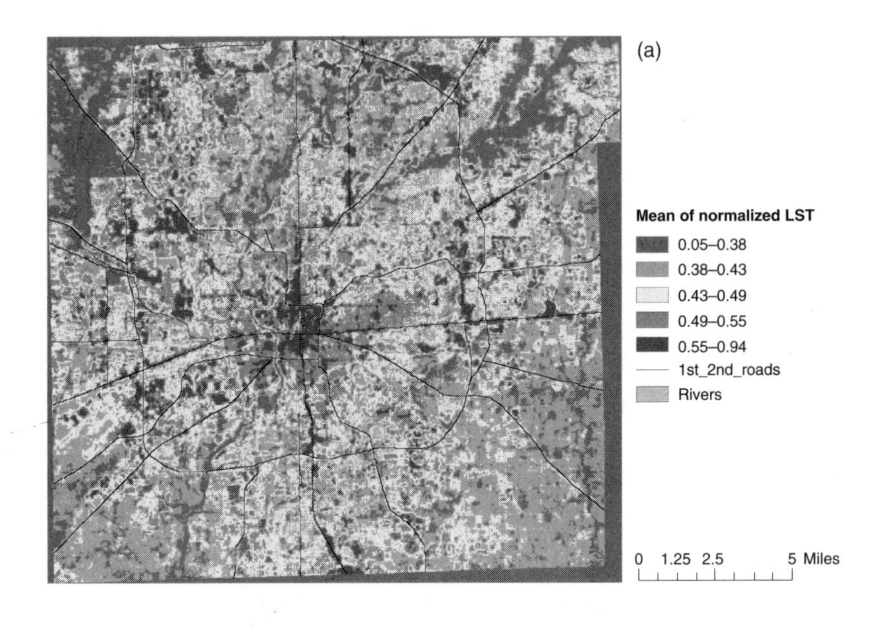

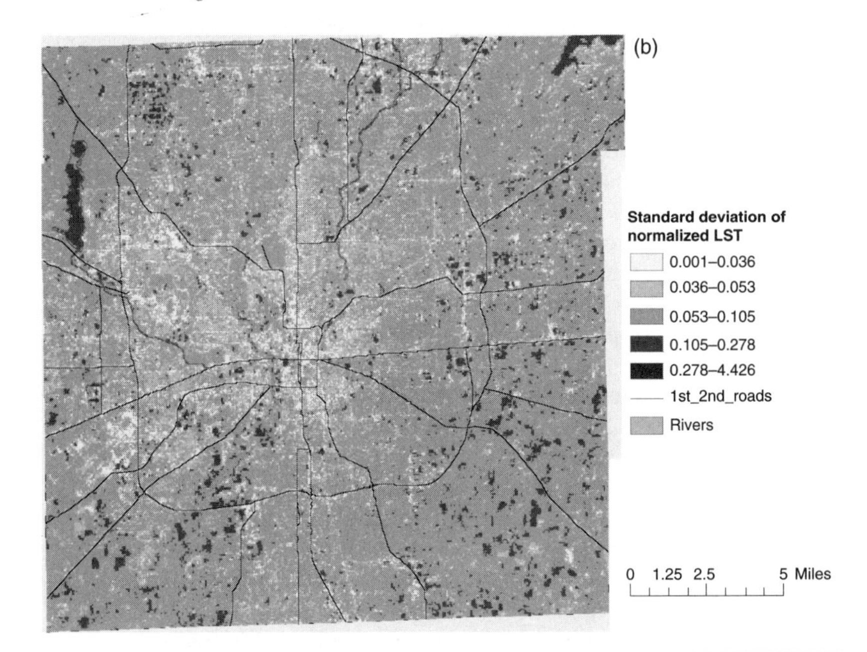

FIGURE 7.2 Normalized land surface temperatures computed based on the four dates of images: (*a*) mean value of T_N and (*b*) standard deviation values of T_N. (*Adapted from Weng et al., 2008.*) See also color insert.

for Indianapolis). It became clear that land-use zones with less human activity generally had more notable seasonal variability in LST regardless of referencing to general or residential zoning. On the other hand, zones with intensive human activity fluctuated less in LST. It can be hypothesized that anthropogenic heat, resulting from the use of air conditioners, automobiles, and others heat-producing devices, had a significant impact on the urban surface energy balance, leading to a reduction in the seasonal variability in LST.

7.3 Discussion and Conclusions: Remote Sensing–GIS Integration in Urban Land-Use Planning

This chapter has examined LST variations within/among land-use zoning polygons and has assessed the relationship between LST values and various GIS- and remote sensing-derived variables in Indianapolis, Indiana. Zoning data are an essential part of GIS data in urban planning. Potential factors for LST variations were grouped into four categories: LULC composition, biophysical conditions, intensity of human activity, and landscape pattern. Results indicate that the biophysical variables were most significant in explaining the spatial variations in LST, which included biophysical variables derived from SMA and NDVI. Regardless of whether residential or general zoning polygons were examined, LST had a weaker relationship with LULC composition than with the biophysical variables. PCA further shows that one of the factors was associated with the percentage of water and the other with the landscape-pattern variable(s). Population variables were found to be strong explanatory variables in LST variability in residential zoning, whereas the percentage of agricultural and/or forest land may be an explanatory variable in general zoning.

Impervious surface coverage is an indicator of human activity intensity and has a great impact on urban surface energy exchange. GIS data layers of building footprints and pavements were obtained to create two imperviousness variables. However, the stepwise regression models for general zoning did not enter either variable. On the other hand, the regression models for residential zoning did enter one or both variables, with exception of the 2004 image. GIS-derived impervious surface variables did not contribute significantly to the principal components (loadings < 0.71). These variables may have correlated highly with some entered variables. It is also worthy to note that remotely sensed data and GIS data were collected with different formats. Remote sensing data had finer resolution than the GIS data. When integrating these two types of data, a common method is to aggregate remotely sensed data to an appropriate level of a geographic entity, such as zoning polygons in this study. This aggregation has the potential to generate the same mean LST values in different zoning polygons despite their differences in biophysical conditions, LULC types, landscape pattern, and intensity of human activity. This problem added to the difficulty in analyzing urban LST variability.

The impact of land-use zoning on urban biophysical conditions in general and LST in particular was further examined. Results show that the cumulative variance was always larger in residential zoning than in general zoning, implying that the factors contributing to LST variations in general zoning were more complex or that the variables examined were more appropriate for explaining residential LST patterns. Our further investigation of the relationship between LST and the landscape metrics of zoning polygons suggests that for the residential zoning, smaller zoning polygons were associated with larger temperature variations and that the more complex in shape a residential zoning category was, the more intrapolygon variation in LST there tended to be. However, these correlations did not exist in the non-residential zoning categories. Given the significance of landscape ecology in the understanding of LST patterns and processes and the implications of these findings in landscape design and urban planning, more studies in this direction are called for. An interesting finding in this study is that Shannon's diversity index did not enter the stepwise regression models, so it was not included in the consequent factor analysis. This is likely due to its close correlation with other variables.

The spatial pattern of LST in Indianapolis may be characterized as concentric in the central part of the city, a hot ring along the city's bypass, and several hot corridors along the radial highways outward to the countryside. The seasonal fluctuation in LST was observed to be less in the central part and increased toward the countryside. Numerous clusters of high LST fluctuations can be clearly identified in the surrounding agricultural areas, which alternated between a growing season and a dry, often plowed season typical of many areas in the Midwest. At the general zoning level, LST fluctuated most in dwellings, agriculture and single-family areas, and parks and least in commercial uses, a few low- to medium-density single/multifamily residential areas, and industrial zones, whereas at the residential zoning level, LST fluctuated most in the planned-unit development areas and least in medium-density single-family, low- or medium-density multifamily, low- or medium-density single-family, urban multiuse residential, and suburban high-rise apartment areas. In sum, zones with less human activity observed a strong seasonal LST variability, whereas zones with intensive human activity fluctuated less in LST. These findings may be interpreted in terms of the amount of anthropogenic heat emittance in different seasons. Apparently, the emittance of anthropogenic heat had a direct link with the reduction in seasonal LST variability in Indianapolis. Whether this finding can be applied to other midlatitude cities in the northern hemisphere remains to be determined.

References

Adams, J. B., Sabol, D.E., Kapos, et al. 1995. Classification of multispectral images based on fractions of endmembers: Application to land cover change in the Brazilian Amazon. *Remote Sensing of Environment* **52,** 137–154.

Aguiar, A. P. D., Shimabukuro, Y. E., and Mascarenhas, N. D. A. 1999. Use of synthetic bands derived from mixing models in the multispectral classification of remote sensing images. *International Journal of Remote Sensing* **20,** 647–657.

Artis, D. A., and Carnahan, W. H. 1982. Survey of emissivity variability in thermography of urban areas. *Remote Sensing of Environment* **12,** 313–329.

Asner, G. P., and Lobell, D. B. 2000. A biogeophysical approach for automated SWIR unmixing of soils and vegetation. *Remote Sensing of Environment* **74,** 99–112.

ASTER Temperature/Emissivity Working Group. 1999. *Temperature/Emissivity Separation Algorithm Theoretical Basis Document,* Version 2.4; available at *http://eospso.gsfc.nasa.gov/eos_homepage/for_scientists/atbd/docs/ASTER/atbd-ast-03.pdf,* Jet Propulsion Laboratory, California Institute of Technology (last date accessed August 10, 2007).

Balling, R. C., and Brazell, S. W. 1988. High resolution surface temperature patterns in a complex urban terrain, *Photogrammetric Engineering and Remote Sensing,* 54:1289–1293.

Bottyán, Z., and Unger, J. 2003. A multiple linear statistical model for estimating the mean maximum urban heat island. *Theoretical and Applied Climatology* **75,** 233–243.

Carlson, T. N., Gillies, R. R., and Perry, E. M. 1994. A method to make use of thermal infrared temperature and NDVI measurements to infer surface soil water content and fractional vegetation cover. *Remote Sensing Review* **9,** 161–173.

Carnahan, W. H., and R. C. Larson, 1990. An analysis of an urban heat sink, *Remote Sensing of Environment,* 33:65–71.

Cendese, A., and Monti, P. 2003. Interaction between an inland urban heat island and a sea-breeze flow: A laboratory study. *Journal of Applied Meteorology* **42,** 1569–1583.

Chavez, P. S., Jr. 1996. Image-based atmospheric corrections: Revisited and improved. *Photogrammetric Engineering and Remote Sensing* **62,** 1025–1036.

Chrysoulakis, N. 2003. Estimation of the all-wave urban surface radiation balance by use of ASTER multispectral imagery and in situ spatial data. *Journal of Geophysical Research* 108, 4582, doi: 10.1029/2003JD003396.

Cochrane, M. and Souza, C. M., Jr. 1998. Linear mixture model classification of burned forests in the eastern Amazon. *International Journal of Remote Sensing* **19,** 3433–3440.

Comrey, A. L. and Lee, H. B. 1992. *A First Course in Factor Analysis,* 2nd ed. Hillsdale, NJ: Erlbuum.

Dash, P., Gottsche, F. M, Olesen, F. S., and Fischer, H. 2002. Land surface temperature and emissivity estimation from passive sensor data: Theory and practice—current trends. *International Journal of Remote Sensing* **23,** 2563–2594.

Dousset, B., and Gourmelon, F. 2003. Satellite multi-sensor data analysis of urban surface temperatures and landcover. *ISPRS Journal of Photogrammetry and Remote Sensing* **58,** 43–54.

Eliasson, I. 1996. Urban nocturnal temperatures, street geometry and land use. *Atmospheric Environment* **30,** 379–392.

Foody, G. M. 2002. Status of land cover classification accuracy assessment. *Remote Sensing of Environment* **80,** 185–201.

Fujisada, H. 1998. ASTER level-1 data processing algorithm. *IEEE Transactions on Geoscience and Remote Sensing* **36,** 1101–1112.

Gallo, K. P., McNab, A. L., Karl, T. R., et al. 1993. The use of NOAA AVHRR data for assessment of the urban heat island effect, *Journal of Applied Meteorology,* **32,** 899–908.

Gallo, K. P., and Owen, T. W. 1998. Assessment of urban heat island: A multi-sensor perspective for the Dallas–Ft. Worth, USA, region. *Geocarto International* **13,** 35–41.

Gillespie, A., Rokugawa, S., Matsunaga, T., et al. 1998. A temperature and emissivity separation algorithm for advanced spaceborne thermal emission and reflection radiometer (ASTER) images. *IEEE Transactions on Geoscience and Remote Sensing* **36,** 1113–1126.

Gillies, R. R., Carlson, T. N., Cui, J., et al. 1997. A verification of the "triangle" method for obtaining surface soil water content and energy fluxes from remote measurements of the normalized difference vegetation index (NDVI) and surface radiant temperature. *International Journal of Remote Sensing* **18,** 3145–3166.

Goward, S. N. 1981. Thermal behavior or urban landscapes and the urban heat island. *Physical Geography* **1,** 19–33.

Kaiser, H. 1960. The application of electronic computers to factor analysis. *Education and Psychology* **52,** 621–652.

Kidder, S. Q., and Wu, H. T. 1987. A multispectral study of the St. Louis area under snow-covered conditions using NOAA-7 AVHRR data, *Remote Sensing of Environment*, 22:159–172.

Kim, H. H., 1992. Urban heat island, *International Journal of Remote Sensing,* **13,** 2319–2336.

Landsberg, H. E.1981. *The Urban Climate*. New York: Academic Press.

Li, F., Jackson, T. J., Kustas, W. P., et al. 2004. Deriving land surface temperature from Landsat 5 and 7 during SMEX02/SMACEX. *Remote Sensing of Environment* **92,** 521–534.

Li, H., and Reynolds, J. F. 1994. A simulation experiment to quantify spatial heterogeneity maps. *Ecology* **75,** 2446–2455.

Liu, H., and Weng, Q. 2008. Seasonal variations in the relationship between landscape pattern and land surface temperature in Indianapolis, U.S.A. Environmental Monitoring and Assessment **144,** 199–219.

Lo, C. P., Quattrochi, D., and Luvall, J. 1997. Application of high-resolution thermal infrared remote sensing and GIS to assess the urban heat island effect. *International Journal of Remote Sensing* **18,** 287–304.

Lu, D., and Weng, Q. 2004. Spectral mixture analysis of the urban landscapes in Indianapolis city with Landsat ETM+ imagery. *Photogrammetric Engineering and Remote Sensing* **70,** 1053–1062.

Lu, D., and Weng, Q. 2006. Spectral mixture analysis of ASTER imagery for examining the relationship between thermal features and biophysical descriptors in Indianapolis, Indiana. *Remote Sensing of Environment* **104,** 157–167.

Lu, D., Mausel, P., Brondizio, E., and Moran, E. 2002. Assessment of atmospheric correction methods for Landsat TM data applicable to Amazon Basin LBA research. *International Journal of Remote Sensing* **23,** 2651–2671.

Markham, B. L., and Barker, J. K. 1985. Spectral characteristics of the LANDSAT Thematic Mapper sensors. *International Journal of Remote Sensing* **6,** 697–716.

McGarigal, K., and Marks, B. J. 1995. FRAGSTATS: Spatial Pattern Analysis Program for Landscape Structure, p. 22. Portland, OR: U.S. Department of Agriculture, Forest Service, Northwest Research Station.

McGwire, K., Minor, T., and Fenstermaker, L. 2000. Hyperspectral mixture modeling for quantifying sparse vegetation cover in arid environments. *Remote Sensing of Environment* **72,** 360–374.

Mustard, J. F., and Sunshine, J. M. 1999. Spectral analysis for earth science: Investigations using remote sensing data. In *Remote Sensing for the Earth Sciences: Manual of Remote Sensing,* 3rd ed., Vol. 3, edited by A. N. Rencz, pp. 251–307. New York: Wiley.

Nichol, J. E. 1994. A GIS-based approach to microclimate monitoring in Singapore's high-rise housing estates, *Photogrammetric Engineering and Remote Sensing,* 60:1225–1232.

Nichol, J. E. 1996. High-resolution surface temperature patterns related to urban morphology in a tropical city: a satellite-based study. Journal of Applied Meteorology **35,** 135–146.

Nichol, J. E. 1998. Visualization of urban surface temperature derived from satellite images. International Journal of Remote Sensing **19,** 1639–1649.

Nichol, J. E. 2005. Remote sensing of urban heat islands by day and night. Photogrammetric Engineering and Remote Sensing **71,** 613–623.

Oke, T. R. 1979. *Technical Note No. 169: Review of Urban Climatology*, World Meteorological Organization, Geneva, Switzerland, 43 p.

Oke, T. R. 1982. The energetic basis of the urban heat island. *Quarterly Journal of the Royal Meteorological Society* **108,** 1–24.

Oke, T. R. 1988. The urban energy balance. *Progress in Physical Geography* **12,** 471–508.

Oke, T. R., Spronken-Smith, R. A., Jauregui, E., and Grimmond, C. S. B. 1999. The energy balance of central Mexico City during the dry season. *Atmospheric Environment* **33,** 3919–3930.

Phinn, S., Stanford, M., Scarth, P., et al. 2002. Monitoring the composition of urban environments based on the vegetation-impervious surface-soil (VIS) model by subpixel analysis techniques. *International Journal of Remote Sensing* **23,** 4131–4153.

Rashed, T., Weeks, J. R., Gadalla, M. S., and Hill, A. G. 2001. Revealing the anatomy of cities through spectral mixture analysis of multisepctral satellite imagery: A case study of the Greater Cairo Region, Egypt. *Geocarto International* **16,** 5–15.

Read, J. M. and Lam, N. S. N. 2002. Spatial methods for characterising land cover and detecting land-cover for the tropics. *International Journal of Remote Sensing* **23,** 2457–2474.

Riitters, K. H., O'Neill, R. V., Hunsaker, C. T., et al. 1995. A factor analysis of landscape pattern and structure metrics. *Landscape Ecology* **10,** 23–39.

Roberts, D. A., Batista, G. T., Pereira, J. L. G., et al. 1998. Change identification using multitemporal spectral mixture analysis: Applications in eastern Amazonia. In *Remote Sensing Change Detection: Environmental Monitoring Methods and Applications,* edited by R. S. Lunetta and C. D. Elvidge, pp. 137–161. Ann Arbor, MI:Ann Arbor Press.

Roth, M., Oke, T. R. and Emery, W. J. 1989. Satellite derived urban heat islands from three coastal cities and the utilisation of such data in urban climatology, *International Journal of Remote Sensing*, 10:1699–1720.

Saitoh, T. S., Shimada, T., and Hoshi, H. 1996. Modeling and simulation of the Tokyo urban heat island. *Atmospheric Environment* **30,** 3431–3442.

Schmugge, T., French, A., Ritchie, J. C., et al. 2002. Temperature and emissivity separation from multispectral thermal infrared observations. *Remote Sensing of Environment* **79,** 189–198.

Small, C. 2001. Estimation of urban vegetation abundance by spectral mixture analysis. *International Journal of Remote Sensing* **22,** 1305–1334.

Smith, M. O., Ustin, S. L., Adams, J. B., and Gillespie, A. R. 1990. Vegetation in deserts: I. A regional measure of abundance from multispectral images *Remote Sensing of Environment* **31,** 1–26.

Smith, R. M. 1986. Comparing traditional methods for selecting class intervals on choropleth maps. *Professional Geographer* **38,** 62–67.

Snyder, W. C., Wan, Z., Zhang, Y., and Feng, Y.-Z. 1998. Classification-based emissivity for land surface temperature measurement from space. *International Journal of Remote Sensing* **19,** 2753–2774.

Streutker, D. R. 2002. A remote sensing study of the urban heat island of Houston, Texas. *International Journal of Remote Sensing*, **23**, 2595–2608.

Streutker, D. R. 2003. Satellite-measured growth of the urban heat island of Houston, Texas. *Remote Sensing of Environment* **85,** 282–289.

Tabachnick, B., and Fidell, L. 1996. *Using Multivariate Statistics,* 3rd ed. New York: Harper Collins College Publishers.

Tong, H., Walton, A., Sang, J., and Chan, J. C. L. 2005. Numerical simulation of the urban boundary layer over the complex terrain of Hong Kong. *Atmospheric Environment* **39,** 3549–3563.

Voogt, J. A. and Oke, T. R. 1998. Effects of urban surface geometry on remotely-sensed surface temperature. *International Journal of Remote Sensing* **19,** 895–920.

Voogt, J. A., and Oke, T. R. 2003. Thermal remote sensing of urban climate. *Remote Sensing of Environment* **86,** 370–384.

Weng, Q. 2001. A remote sensing–GIS evaluation of urban expansion and its impact on surface temperature in the Zhujiang Delta, China. *International Journal of Remote Sensing* **22,** 1999–2014.

Weng, Q. 2003. Fractal analysis of satellite-detected urban heat island effect. Photogrammetric Engineering and Remote Sensing **69,** 555–566.

Weng, Q., Lu, D., and Schubring, J. 2004. Estimation of land surface temperature-vegetation abundance relationship for urban heat island studies. *Remote Sensing of Environment* **89,** 467–483.

Weng, Q., and Lu, D. 2008. A sub-pixel analysis of urbanization effect on land surface temperature and its interplay with impervious surface and vegetation coverage in Indianapolis, United States. International Journal of Applied Earth Observation and Geoinformation **10,** 68–83.

Weng, Q., Lu, D., and Liang, B. 2006. Urban surface biophysical descriptors and land surface temperature variations. Photogrammetric Engineering & Remote Sensing **72,** 1275–1286.

Weng, Q., Liu, H., and Lu, D. 2007. Assessing the effects of land use and land cover patterns on thermal conditions using landscape metrics in city of Indianapolis, United States. Urban Ecosystem **10,** 203–219.

Weng, Q., Liu, H., Liang, B., and Lu, D. 2008. The spatial variations of urban land surface temperatures: pertinent factors, zoning effect, and seasonal variability. *IEEE Journal of Selected Topics in Applied Earth Observations and Remote Sensing* **1,** 154–166.

Wilson, J. S., Clay, M., Martin, E., Stuckey, D., and Vedder-Risch, K. 2003. Evaluating environmental influences of zoning in urban ecosystems with remote sensing. *Remote Sensing of Environment* **86,** 303–321.

Wu, C., and Murray, A. T. 2003. Estimating impervious surface distribution by spectral mixture analysis. *Remote Sensing of Environment* **84,** 493–505.

CHAPTER 8

Surface Runoff Modeling and Analysis

Land-use and land-cover (LULC) changes may have four major direct impacts on the hydrologic cycle and water quality. They can cause floods, droughts, and changes in river and groundwater regimes, and they can affect water quality (Rogers, 1994). In addition to these direct impacts, there are also indirect impacts on climate and the altered climate's subsequent impact on the waters. Urbanization, the conversion of other types of land to uses associated with population and economic growth, is a main type of LULC change, especially in recent human history. The process of urbanization has a considerable hydrologic impact in terms of influencing the nature of runoff and other hydrologic characteristics, delivering pollutants to rivers, and controlling rates of erosion (Goudie, 1990).

At different stages of urban growth, various impacts can be observed (Kibler, 1982). In the early stage of urbanization, removal of trees and vegetation may decrease evapotranspiration and interception and increase stream sedimentation. Later, when construction of houses, streets, and culverts begins, the impacts may include decreased infiltration, lowered groundwater table, increased storm flows, and decreased base flows during dry periods. After the development of residential and commercial buildings has been completed, increased imperviousness will reduce time of runoff concentration so that peak discharges are higher and occur sooner after rainfall starts in basins. The volume of runoff and flood damage potential will increase greatly. Moreover, the installation of sewers and storm drains accelerates runoff (Goudie, 1990). As a result, the rainfall-runoff process in an urban area tends to be quite different from that in natural conditions depicted in classic hydrologic cycles. This effect of urbanization, however, varies according to the size of a flood. As the size of the flood becomes larger and its recurrence interval increases, the effect of urbanization decreases (Hollis, 1975).

The integration of remote sensing and geographic information systems (GIS) has been applied widely and has been recognized as a powerful and effective tool in detecting urban growth (Ehlers et al., 1990; Harris and Ventura, 1995; Treitz et al., 1992; Yeh and Li, 1996, 1997). Remote sensing collects multispectral, multiresolution, and multitemporal data and turns them into information that is valuable for understanding and monitoring urban land processes and for building urban land-cover data sets. GIS technology provides a flexible environment for entering, analyzing, and displaying digital data from various sources necessary for urban feature identification, change detection, and database development. In hydrologic and watershed modeling, remotely sensed data are found to be valuable for providing cost-effective data input and for estimating model parameters (Drayton et al., 1992; Engman and Gurney, 1991; Mattikalli et al., 1996). The introduction of GIS to the field makes it possible for computer systems to handle the spatial nature of hydrologic parameters. The hydrologic community now increasingly adopts GIS-based distributed modeling approaches (Berry and Sailor, 1987; Drayton et al., 1992; Mattikalli et al., 1996). However, little attempt has been made to relate urban growth studies to distributed hydrologic modeling, although both studies share LULC data. This chapter attempts to develop an integrated approach of remote sensing and GIS for examining the effect of urban growth on surface runoff at the local and regional levels by using the Zhujiang Delta of South China as a case study.

8.1 The Distributed Surface Runoff Modeling

The model used for estimating surface runoff in this study was developed by the U.S. Soil Conservation Service (SCS). It has been applied widely to estimate storm-runoff depth for every patch within a watershed based on runoff curve numbers (CN) (Soil Conservation Service, 1972). The SCS equation for storm-runoff depth is mathematically as

$$Q = \frac{(P - 0.2S)^2}{P + 0.8S} \tag{8.1}$$

where Q = storm runoff
P = rainfall
S = potential maximum storage [$S = (1000/CN) - 10$

where CN is the runoff curve number of the hydrologic soil group-land cover complex].

To solve this equation, two input values are needed: P and CN. Precipitation data are often available from meteorologic observations. A runoff curve number is a quantitative description of land-cover and soil conditions that affect the runoff process. The CN values are normally estimated using field survey data with reference to U.S. Department of Agriculture's (USDA's) SCS tables (Table 8.1). From Table 8.1, it is

Cover			Runoff Curve Number for Hydrologic Soil Group			
Land Use	Treatment or Practice	Hydrologic Condition	A	B	C	D
Fallow	Straight row	—	77	86	91	94
Row crops	Straight row	Poor	72	81	88	91
		Good	67	78	85	89
	Contoured	Poor	70	79	84	88
		Good	65	75	82	86
	Contoured and terraced	Poor	66	74	80	82
		Good	62	71	78	81
Small grain	Straight row	Poor	65	76	84	88
		Good	63	75	83	87
	Contoured	Poor	63	74	82	85
		Good	61	73	81	84
	Contoured and terraced	Poor	61	72	79	82
		Good	59	70	78	81
Close-seeded legumes*	Straight row	Poor	66	77	85	89
		Good	58	72	81	85
	Contoured	Poor	64	75	83	85
		Good	55	69	78	83
	Contoured and terraced	Poor	63	73	80	83
		Good	51	67	76	80
Pasture or range		Poor	68	79	86	89
		Fair	49	69	79	84
		Good	39	61	74	80
	Contoured	Poor	47	67	81	88
		Fair	25	59	75	83
		Good	6	35	70	79
Meadow		Good	30	58	71	78
Woods		Poor	45	66	77	83
		Fair	36	60	73	79
		Good	25	55	70	77

*Close drilled or broadcast.
From Soil Conservation Service, 1972.

TABLE 8.1 Runoff Curve Numbers for Hydrologic Soil-Cover Complexes (Antecedent Moisture Condition II and $I_a = 0.2S$)

apparent that *CN* values approaching 100 are associated with high runoff from cultivated agricultural land, whereas low to moderate *CN* values indicate the reduced runoff from heavily vegetated areas (Slack and Welch, 1980). A hydrologic soil group code A, B, C, or D was set up by the SCS for over 4000 soils in the United States based on permeability and infiltration characteristics. Group A soils are coarse, sandy, well-drained soils with the highest rate of infiltration and the lowest potential for runoff. Group D soils, on the other hand, are heavy, clayey, poorly drained soils with the lowest rate of infiltration and highest potential for runoff. In between groups A and D are groups B and C soils.

Since the availability of Landsat data in the 1970s, several attempts have been made to use these satellite data to determine curve numbers because of the cost-effective nature of these data. Mintzer and Askari (1980) employed Landsat MSS data, in combination with color infrared photography, to derive the runoff coefficients for the watersheds of Mill Creek, Ohio, by using polynomial regression modeling. Ragan and Jackson (1980) compared the estimation of curve numbers from Landsat MSS with those from field surveys and from high-level aerial photography. It was suggested that there was no significant difference among the three sources but that a modified curve-number classification system compatible with Landsat data needs to be established. Slack and Welch (1980) conducted a similar study for the Little River watershed near Tifton, Georgia. They generated four hydrologically important land classes—agricultural vegetation, forest, wetland, and bare ground—from Landsat MSS data and found that curve numbers for six subwatersheds and for the entire watershed were estimated within two curve-number units. Rango and colleagues (1983) claimed only a 5 percent error in land-cover estimation by Landsat data at the basin level but a much greater error at the cell level.

The development and maturity of GIS technology in the late 1980s have made it possible to combine various data sources for the derivation of model input parameters and have automated the SCS modeling process. Berry and Sailor (1987) used Map Analysis Package (MAP) to automate the procedures for estimating input parameters for the SCS method of predicting storm-runoff volume and lag to peak timing. Drayton and colleagues (1992) used GIS to develop the runoff model for the Tywi catchment of West Wales based on a rectangular grid cell network. The incorporation of topographic information allows them to further derive flow routing and to generate a hydrograph. Mattikalli and colleagues (1996) conducted the modeling in a vector GIS environment using Arc/Info and evaluated the effect of land-use change on river discharge by comparing rainfall-runoff curves over time and on water quality as indexed by nitrogen loading.

8.2 Study Area

The specific objectives of this study were (1) to examine the effect of urban growth on surface runoff and (2) to evaluate the impact of urban growth on the rainfall-runoff relationship. The study area, the Zhujiang (literally "Pearl River") Delta, is located between latitudes 21°40′N and 23°N and longitudes 112°E and 113°20′E (Fig. 8.1). It is the third biggest river delta in the nation and has an area of 17,200 km^2. Because of the constraints of satellite imagery coverage, this research focuses on the core area of the delta, which includes 15 cities/counties shown in Fig. 8.1. Geomorphologically, the Zhujiang Delta consists of three subdeltas, the Xijiang (West River), Beijiang (North River), and Dongjiang (East River) deltas, which originated approximately 40,000 years ago (Department of Geography, Zhongshan University, 1988). The process of sedimentation still continues today, extending seaward at a rate of 40 m/year (Gong and Chen, 1964). The delta has a subtropical climate with an average annual temperature of between 21 and 23°C and an average precipitation ranging from 1600 to 2600 mm. Because of the impact of the East Asian monsoonal circulation, about 80 percent of the rainfall comes in the period from April to September, with a concentration in the months of May, June, and July, when the delta is most prone to flooding (Ditu Chubanshe, 1977). Another hazard is typhoons, which occur most frequently from June to October.

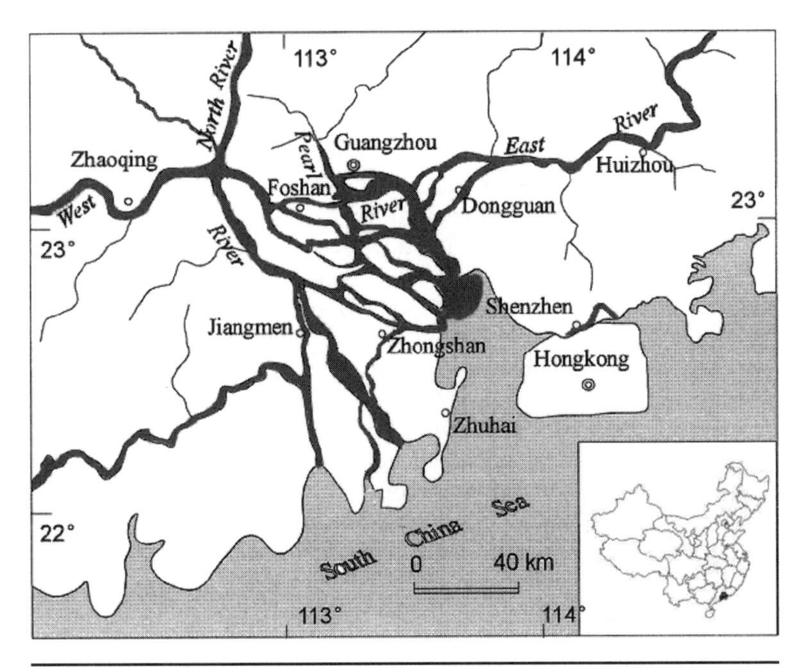

FIGURE 8.1 Location of the study area.

The drainage system in the delta is well developed as a result of abundant rainfall. Of the total annual runoff of 341.2 billion m^3, the Xijiang contributes the most (72.10 percent), followed by the Beijiang (14.13 percent) and the Dongjiang (9.14 percent). The annual mean discharge for the Xijiang ranges from 848 to 48,800 m^3/s, for the Beijiang it ranges from 139 to 14,900 m^3/s, and for the Dongjiang it ranges from 31.4 to 12,800 m^3/s (Department of Geography, Zhongshan University, 1988). The load discharge of the river system is large, with an annual total silt discharge of 83.36 million tons, of which the Xijiang is the major contributor (87 percent) (Huang et al., 1982). About 20 percent of the silt discharge is deposited in the delta region, whereas 80 percent enters the sea, thus causing a seaward extension in the mouth region of the Zhujiang. The rivers of the system arrive at the South China Sea through eight estuaries ("gates" in Chinese), namely, Humen, Jiaomen, Hongqili, Hengmen, Modaomen, Jitimen, Hutiaomen, and Yamen from north to south. Perhaps the most distinguishing characteristic of the river system is its numerous tributaries. There are 100 main branches in the system, with a total length over 1700 km. The drainage density in the Xijiang and the Beijiang deltas is 0.81 km/km^2, and it is 0.88 km/km^2 in the Dongjiang Delta. The average channel width-to-depth ratio is 1.8 to 11.5. However, 68 percent of all channels have a ratio of less than 6.0, which indicates that most river channels in the delta are stable. The average channel gradient in all major rivers is low: 0.0023 percent in the Xijiang, 0.0037 percent in the Beijiang, and 0.026 percent in the Dongjiang.

The delta's fertile alluvial deposits, in combination with the subtropical climate, make it one of the richest agricultural areas in China. A variety of crops, vegetation, cash crops, and fruit trees are the major agricultural products. Agricultural land accounts for almost half the total area. Known as an ecologically well-integrated agriculture-aquaculture system (Ruddle and Zhong, 1988; Zhong, 1980), the dike-pond land is a unique landscape feature of the delta, accounting for 4 percent of the area. Inside the delta, there are over 160 hills and terraces located at heights between 100 and 300 m above sea level (Huang et al., 1982). These hills and terraces are largely occupied by forest (28 percent) and grassland (15 percent) (Department of Geography, Zhongshan University 1988). Only a small share of the land was urban or built up in 1978. However, urban growth has been speeded up owing to accelerated economic development after 1978, when the Chinese government initiated economic reform policies. Massive parcels of agricultural land are disappearing each year for urban or related uses. Because of the lack of appropriate land-use planning and measures for sustainable development, rampant urban growth has created severe environmental consequences.

Economically, the Zhujiang Delta is the largest area of economic concentration in South China. Hong Kong and Macau are located here. Since 1978, the delta has become a rising star owing to its dramatic economic expansion and therefore has been regarded as a model for Chinese regional development. The establishment of the Shenzhen and Zhuhai Special Economic Zones in 1979 and the Zhujiang Delta Economic Open Zone in 1985 has stimulated Hong Kong and foreign firms to locate their factories there as village-township enterprises. The labor-intensive industries, in association with the cash-crop production, have transformed the spatial economy of the delta (Lin, 1997; Lo, 1989; Weng, 1998).

8.3 Integrated Remote Sensing–GIS Approach to Surface Runoff Modeling

Integration of GIS and remote sensing in runoff modeling involves two processes: (1) hydrologic parameter determination using GIS and (2) hydrologic modeling within GIS. Hydrologic parameter determination using GIS entails preparing land-cover, soil, and precipitation data that go into the SCS model, whereas hydrologic modeling within GIS automates the SCS modeling process using generic GIS functions. Remote sensing is used for obtaining land-cover data each year and for obtaining information about the nature, rate, and location of LULC changes. Urban growth analysis then is carried out by superimposing administrative boundaries on the LULC change map. After a surface runoff image is obtained from the hydrologic modeling, the technique of image differencing is applied to evaluate the changes in surface runoff over time. The urban-expansion map is overlaid with the runoff-change map to analyze the impact of LULC change on the environment. Figure 8.2 shows the flowchart of image processing procedures. These methods and implementation procedures are discussed below.

8.3.1 Hydrologic Parameter Determination Using GIS

Derivation of Land-Cover Data and Change Detection

LULC patterns for 1989 and 1997 were mapped using Landsat Thematic Mapper (TM) data (dates: December 13, 1989 and August 29, 1997). A modified version of the Anderson scheme of LULC classification that takes into account local conditions (Anderson et al., 1976) was adopted. The categories include (1) urban or built-up land, (2) barren land, (3) cropland, (4) horticulture farms, (5) dike-pond land, (6) forest, and (7) water. Using ERDAS IMAGINE computer software, each Landsat image was enhanced using linear contrast stretching and histogram equalization to increase the volume of visible information. All images were rectified to a common Universal Transverse Mercator (UTM) coordinate system based on the 1:50,000 scale topographic maps of Guangdong Province produced by the Chinese government. Each image then was corrected radiometrically using a relative radiometric

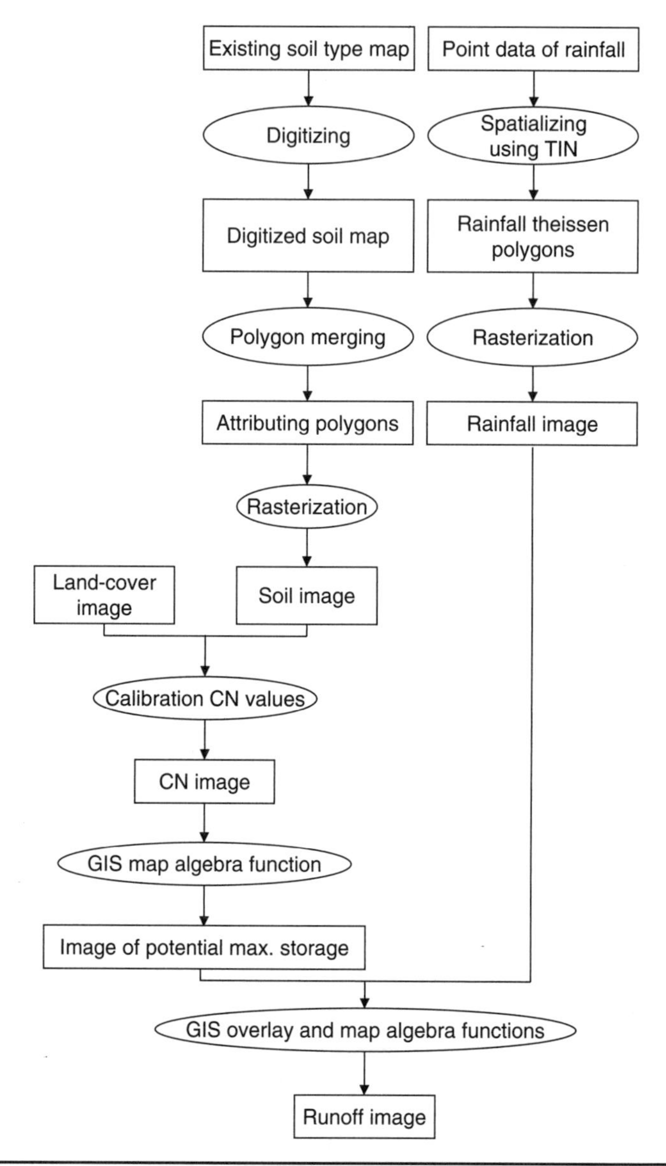

FIGURE 8.2 The implementation procedure for GIS-based surface runoff modeling.

correction method (Jensen, 1996). A supervised classification with the maximum likelihood algorithm was conducted to classify the Landsat images. The accuracy of the classification was verified by field checking or comparing with existing LULC maps that have been field checked. In Fig. 8.3 are the maps showing urban or built-up land in the two years, with nonurban land as the background.

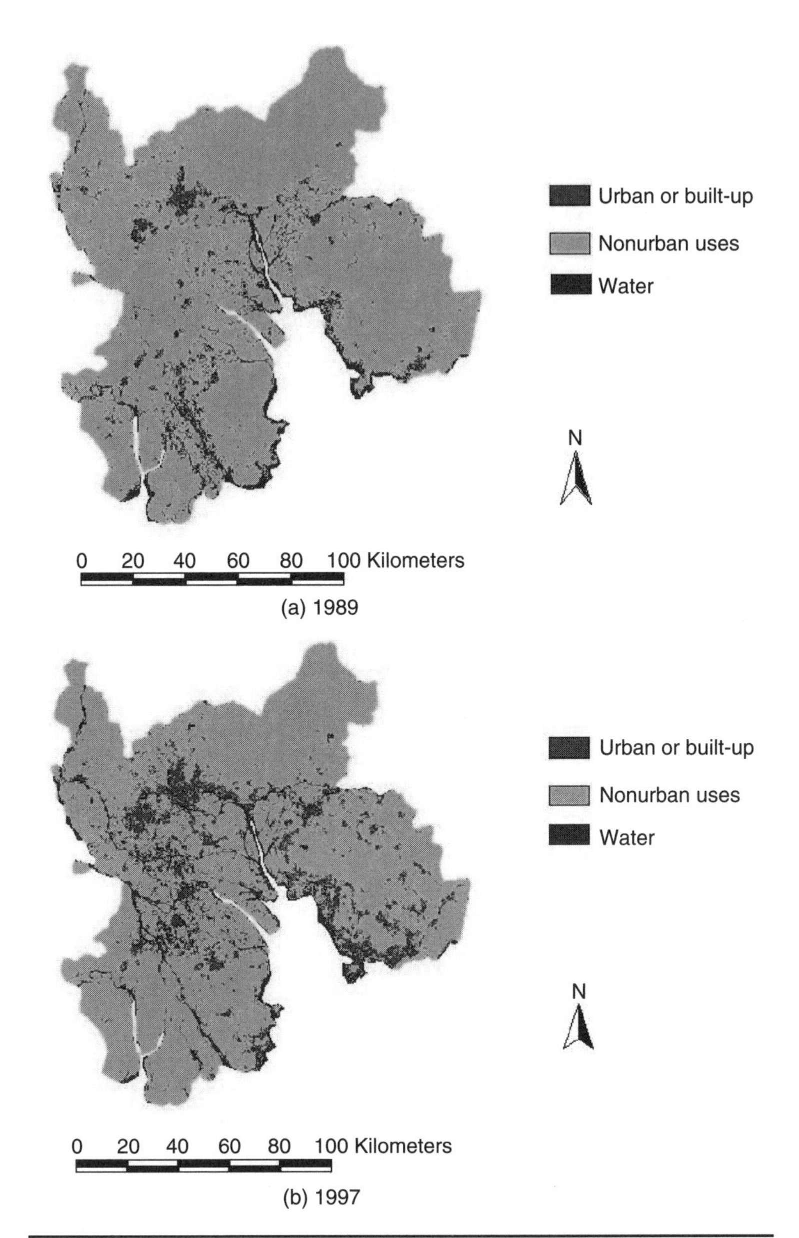

FIGURE 8.3 Urban or built-up land of the Zhujiang Delta in (*a*) 1989 and (*b*) 1997. See also color insert.

In performing LULC change detection, a cross-tabulation detection method was employed. A change matrix was produced with the help of the ERDAS IMAGINE software. Quantitative areal data of the overall LULC changes, as well as gains and losses in each category, can be compiled. In order to analyze the nature, rate, and location of urban LULC change, an image of urban and built-up land was extracted from each original land-cover image. The two extracted images then were overlaid and recoded to obtain an urban LULC change (growth) image.

This urban growth image was further overlaid with several geographic reference images to help analyze the patterns and processes of urban expansion, including an image of the county/city boundary, major roads, and major urban centers. These layers were constructed in a vector geographic information system (GIS) environment and converted into a raster format (grid size = 30 m). The county/city boundary image was used to find urban LULC change information within each county/city.

Derivation of Soil Data

The soil data are available in a book entitled, *The Soils of Guangdong* (Liu, 1993), which results from the Second National Soils Survey between 1979 and 1990. The soil types were extracted from the 1:2,800,000 provincial soil map and digitized into a polygon coverage and registered to the UTM coordinate system. Ten types of soils are found in the study area and they can be grouped into two major categories: (1) podzolized old and young red earths (39.17 percent) and (2) non-calcareous alluvium and paddy soils (60.83 percent). The former is seen in the uplands with clay accumulation and low base supply known as *udults* in soil taxonomy (Soil Conservation Service, 1975) and the latter in the flood plains and deltas. Red earths normally are permeable and well drained and can be related to class A or B in the Hydrological Soil Group codes of the SCS classification. Paddy soils have been modified by intensive agricultural activities, and their hydrologic properties are subject to human influence. As the fields were reclaimed from the sea at various periods of time, a distinction can be made according to their distance from the sea (Lo and Pannell, 1985). Inland fields, enclosed with irrigation dikes, usually were developed earlier than those found along the coast and are more fertile and higher yielding. The fields found near the coast are susceptible to flooding, and their soils tend to be more saline and less suitable for agricultural purposes. Most of the paddy soils consist of loam or silt loam and can be classified into hydrologic soil class C. In most of Shunde County in the central delta, however, a much larger proportion of clay may be found in the soils. These soils are grouped into the class D given their relative weak permeability and infiltration. The Hydrological Soil Group classes (i.e., A, B, C, and D) were associated with each polygon in the soil coverage. This coverage was converted into a raster layer with a resolution of 30 m. After the vector-to-raster conversion, a 3 × 3 mode filter was passed over the data layer to eliminate any slivers (Lo and Shipman, 1990).

Derivation of Precipitation Data

Rainfall data are available for all cities and counties of the delta in *Guangdong Statistical Yearbooks* (Guangdong Statistical Bureau, 1990, 1998) and the *Guangdong Province Gazetteer of Geography* (Liu, 1998).

These rain gauges usually were located in the urban center of a county seat or a city proper and recorded continuous data from the early 1950s. Daily, monthly, and yearly rainfall totals are available for every year. The gauge stations were digitized and registered to the UTM coordinate system. A Theissen polygon coverage (in which two neighboring stations have an equal distance to the boundary) was built using Arc/Info (a vector GIS program) commands. By assigning average yearly rainfall totals to each polygon in the coverage, a rainfall data layer was generated. The data layer then was converted into raster format with a resolution of 30 m.

8.3.2 Hydrologic Modeling within the GIS

To start modeling, a land-cover image and the soil layer were combined and recoded to calibrate curve number values with the aid of a standard SCS table (Soil Conservation Service, 1972), and a curve number image thus was created. By using the map algebra function of the GIS, a potential maximum storage S can be computed for each pixel. A layer of potential maximum storage then was created for each year. This layer was further overlaid with the rainfall layer to create a runoff image (see Fig. 8.2).

Image differencing was performed between the 1989 and 1997 runoff layers. The resulting differencing image was reclassified into runoff change zones. The areal extent and spatial occurrence of these zones were studied in reference to the spatial patterns of urban growth in order to understand the effects of LULC changes.

Urban growth alters the relationship between rainfall and runoff through potential maximum storage. An average value of potential maximum storage for each city/county was computed by superimposing city/county boundaries on a potential maximum storage layer. Assuming that uniform rainfall events from 10 to 100 mm with increments of 10 mm occurred in each city/county, runoff depths for these events can be calculated based on the SCS model. Runoff coefficients, defined as the ratio of runoff to rainfall, also can be computed for each event. A runoff coefficient curve was constructed as a function of the size of the flood. By comparing the curves in 1989 and 1997, the effect of urbanization can be examined, showing how it varies according to the size of the flood. By relating runoff coefficient curve patterns and changes with urban growth patterns in each city/county, the effect of urbanization was further studied.

8.4 Urban Growth in the Zhujiang Delta

The remote sensing–GIS analysis indicates that urban or built-up land has expanded by 47.68 percent (65,690 ha) in the delta from 1989 to 1997. Overlaying the 1989 and 1997 LULC maps reveals that most urban or built-up land increases were at the expense of cropland (37.92 percent) and horticulture farms (16.05 percent). The overlay of

City/ County	Total Area (ha)	Urban Area 1989 (ha)	Urban Area 1997 (ha)	Change (ha)	Change (%)
Bao'an	159,752	6,403	21,344	14,941	233.33
Dongguan	230,202	18,676	42,155	23,479	125.71
Doumen	55,566	2,134	3,735	1,601	75.00
Foshan	9,922.5	6,403	6,937	534	8.33
Guangzhou	140,900	23,479	28,281	4,802	20.45
Jiangmen	9,923	1,601	3,735	2,134	133.33
Nanhai	106,171	13,340	21,344	8,004	60.00
Panyu	79,380	7,471	8,538	1,067	14.29
Sanshui	91,287	2,134	2,134	0	0.00
Shenzhen	27,783	1,050	4,269	3,219	306.65
Shunde	78,388	6,403	10,139	3,736	58.33
Xinhui	151,814	5,870	7,471	1,601	27.27
Zengcheng	176,621	5,870	5,870	0	0.00
Zhongshan	169,675	13,340	16,542	3,202	24.00
Zhuhai	23,814	534	6,403	5,869	1100.00

TABLE 8.2 Satellite-Detected Urban Expansion in the Zhujiang Delta, 1989–1997

this map with a city/county mask reveals the areal extent and spatial occurrence of urban expansion within administrative regions. Table 8.2 and Fig. 8.4 illustrate the result of this GIS overlay. It is shown that in absolute terms, the greatest urban expansion occurred in Dongguan (23478.90 ha), Bao'an (14941.08 ha), Nanhai (8004.1 ha), and Zhuhai (5869.71 ha). However, in percentage terms, the largest increase in urban or built-up land occurred in Zhuhai (1100.00 percent), followed by Shenzhen (306.65 percent), Bao'an (233.33 percent), and Dongguan (125.71 percent). Massive urban sprawl in these areas can be ascribed to rural urbanization, a common phenomenon in post-reform China. Rapid urban development in the form of small towns on the east side of the delta is highly influenced by investment from Hong Kong. In contrast, the old cities, such as Guangzhou and Foshan, do not show a rapid increase in urban or built-up land because they have no land to expand further (because they have already expanded fully in the past) and the concentration of urban enterprises in the city proper. Shenzhen and Zhuhai were designated as special economic zones at the same time, but the pace of urbanization in the two cities is quite different. Urban development in Shenzhen was mostly completed in

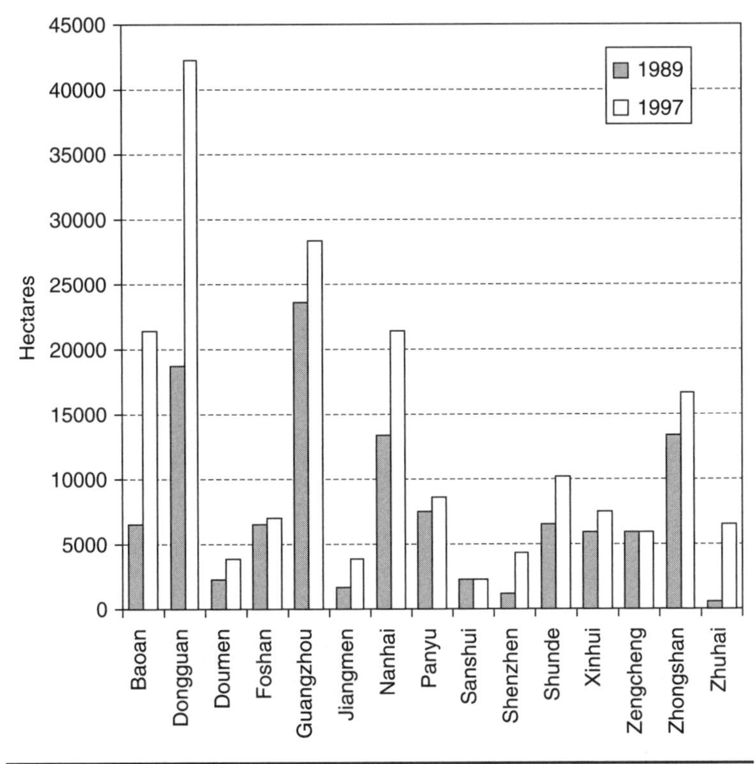

FIGURE 8.4 Urban growth among cities in the Zhujiang Delta, 1989–1997.

the 1980s, whereas Zhuhai's urban expansion appeared primarily during the period 1989–1997 (5869.71 ha).

8.5 Impact of Urban Growth on Surface Runoff

The impacts of LULC change on surface runoff were examined by comparing predicted runoff volumes in 1989 with those in 1997. The runoff image of 1997 was subtracted from that of 1989. The resulting image of "change" indicated that the annual runoff volume had increased by 8.10 mm during the 8-year period owing to LULC changes. This number refers to a uniform runoff depth for the whole delta, and it has a standard deviation of 9.57 mm.

To understand the spatial pattern of the surface runoff changes, the change image was reclassified into 10 categories (Table 8.3). Each of these categories is a set of contiguous or discontinuous locations that exhibit the same value and is conventionally called a *zone* in raster GIS. Zone 1 has the largest negative value (less than –3.9 mm), indicating a decrease in runoff change, whereas zone 10 has the largest positive value (4.1 mm or more), indicating an increase in surface runoff. A visual interpretation of the areal extent and spatial occurrence of these zones (Fig. 8.5)

Runoff-Change Value	Runoff Zone
Less than –3.9	1
–3.9 to –2.9	2
–2.9 to –1.9	3
–1.9 to –0.9	4
–0.9 to 0.1	5
0.1 to 1.1	6
1.1 to 2.1	7
2.1 to 3.1	8
3.1 to 4.1	9
4.1 or more	10

Table 8.3 Class Assignments of the Runoff-Change Image

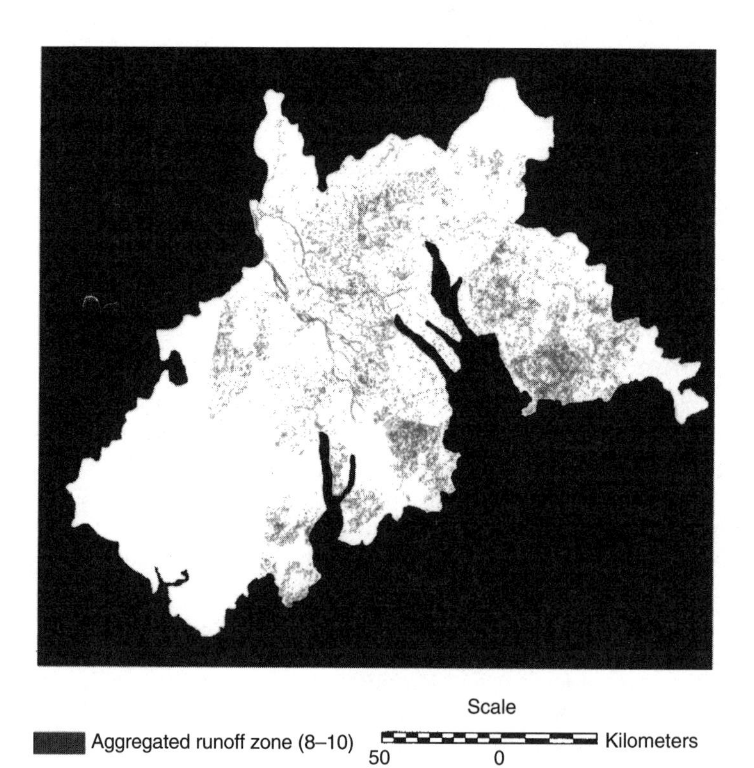

Figure 8.5 Surface runoff changes in the Zhujiang Delta, 1989–1997. See also color insert.

implies a similarity between the urban expansion pattern and the spatial pattern created by aggregating zones 8 through 10. These three zones have an increase in the value of surface runoff volume ranging from 2.10 to 24.59 mm, occupying 3.33 percent of the total area of the delta. By superimposing a city/county map onto these changed-runoff zones, the percentages of aggregated runoff zones in the total city/county area were computed. A correlation analysis then was carried out to examine the relationship between the distribution of the aggregated runoff zone and that of urban expansion within each city and county. The result showed a relatively strong positive correlation between the two mapped patterns with a multiple r value of 0.67 (significant at 0.05 level). This correlation suggests that the more urban growth a city or county experienced, the greater potential it had to increase surface runoff.

8.6 Impact of Urban Growth on Rainfall-Runoff Relationship

Figure 8.6 shows the runoff coefficient as a function of rainfall from 10 to 100 mm in Shenzhen and Zengcheng in 1989 and 1997. The cities/counties with greater urban growth, such as Shenzhen, Zhuhai, and Bao'an, have a curve for 1989 that is distinct from that for 1997. In contrast, those cities/counties with less urban growth have two curves that match well. This is particularly true in Zengcheng, Sanshui, Panyu, and Xinhui, where the two curves are so similar that visual differentiation is nearly impossible. According to the SCS model, the rainfall-runoff relationship is controlled by potential maximum storage. Therefore, the effect of urban growth on the relationship can be studied by relating the following two variables: change in potential maximum storage value and urban growth rate (percentage of urban growth). A correlation analysis between them gives a multiple r value of 0.6 (significant at 0.05 level). This result suggests that urban growth is a major contributor to the changes in potential maximum storage and thus to the relationship between rainfall and runoff.

A city or county with a higher degree of urbanization (ratio of urbanized area to total area) generally has a lower average value for potential maximum storage, and vice versa. A correlation between the two sets of variables gives $r = -0.56$ in 1989, and the coefficient increases to -0.68 by 1997. This increase is a good indication that urbanization has played an increasingly important role in shaping the relationship between rainfall and runoff. Furthermore, highly urbanized areas (such as Foshan, Jiangmen, and Zhuhai) are more prone to flooding than less urbanized areas (such as Sanshui, Zengcheng, Xinhui, and Doumen). This is so because lower values in potential maximum storage often imply that the same amount of rainfall will generate more runoff. Foshan City in the central delta, for example, had a 65 percent degree of urbanization in 1989 and 70 percent in 1997. Its average runoff

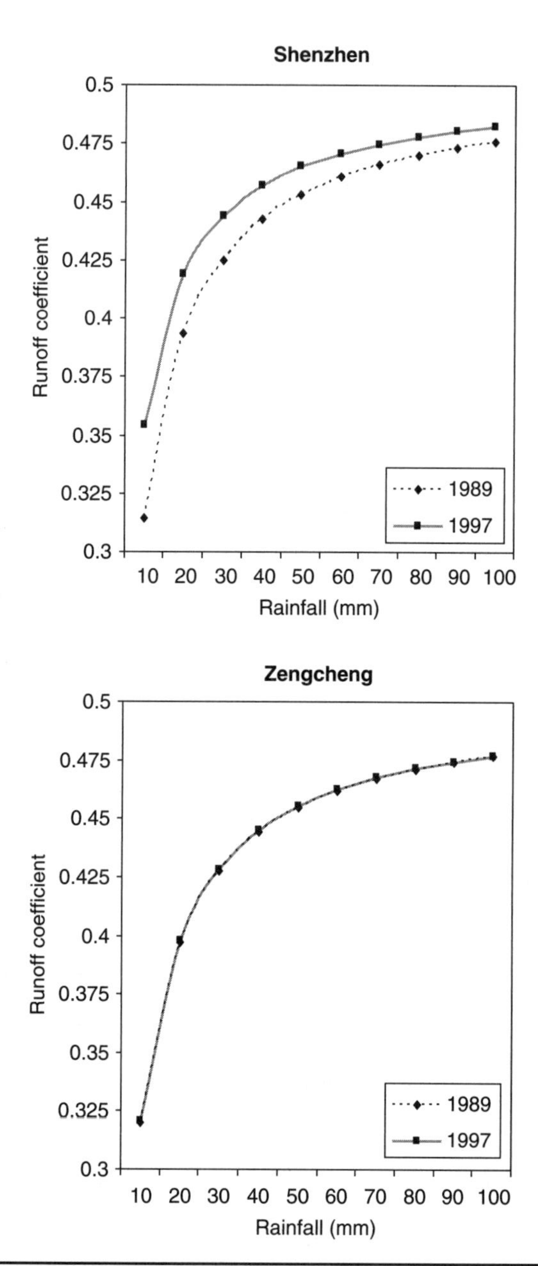

FIGURE 8.6 Runoff coefficient as a function of rainfall from 10 to 100 mm.

coefficient reached 0.46 (standard deviation = 0.034) in 1989 and 0.47 (standard deviation = 0.028) in 1997. The impact of urbanization on runoff can be examined further by comparing runoff coefficient curves. Foshan and Jiangmen have two highly similar curves, indicating that the degree of urbanization was similar in both 1989 and 1997.

Indeed, these cities have long been designed to function as pure urban centers, supported by secondary and tertiary production. Most of the land near the urban centers was filled before 1978. Recent urban development in these cities has had to seek spare land in the suburban areas and is limited to a small scale. Similar runoff coefficient curves also can be identified between Shenzhen and Zhuhai. Both cities were small towns before 1978. Urban development in these cities therefore has had much more freedom and is subject to the influence of the economic reform policies. From satellite imagery, a scattered pattern can be detected in these cities, in which urban development spread over to the suburban and surrounding rural areas.

8.7 Discussion and Conclusions

This chapter has focused on the development of an integrated approach of remote sensing and GIS for the study of urban growth and for distributed hydrologic modeling. It also has established the linkage between the two subjects through spatial analysis. By applying this methodology to the Zhujiang Delta of China, urban growth, which resulted from a rapid industrialization, and its relationship with surface runoff have been examined.

The combined use of remote sensing and GIS proves to be an effective tool for urban growth analysis. The technique of LULC change detection can be refined to find out the nature, rate, and location of urban growth. GIS allows for determining the magnitude of satellite-derived urban growth rates within administrative units. Results show that there was a remarkable expansion in urban land cover in the delta between 1989 and 1997 and that urban land development was extremely uneven among administrative units.

The integration of GIS and remote sensing has been applied successfully to surface runoff modeling. This study uses GIS to derive two key parameters: rainfall and hydrologic soil groups. Based on these data and land-cover digital data, surface runoff images may be obtained through the map algebra and overlay functions of GIS. Thus the integration has automated SCS modeling. Results indicate that annual runoff depth had increased by 8.10 mm between 1989 and 1997. The more urban growth a city or county experienced, the greater potential it had to increase surface runoff. Urban growth played a critical role in the changing relationships between rainfall and surface runoff.

The missing link between urban growth analysis and surface runoff modeling has hindered modeling and assessing the dynamics of LULC change and significantly impeded progress toward understanding earth-atmosphere interactions and global environmental change. This chapter demonstrates that the effects of urban growth on surface runoff can be modeled at local levels using an integrated approach of remote sensing and GIS. This linkage is based on the fact that LULC data are the main input parameter to both urban growth analysis and surface runoff modeling and that remote sensing and GIS are able techniques for spatial data acquisition and handling. The methodology developed in this chapter provides an alternative to traditional empirical observations and analysis using in situ (field) data for environmental studies. Future research efforts should validate these spatial modeling results and investigate the possibility and feasibility that the integration of remote sensing and GIS can be applied in a regional and global context.

References

Anderson, J. R., Hardy, E. E., Roach, J. T., and Witmer, R.E. 1976. A land use and land cover classification systems for use with remote sensing data. USGS Professional Paper 964, Washington, DC.

Berry, J. K., and Sailor, J. K. 1987. Use of a geographic information system for storm runoff prediction from small urban watersheds. *Environmental Management* **11,** 21–27.

Department of Geography, Zhongshan University. 1988. *The Land and Water Resources in the Zhujiang Delta.* Guangzhou, China: Zhongshan University Press.

Ditu Chubanshe (The Map Production Society). 1977. *Provincial Atlas of the People's Republic of China.* Beijing: People's Printer.

Drayton, R. S., Wilde, B. M., and Harris, J. H. K. 1992. Geographic information system approach to distributed modeling. *Hydrological Processes* **6,** 361–368.

Ehlers, M., Jadkowski, M. A., Howard, R. R., and Brostuen, D. E. 1990. Application of SPOT data for regional growth analysis and local planning. *Photogrammetric Engineering and Remote Sensing* **56,** 175–180.

Engman, E. T., and Gurney, R. J. 1991. *Remote Sensing in Hydrology.* London: Chapman & Hall.

Gong, Z., and Chen, Z. 1964. The soils of the Zhujiang Delta. *Journal of Soils* **36,** 69–124 (in Chinese).

Goudie, A. 1990. *The Human Impact on the Natural Environment,* 3rd ed. Cambridge, MA: MIT Press.

Guangdong Statistical Bureau. 1990. *Statistical Yearbook of Guangdong, 1989.* Beijing: China Statistics Press.

Guangdong Statistical Bureau. 1998. *Statistical Yearbook of Guangdong, 1997.* Beijing: China Statistics Press.

Huang, Z., Li, P., Zhang, Z., et al. 1982. *The Formation, Development, and Evolution of Zhujiang Delta.* Guangzhou, China: Kexue Puji Press Guangzhou Branch (in Chinese).

Harris, P. M., and Ventura, S. J. 1995. The integration of geographic data with remotely sensed imagery to improve classification in an urban area. *Photogrammetric Engineering and Remote Sensing* **61,** 993–998.

Hollis, G. E. 1975. The effects of urbanization on floods of different recurrence interval. *Water Resources Research* **11,** 431–435.

Jensen, J. R. 1996. *Introductory Digital Image Processing: A Remote Sensing Perspective,* 2nd ed. Upper Saddle River, NJ: Prentice-Hall.

Kibler, D. F. (ed.) 1982. *Urban Stormwater Hydrology.* Washington: American Geophysical Union.

Lin, G. C. S. 1997. *Red Capitalism in South China: Growth and Development of the Pearl River Delta.* Vancouver, WA: UBC Press.

Liu, A. 1993. *The Soils of Guangdong.* Beijing: Science Press (in Chinese).

Liu, N. 1998. *Guangdong Province Gazetteer of Geography.* Guangzhou, China: Department of Geography, South China Normal University (in Chinese).

Lo, C. P. 1989. Recent spatial restructuring in Zhujiang Delta, South China: A study of socialist regional development strategy. *Annals of the Association of the American Geographers* **79,** 293–308.

Lo, C. P., and Pannell, C. W. 1985. Seasonal agricultural land use patterns in China's Pearl River Delta from multi-date Landsat images. *GeoJournal* **10,** 183–195.
Lo, C. P., and Shipman, R. L. 1990. A GIS approach to land-use change dynamics detection. *Photogrammetric Engineering and Remote Sensing* **56,** 1483–1491.
Mattikalli, N. M., Devereux, B. J., and Richards, K. S. 1996. Prediction of river discharge and surface water quality using an integrated geographic information system approach. *International Journal of Remote Sensing* **17,** 683–701.
Mintzer, O., and Askari, F. 1980. *A Remote Sensing Technique for Estimating Watershed Runoff.* Washington: U.S. Department of Commerce.
Ragan, R. M., and Jackson, T. J. 1980. Runoff synthesis using Landsat and SCS model. *Journal of Hydraulic Division of the American Society of Civil Engineers* **106,** 667–678.
Rango, A., Feldman, A., George, T. S., and Ragan, R. M. 1983. Effective use of Landsat data in hydrological models. *Water Resources Bulletin* **19,** 165–174.
Rogers, P. 1994. Hydrology and Water Quality. In *Changes in Land Use and Land Cover: A Global Perspective,* edited by W. B. Meyer and B. L. Turner II, pp. 231–258 Cambridge, England: Cambridge University Press.
Ruddle, K., and Zhong, G. 1988. *Integrated Agriculture-Aquaculture in South China: The Dike-Pond System of the Zhujiang Delta.* Cambridge, England: Cambridge University Press.
Slack, R. B., and Welch, R. 1980. Soil conservation service runoff curve number estimates from LANDSAT data. *Water Resources Bulletin* **16,** 887–893.
Soil Conservation Service, USDA. 1972. *National Engineering Handbook,* Section 4, *Hydrology.* Washington: U.S. Government Printing Office.
Soil Conservation Service, USDA. 1975. *Soil Taxonomy.* Washington: U.S. Government Printing Office.
Treitz, P. M., Howard, P. J., and Gong, P. 1992. Application of satellite and GIS technologies for land-cover and land-use mapping at the rural-urban fringe: A case study. *Photogrammetric Engineering and Remote Sensing* **58,** 439–448.
Weng, Q. 1998. Local impacts of the post-Mao development strategy: The case of the Zhujiang Delta, southern China. *International Journal of Urban and Regional Studies* **22,** 425–442.
Yeh, A. G. O., and Li, X. 1996. Urban growth management in the Pear River Delta: An integrated remote sensing and GIS approach. *ITC Journal* **1,** 77–85.
Yeh, A. G. O., and Li, X. 1997. An integrated remote sensing–GIS approach in the monitoring and evaluation of rapid urban growth for sustainable development in the Pearl River Delta, China. *International Planning Studies* **2,** 193–210.
Zhong, G. 1980. The Mulberry dike-fish pond system in the Zhujiang Delta: A man-made ecosystem of land-water interaction. *Acta Geographica Sinica* **35,** 200–212.

CHAPTER 9

Assessing Urban Air Pollution Patterns

Urban air pollution, land-use, and thermal environmental characteristics are all important issues to be considered in urban environmental management and planning practices. The relationship between air pollution and urban thermal conditions is not clearly understood, although both relate to the pattern of urban land use and land cover. Remote sensing and geographic information services (GIS) have been proven to be most appropriate and effective techniques for handling spatial data that describe urban air pollution, urban thermal environment, and land use and land cover (LULC) patterns. Current remote sensing technology allows researchers to obtain reasonably high-quality land surface temperature (LST) estimates through various correction processes (Voogt and Oke, 2003). GIS technology provides a flexible environment for entering digital data from various sources and is a powerful tool in analyzing numerical relationships within and among various data layers. It is advantageous that an integrated approach of remote sensing and GIS can be developed and applied when examining the relationships among spatial variables such as urban land use, pollution, and thermal variation within an urban context.

This chapter aims to examine the relationship between air pollution patterns and LULC changes and urban thermal landscapes based on a case study of Guangzhou, China. This coastal city in South China has undergone a fundamental change in LULC owing to accelerated economic development under the reform policies in China since 1978. Because of less attention paid onto urban air quality and a lack of appropriate land-use planning and the measures for sustainable development, economic development, and rampant urban growth have created severe negative environmental consequences. By investigating air pollution pattern in the city from 1980 to 2000 and its interplay with land use and the thermal landscape, an integrated approach of remote sensing and GIS will be demonstrated and used in urban environmental studies.

9.1 Relationship between Urban Air Pollution and Land-Use Patterns

Urban areas are associated with sources of a variety of air pollutants and such regional air pollution problems as acid rain and photochemical smog. Cities are also major contributors to global air pollution related to ozone depletion and carbon dioxide (CO_2) warming. Within an urban area, the level of pollution varies with the distance to pollution sources, including both stationary and mobile sources (e.g., vehicles). Local pollution patterns in cities mainly relate to the distribution of different LULC categories, the occurrence of water bodies and parks, building and population densities, the division of functional districts, the layout of the transportation network, and air flushing rates. It is well know that pollution levels rise with land-use density, which tends to increase toward a city center (Marsh and Grossa, 2002, p. 224). Therefore, there is generally an urban-rural gradient in the concentrations of air pollutants. For example, the concentrations of particulates, carbon dioxide, and nitrate ions (an oxide, as in acid rain) in the inner city typically are two to three times higher than in the suburban area and five times higher than in the rural areas (Marsh and Grossa, 2002, p. 224).

Furthermore, urban areas experience another type of pollution—heat pollution. Urban development results in a dramatic alteration in the energy balance at the earth's surface as natural vegetation is removed and replaced by nonevaporating, nontranspiring surfaces. Under such alterations, the partitioning of incoming solar radiation into fluxes of sensible and latent heat is skewed in favor of increased sensible heat flux as evapotranspirative surfaces are reduced. A higher level of latent heat exchange is found in more vegetated areas, whereas sensible heat exchange is more favored by sparsely vegetated surfaces such as urban impervious areas (Oke, 1982). Therefore, urban areas generally have higher solar radiation absorption and a greater thermal capacity and conductivity. This thermal difference, in conjunction with waste heat released from urban houses, transportation, and industry, contributes to the development of an urban heat island (UHI). Because of the construction of tall and closely spaced buildings, the flushing capability of the air at ground level is largely reduced.

Thermal variations within an urban area mainly relate to different LULC classes, surface materials, and air flushing rates (Marsh and Grossa, 2002, p. 227). Each component surface in urban landscapes (e.g., lawns, parking lots, roads, buildings, cemeteries, and gardens) exhibits unique radiative, thermal, moisture, and aerodynamic properties, and these properties relate to their surrounding site environment (Oke, 1982). Because of its significance, recent literature has witnessed a growing interest in the relationship between LULC (vegetation) and LST (e.g., Carlson et al., 1994; Gallo and Owen, 1998; Gillies and Carson, 1995; Goward et al., 2002; Weng, 2001; Weng et al., 2004). Remote sensing data have been employed to measure LSTs and study urban thermal variations, including National Oceanographic and Atmospheric Administration (NOAA) Advanced Very High Resolution Radiometer (AVHRR) data (Balling and Brazell, 1988; Gallo et al., 1993; Kidder and Wu, 1987; Roth et al., 1989; Streutker, 2002, 2003), Landsat thematic Mapper (TM) and Enhanced Thematic Mapping Plus (ETM+) (Carnahan and Larson, 1990; Kim, 1992; Nichol, 1994, 1996; Weng, 2001, 2003; Weng et al., 2004), and ATLAS data (Quattrochi and Luvall, 1999; Quattrochi and Ridd, 1994). Studies using satellite-derived radiant temperatures have been termed as the *surface temperature heat islands* (Streutker, 2002). LST is believed to correspond more closely with the urban canopy-layer heat islands, although a precise transfer function between LST and the near-ground air temperature is not yet available (Nichol, 1994). Byrne (1979) has observed a difference of as much as 20°C between the air temperature and the warmer surface temperature of dry ground.

UHIs favor the development of air pollution problems but are not an indicator of air pollution (Ward and Baleynaud, 1999). It is known that higher urban temperatures generally result in higher ozone levels owing to increased ground-level ozone production (DeWitt and Brennan, 2001). Moreover, higher urban temperatures mean increased energy use, mostly owing to a greater demand for air conditioning. As power plants burn more fossil fuels, the pollution level is driven up. A few studies have so far examined the correlation between LST and air pollution measurements. Poli and colleagues (1994) investigated the relationship between satellite-derived apparent temperatures and daily sums of total suspended particulates (TSPs) and sulfur dioxide (SO_2) in the winter season in five locations in Rome, Italy. It was found that apparent temperatures had a strong negative correlation with TSPs but a weak correlation with SO_2. Brivio and colleagues (1995) used three AVHRR images of Milan, Italy, acquired on February 12 to 14, 1993, to study the correlation of apparent temperatures with air-quality parameters, including TSPs and SO_2. A weak correlation was discovered with both TSPs and SO_2 that could be explained by the large pixel size of the image. Ward and Baleynaud (1999) explored the correlation between Landsat TM band 6 digital counts and the concentrations of pollutants, including black particulates (BPs), SO_2, nitrogen dioxide (NO_2), nitrogen monoxide (NO), and strong acidity (AF) in Nates, France, based on measurements of daily sums, individual measurements every 15 minutes, and daily mean values taken on May 22, 1992. It was found that apparent temperatures were highly positively correlated with BPs; moderately correlated with SO_2 and daily means of NO_2, NO, and AF; but weakly correlated with instantaneous measurements of NO_2 and NO. These studies contribute to the current literature by adding more evidence of the correlation between air pollution and urban thermal patterns.

9.2 Case Study: Air Pollution Pattern in Guangzhou, China, 1980–2000

Specific questions that will be addressed in this chapter include: (1) How has the local air pollution pattern changed over the past two decades? (2) How has urban growth altered the air pollution pattern and promoted the UHI effect, as indicated by satellite-derived LST measurements? and (3) How did the pollution patterns relate to urban land use and urban thermal patterns? Detailed information on the relationships among these three spatial variables in an urban area would be valuable to urban and environmental planners. Knowledge of their relationships can be used by planners to evaluate the need for new or revised urban design and landscaping policies for mitigating the adverse environmental effects of building mass, transportation networks, and poor landscape layouts.

9.2.1 Study Area: Guangzhou, China

Guangzhou (also known as Canton) is located at latitude 23°08′N and longitude 113°17′E and lies at the confluence of two navigable branches of the Zhujiang River (literally the "Pearl River") system (Fig. 9.1). With a population of 3.99 million and a total area of 1444 km^2 (Guangdong Statistical Bureau, 1999), it is the sixth most populous city in China. It has been the most important political, economic, and cultural center in South China. The Guangzhou municipality consists of eight administrative districts in the city proper (see Fig. 9.1) and four rural counties. This study focuses on the city proper. It has a subtropical climate with an average annual temperature of 22°C, with the lowest temperature in January (averaging 18.3°C) and the highest temperature in July (averaging 32.6°C) (computed based on averages of the daily mean temperatures). The annual average precipitation ranges from 1600 to 2600 mm. Because of the impact of the East Asian and Indian monsoonal circulation, about 80 percent of the rainfall comes in the period of April to September, with highest concentration in spring. The humidity is high in all seasons, and the average relative humidity is 80 percent. This stems principally from the wetland environment and the basin-like terrain of the city (Fig. 9.2). Guangzhou has a history of urban development spanning 2000 years (Xu, 1990). The urban development has progressed at an unprecedented pace, especially after implementation of economic reform in 1978, with dominant expansions toward the east and the north. The

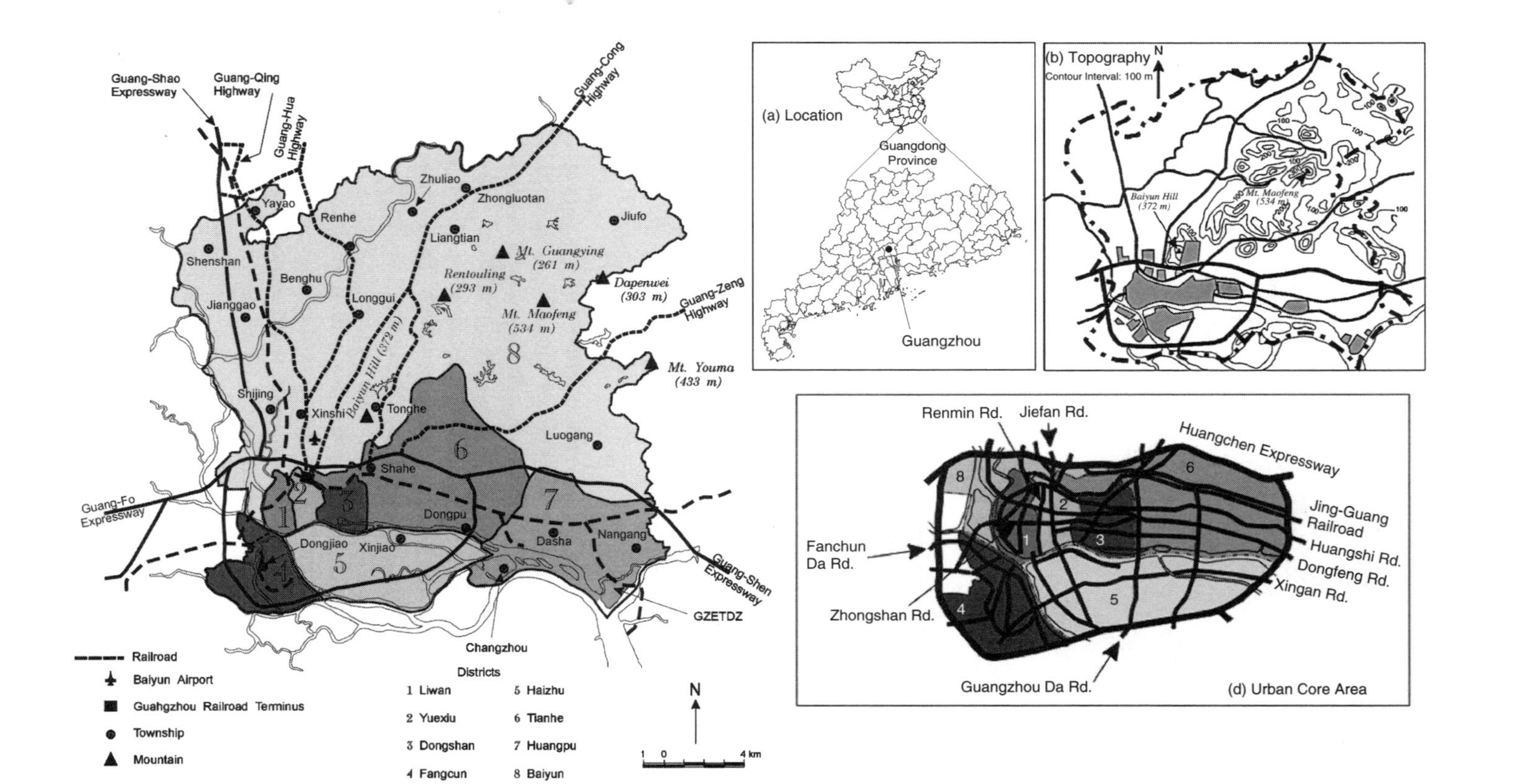

Figure 9.1 Study-area map showing major geographic features and the eight districts, with inset maps of location and the urban core area.

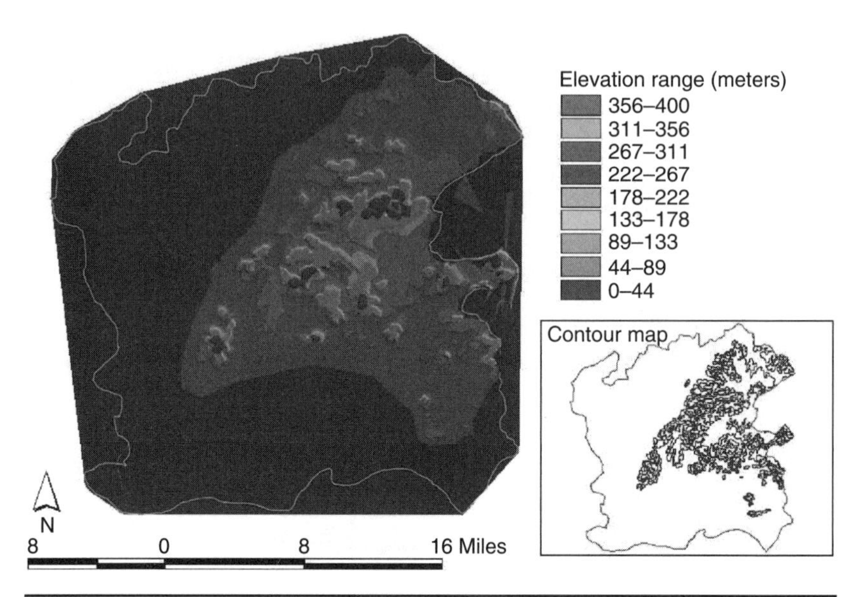

Figure 9.2 Digital elevation model of Guangzhou. (Inset: Contour map.) See also color insert.

total area of finished housing covered only 12.3 km^2 in 1949 when the People's Republic of China was founded but had increased to 127.9 km^2 in 1999.

9.2.2 Data Acquisition and Analysis

Ambient Air Quality

In Guangzhou, ambient air quality was monitored through six municipal monitoring stations (the number of monitoring stations was increased to nine after 1996). Concentrations of the pollutants sulfur dioxide (SO_2), nitrogen oxides (NO_x, consisting of NO and NO_2), carbon monoxide (CO), and total suspended particles (TSPs), as well as dust level, are measured routinely in the stations. These measurements are analyzed with the standard methods stipulated by the State Environmental Protection Administration (SEPA) of China. Concentrations of SO_2 were collected using the designated impinger (GS-3) that was calibrated for each measurement with bubble flowmeters and were analyzed by the method of pulse fluorescence. Concentrations of NO_x were measured using the NO_x impinger and then analyzed by the method of chemical fluorescence. TSP concentrations were sampled using KB-120E samplers that were calibrated quarterly with orifice plate meters and were determined by the method of filter lightening. Concentrations of CO were measured by the method of infrared absorption. The six monitoring stations are nationally regulated ones located in various functional districts of land use:

Pollutant	1980	1985	1990	1995	2000	NAQS for Residential Areas
SO_2 (mg/m^3)	0.09	0.089	0.097	0.057	0.045	0.06
NO_x (mg/m^3)	0.04	0.073	0.137	0.123	0.061	0.10
TSPs (mg/m^3)	0.24	0.226	0.22	0.31	0.157	0.20
Dust (ton/km^2/ month)	15.94	12.7	9.56	9.16	7.34	6–8
CO (mg/m^3)	No data	2.8	3.16	2.91	No data	4.00

Note: NAQS = national air quality standard.

Table 9.1 Annual Means of Concentrations of the Air Pollutants Observed in Major Years

industrial, commercial, mixed urban uses, dense transportation, residential, and sanitary suburban areas. Data for the monitoring stations were strictly restricted to internal use, but this research was able to collect the city's mean measurements of pollutants since the 1980s through compiling a time series of environmental quality reports in an attempt to analyze the temporal change in air quality. Table 9.1 shows annual means of concentrations of the designated pollutants every 5 years and the national air-quality standards for those pollutants. Moreover, geographic distribution maps of SO_2, NO_x, TSPs, and dust level included in the *Guangzhou Natural Resources Atlas* (Guangzhou Municipality Planning Committee, 1997) were manually digitized and converted into GIS data layers so that the spatial patterns of air pollution also could be examined. The maps were produced by the city's environmental monitoring authority based on data collected in the fixed monitoring stations between 1983 and 1992 in conjunction with the results of several mobile studies. The placements of contours in these maps have a fairly high accuracy, and thus the maps are appropriate for this type of analysis.

Land Use and Land Cover

The source data for mapping urban and built-up areas include a topographic map, an LULC map, and satellite imagery. The topographic map was produced by the Chinese government in 1960 based on 1:25,000 aerial photographs taken in 1958. The LULC map was prepared by the Department of Geography, Zhongshan University, based on aerial photographs taken in 1984. The two maps have the same scale of 1:50,000. Urban and built-up areas were delineated from the two maps and manually digitized into a computer with the ArcInfo computer program.

The satellite images are two Landsat TM images, dated December 13, 1989, and August 29, 1997, respectively. Each Landsat image was rectified to a common Universal Transverse Mercator (UTM) coordinate system based on 1:50,000 scale topographic maps. These images were resampled using the nearest-neighbor algorithm with a pixel size of 30 m for all bands. The resulting root-mean-squared error (RMSE) was found to be 0.77 pixel (or 23.1 m on the ground) for the 1989 image and 0.58 pixel (or 17.4 m on the ground) for the 1997 image.

LULC patterns for 1989 and 1997 also were mapped using Landsat TM data. A modified version of the Anderson scheme of LULC classification was adopted (Anderson et al., 1976). The categories include (1) urban or built-up land, (2) barren land, (3) cropland (rice), (4) horticulture farms (primarily fruit trees), (5) dike-pond land, (6) forest, and (7) water. A supervised signature extraction with the maximum likelihood algorithm was employed to classify the Landsat images. Both statistical and graphic analyses of feature selection were conducted, and bands 2, 3, and 5 were found to be most effective in discriminating each class and therefore were used for classification. The accuracy of the classified maps was checked with a stratified random sampling method, by which 50 samples were selected for each LULC category. The reference data were collected from a field survey or from existing LULC maps that have been field checked. Large-scale aerial photographs also were employed as reference data in accuracy assessment when necessary. The overall accuracy of classification was determined to be 90.57 and 85.43 percent, respectively. The kappa index was 0.8905 for the 1989 map and 0.8317 for the 1997 map. A detailed accuracy-assessment result, including the error matrices and the user's and producer's accuracy of the LULC maps, can be found in Weng (2001). The urban and built-up areas were extracted from each LULC map to create the urban maps.

Land Surface Temperatures

LSTs were derived from geometrically corrected Landsat TM thermal infrared bands dated on December 13, 1989, and August 29, 1997. The TM thermal band has a spatial resolution of 120 m and a noise level equivalent to a temperature difference of 0.5°C (Gibbons and Wukelic, 1989). The local time of satellite overpass was in the morning (approximately 10:00 a.m.), so the chance for detecting a weak UHI is maximized. Since both images were acquired at approximately the same time, a comparative study is feasible. A quadratic model was used to convert the digital number (DN) into radiant temperatures ($T_{(k)} = 209.831 + 0.834DN - 0.00133DN^2$) (Malaret et al., 1985). The temperature values obtained then were corrected with emissivity ε according to the nature of the land cover. Vegetated areas were assigned a value of 0.95 and nonvegetated areas 0.92 (Nichol, 1994). The differentiation between vegetated and nonvegetated areas was made according to normalized difference vegetation index (NDVI) values, which were computed from visible (0.63 to 0.69 μm) and near-infrared (0.76 to 0.90 μm) data of TM images. The emissivity-corrected LSTs were computed using the equation developed by Artis and Carnhan (1982):

$$\mathrm{LST} = \frac{T_{(k)}}{1+(\lambda \times T_{(k)}/\rho)\ln \varepsilon} \tag{9.1}$$

where λ = wavelength of emitted radiance [for which the peak response and the average of the limiting wavelengths (λ = 11.5 μm) (Markham and Barker, 1985) will be used]
$\rho = h \times c/s$ (1.438 × 10–2 mK)
s = Boltzmann's constant (1.38×10^{-23} J/K)
h = Planck's constant (6.626×10^{-34} J · s)
c = the velocity of light (2.998×10^{8} m/s)

9.2.3 Air Pollution Patterns

SO_2

Guangzhou's air quality was getting worse after the 1970s and became a major concern after the 1980s with the accelerated urbanization and industrialization. Figure 9.3 shows interyear variations in annual mean concentrations of SO_2, NO_x, TSPs, dust level, and CO from 1981 to 2000. The concentration of SO_2 continued to go up after 1981, peaked at 1989, and then declined gradually. The year 1992 witnessed a long-awaited victory, when the Chinese National Air Quality Standard (NAQS) for SO_2 for residential areas, 0.06 mg/m^3, was not exceeded for the first time since 1972 (GZCLCCC, 1995, pp. 612–613). Figure 9.3*a* indicates that there was an elevated peak of SO_2 concentration between 1996 and 1998, however. All three years had exceeded the NAQS, only marginally in 1996 (0.064 mg/m^3) and 1998 (0.061 mg/m^3) but remarkably in 1997 (0.07 mg/m^3). The intrayear variation was characterized as a higher level of concentration in the winter months and a lower level in the summer months (Xie and Chen, 2001). This seasonal pattern was not evident owing to slight seasonal variations in temperature.

The SO_2 concentration was largely related to the industrial use of coal as a major energy source and the use of coal combustion for cooking and heating in the catering industry and households. As a result, areas with industrial plants, high population density, and clustering of the catering industry tended to have a high concentration of SO_2. Figure 9.4*a* shows two focal points of contamination, one

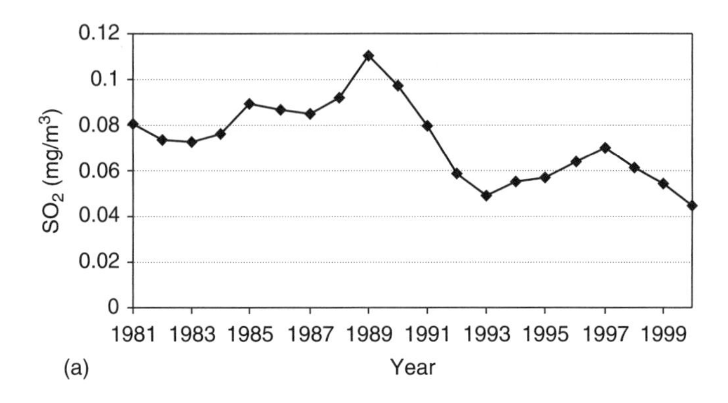

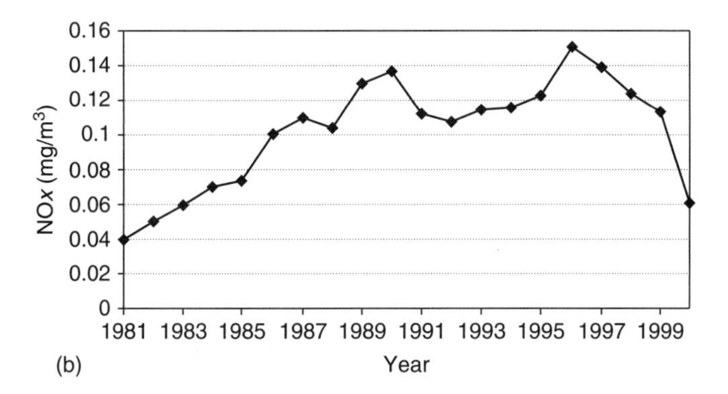

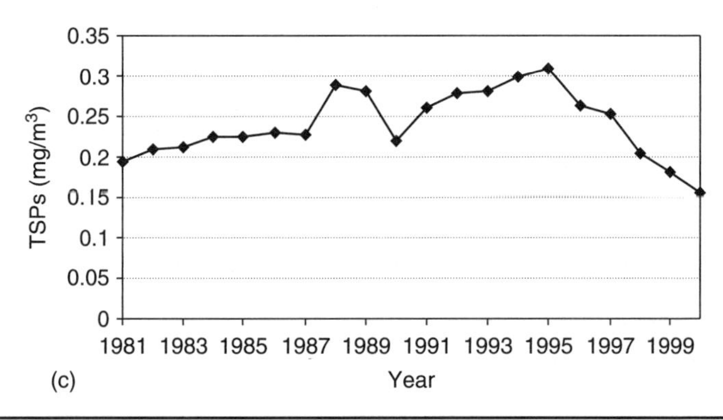

Figure 9.3 Yearly variations in annual mean concentrations of (*a*) SO_2, (*b*) NO_x, (*C*) TSPs, (*d*) dust level, and (*e*) CO, 1981–2000. (*Adapted from Weng and Yang, 2006.*)

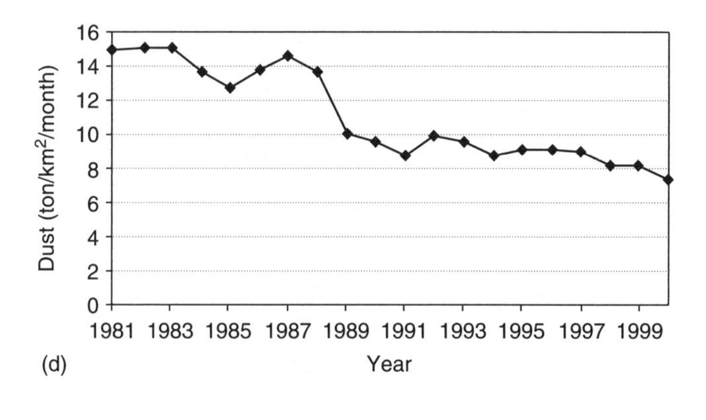

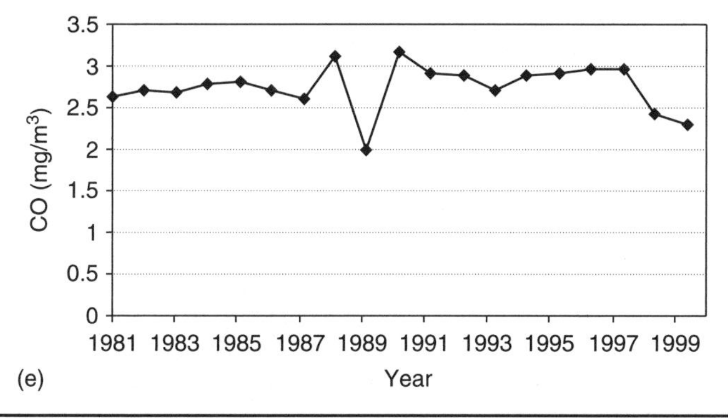

Figure 9.3 *(Continued)*

locating in Liwan District and the other around Yuanchun township of the Tianhe District. Liwan was home to the city's electricity generation plant, and the abnormally high SO_2 concentrations also can be attributed to the biggest concentration of catering enterprises in the city and extremely high building and population densities. In contrast, the Tianhe District did not have a high population density nor a clustering of catering enterprises. However, it possessed sparsely distributed but highly polluted industries, especially chemical plants. Away from these "pollution hubs," SO_2 concentration decreased gradually. An intracity gradient was detected if we looked beyond the old city core (i.e., Yuexiu, Dongshan, and Liwan Districts) and Tianhe toward the northern part of the city. Most of the Baiyun District had kept relatively clean, with a concentration below 0.02 mg/m^3. Research conducted by Xie and Chen (2001) at the Guangzhou Environmental Monitoring Station suggests that SO_2 measurements at the suburban sanitary station, which is located on the southern slope of Baiyun Hill, showed a highly similar temporal variation pattern to those seen in other functional districts. The urban-rural disparity was narrowed down during the period of 1985 to 1997 from 0.09 mg/m^3

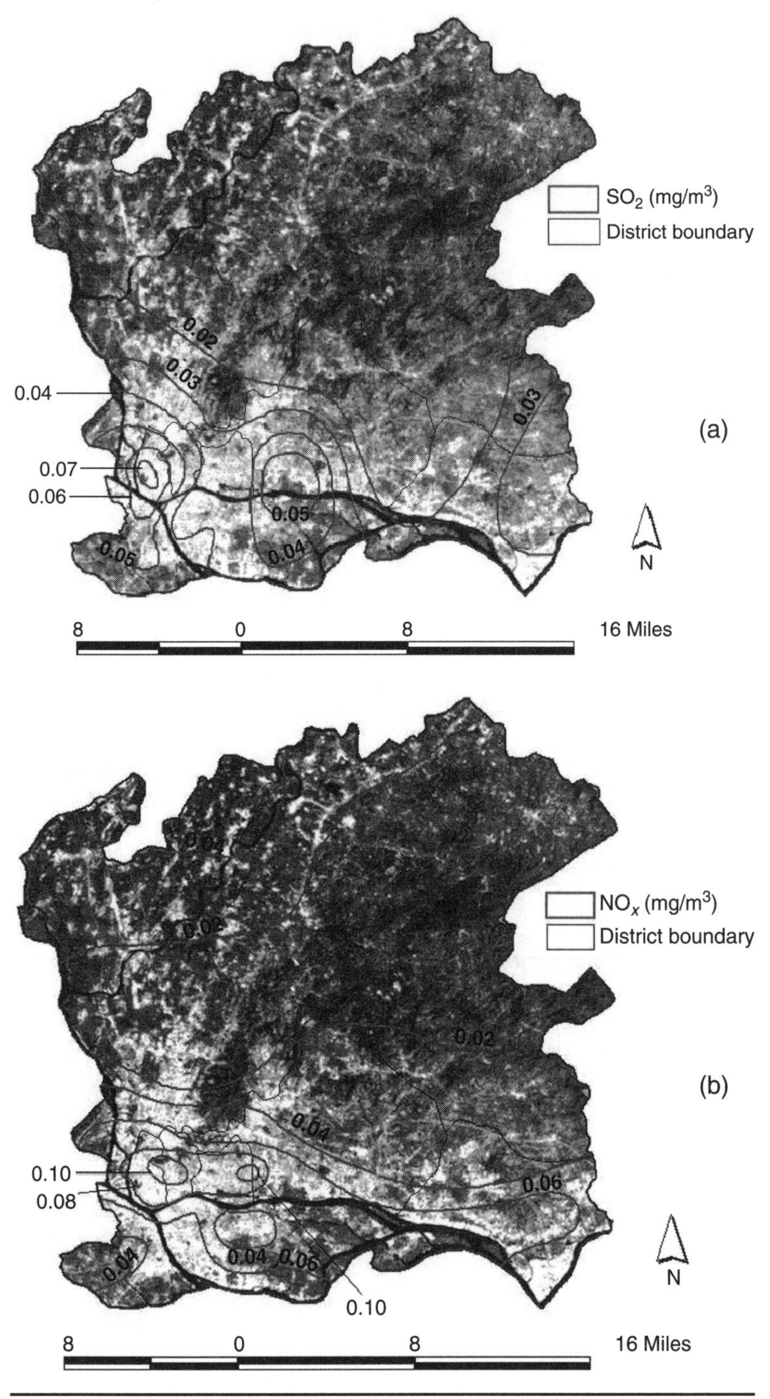

Figure 9.4 Geographic distribution of the concentrations of major air pollutants in Guangzhou. Contours are interpolated from point data observed in the monitoring stations. Data represent annual averages of pollutant measurements. The base map is the LST map of August 29, 1997, derived from Landsat TM thermal infrared data. See also color insert.

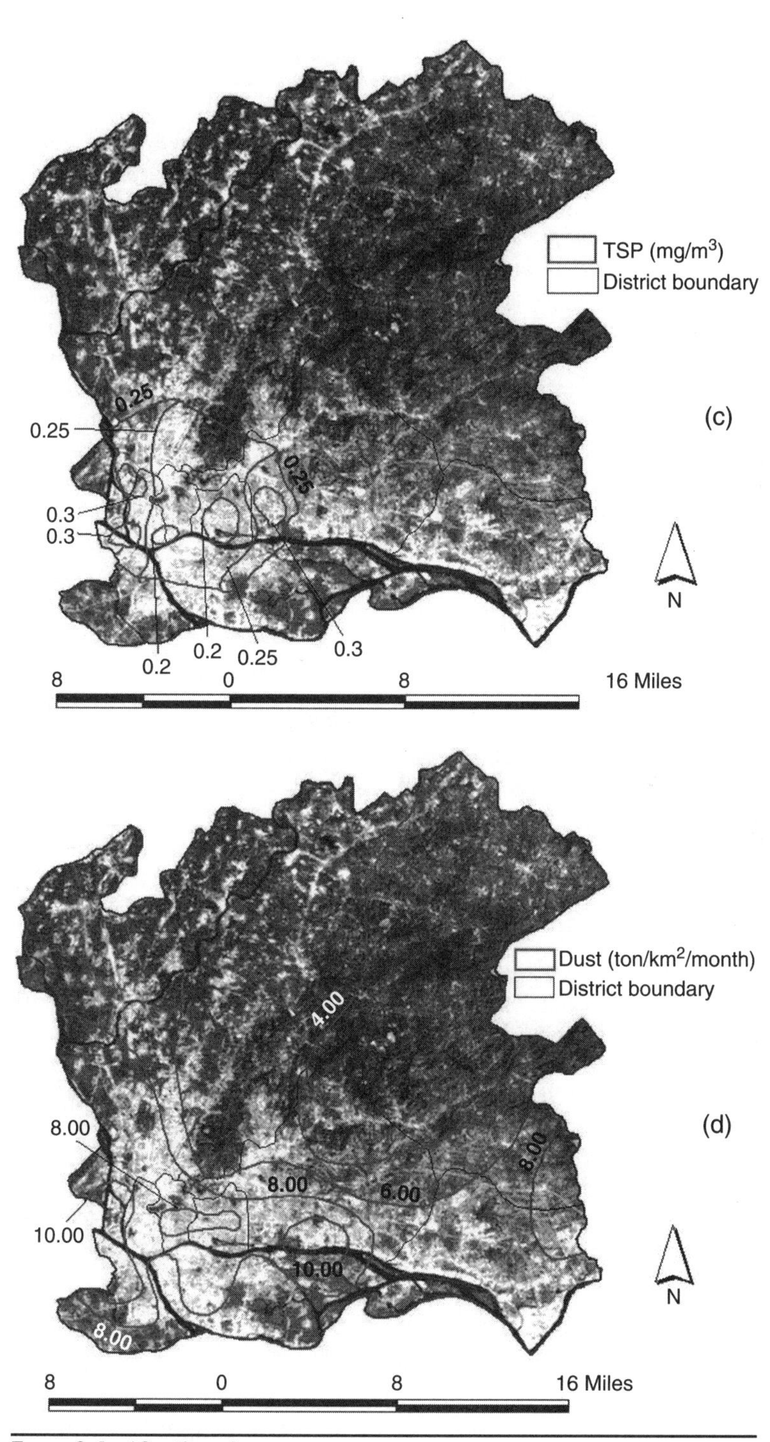

Figure 9.4 (*Continued*)

in 1985 to 0.067 mg/m^3 in 1989, with a sharp decrease in 1993 to 0.026 mg/m^3, leveling off in 1997 to 0.018 mg/m^3.

Since 1990, urban Guangzhou had been able to gradually lower the overall level of SO_2 contamination for to the following three reasons: First, a substantial number of high polluting enterprises were forced to close down or relocate away from the urban core area. Second, new technologies were applied to reduce SO_2 emissions. Finally, use of liquefied petroleum gas increased rapidly, replacing coal as the nearly sole source of energy for catering and home uses. These improvements, however, were somewhat counterbalanced by a sizable increase in the industrial use of coal resulting from an accelerated industrial development in the 1990s. Furthermore, many newly built electricity generation plants in the Zhujiang Delta, where Guangzhou is located, became new sources of air pollution owing to the dominant southeasterly winds in the summer. As a result, although the overall concentration level dropped, the affected area of SO_2 pollution showed an increasing tendency toward dispersal (Xie and Chen, 2001).

NO_x

Ambient NO_x concentrations showed a trend of wavy increase to the peak in 1996 and then decreased gradually (see Fig. 9.3*b*). Based on the Chinese NAQS for NO_x (i.e., 0.1 mg/m^3 for residential areas), the city as a whole exceeded the standard for all the years from 1986 through 1999. In many of the years observed, the city had the highest concentration level among all Chinese cities (Wang et al., 2001; Zhang et al., 1999). The intrayear variation was not as large as that of SO_2, although a slightly higher concentration level were observed in the winter months (Xie and Chen, 2001). Research shows that this pollution problem is strongly related to the rapid growth of vehicular exhaust emissions (Qian et al., 2001; Wang et al., 2001; Zhang et al., 1999). Between 1981 and 1999, the number of cars in the city increased from 26,153 to 273,036, yielding an annual rate of increase of 13.92 percent. The number of motorcycles exhibited a similar rate of increase. The city had 157,677 motorcycles in 1990, but this increased to 385,508 in 1999, yielding an annual rate of increase of 10.44 percent. Traditional dependence of civil transportation on bicycles had been impaired as household possession of motorcycles increased drastically (from 4 per 100 households in 1990 to 23.2 in 1999), whereas the number of bicycles per 100 households decreased from 191.7 to 160.6 during the same period of time. The massive number of vehicles, in conjunction with poor vehicle and fuel quality with no catalytic converters, produced a high volume of pollutants including NO_x and CO. Moreover, the city's road construction and pavement management did not keep up with the speedy growth of the vehicle population. The total area of paved roads was 3.6 km^2 in 1981 and rose to 25.56 km^2 in 1999 (Guangzhou Statistical Bureau, 1999), with an annual rate of increase of 11.5 percent. This increase, however, clearly was lower than that of vehicles, which contributed to heavy traffic congestion most of the time. Low driving speeds, high idling, and frequent acceleration caused automobile engines to be less efficient combustors and to become a main source of air pollution. Consequently, NO_x concentrations were much higher in dense traffic areas. Highway corridors and urban street canyons were notorious for severe levels of NO_x pollution, such as Dongfeng Road, Jiefang Road, and Huangshi Road (see Fig. 9.1*d*) (Xie and Chen, 2001; Zhang et al., 1999).

Figure 9.4*b* shows that two areas exhibited the highest level of NO_x concentration (= 1.0 mg/m^3), one in Yuexiu District and the other smaller one in the eastern part of Dongshan District. Both were directly related to extremely heavy traffic and impaired air circulation. As distance increased from the pollution hubs, the level of NO_x concentration went down gradually. A comparison of Fig. 9.4*b* with Fig. 9.1*c* reveals that the spatial pattern of NO_x concentration was highly correlated with the pattern of the city's transportation network. This is particularly true when both patterns were viewed and referenced to the eastward expansion of urban development. A south-north intracity gradient was observed, and this reflected changes in the density of the transportation network and building density. The old city core (Liwan, Yuexie, and Dongshan Districts) in the south had the highest NO_x concentration, leveling down to Tianhe District and reaching the lowest concentration in Baiyun District in the north. The old city core not only held an extremely high density of the transportation network but also had the highest building density. High and closed-spaced buildings in these districts caused flushing of polluted surface air to be slow and incomplete. Haizhu and Huangpu Districts found an intermediate level of NO_x owing to high building density and moderate traffic. Similar to ambient SO_2 concentrations, NO_x concentrations in the sanitary rural areas rose quicker than those in the urban areas, leading to a smaller urban-rural gradient (Xie and Chen, 2001).

There was a general decline in the NO_x concentrations since 1996 (see Fig. 9.3*b*). Numerous laws and regulations were implemented to tighten up vehicular emission control and to mandate road checks and installation of purification equipment. A series of emission-reduction products was introduced, as well as mandatory use of unleaded gasoline after July 2000. Motorcycles and smoky cars were seen as especially problematic; hence the city government stepped up to set up a series of regulations for emission control. Finally, the city government advocated the adoption of environmentally conservative vehicles that used liquefied petroleum gas, electricity, or other sources of "clean" energy. The data we collected for this study indicate that the preceding measures were effective.

TSPs

Aside from the two major pollutants (SO_2 and NO_x), TSPs are another regulated pollutant measured in the monitoring stations. Figure 9.3*c* illustrates the temporal changes in the concentration of TSPs between 1981 and 2000. It is clear that TSP concentration kept increasing until it reached a peak in 1995 and then declined. Based on the NAQS for residential areas of 0.2 mg/m^3, Guangzhou had 17 years (from 1982 through 1998) out of the 20 in which the standard was exceeded. Only the years 1981, 1999, and 2000 were below the NAQS. Seasonal variations in TSPs were stronger than those in SO_2 and NO_x. TSP levels, as indicated by monthly arithmetic means, were higher in the wintertime (November through March), which may be attributed to the slight increase in domestic fuel combustion (Qian et al., 2001).

Spatially, TSP concentrations were remarkably uneven, showing a great urban-rural gradient. Figure 9.4*c* shows that two north-south corridors of high concentration can be observed. The western corridor extended from Xinshi Township of the Baiyun District, crossed the Pearl River, and went on to near Dongdun Township of the Fangchun District. The eastern corridor set out from the Guangzhou East Railroad Station, passed the Pearl River, and extended to Xinjiao Township of the Haizhu District. Within these two corridors, TSP concentration always was higher than 0.25 mg/m^3, with some spots bearing even higher levels of TSPs. In between the corridors, TSP concentration ranged from 0.15 to 0.25 mg/m^3, but some areas may be cleaner than others. Because no measurements were available to map TSP distribution for the whole city, Huangpu District and most areas in Baiyun, Tianhe, Haizhu, and Fangcun Districts were left out. Presumably, TSP concentrations in these areas were lower than in the mapped areas because there was a higher density of transportation network and population and several high polluting plants in the urban core area.

Dust

The dust level appeared to climb continually in the 1970s. In 1971, the city's average dust level was 16.3 tons/km^2 per month (GZCLCCC, 1995, p. 612). The level escalated to 20.4 tons between 1972 and 1978. Large spatial disparity existed among different functional districts of industrial, commercial, residential, and sanitary rural areas, with an average dust level of 33.4, 20.2, 20.4, and 7 tons, respectively (GZCLCCC, 1995, p. 612). In 1979, the monthly dust level decreased to 19 tons/km^2 (GZCLCCC, 1995, p. 613). Figure 9.3*d* reveals that this declining trend persisted throughout the study period from 1981 through 2000 despite of some ups and downs in the plot. This general declining trend did not mean that Guangzhou had become a "clean" city. Instead, all the years observed, except for 2000, exceeded the Chinese NAQS for dust level, which is 6.0 to 8.0 ton/km^2 for residential areas. The main sources of dust were plants and thousands of construction sites (over 3000 in the early 1990s) in the city (Guangzhou Bureau of City Planning, 1996).

Figure 9.4*d* shows the spatial distribution of dust concentration in 1992. There were two dust pollution centers (both exceeded 10 ton/km^2 per month), with one located in Liwan District and the other located in Yuanchun Township (industrial) of the Tianhe District. Both the Guangzhou Concrete Plant and the Xichun Electric Power Plant had a great impact on the high dust level in the Liwan District. Yuexiu, Dongshan, Haizhu, and Tianhe Districts were next to Liwan in terms of the concentration level. Baiyun District detected the lowest dust level. Contour lines in Fig. 9.4*d* illustrate an overall spatial pattern; that is, the dust level decreased from the southwestern to the northeastern part of the city.

CO

Figure 9.3*e* shows that CO concentration remained relatively stable during the observed years, ranging from 2.0 to 3.16 mg/m^3. The CO concentrations had not exceeded the Chinese NAQS for residential areas, which is 4.0 mg/m^3. The CO concentrations were related mainly to vehicular emissions. Research indicates that 84.8 percent of the total CO in the atmosphere in 1994 resulted from vehicular emissions (Zhang et al., 1999). Accordingly, the geographic distribution of CO concentration was traffic related. The concentrations were extremely high in dense traffic areas such as urban street canyons and highway corridors. A diffusion experiment conducted by Zhang and colleagues from July 14 to August 10, 1998, suggested that CO concentration in Dongfeng Street reached as high as 72.1 mg/m^3, with an average of 9.0 mg/m^3. The concentration level fluctuated with traffic flow and was higher between 7:00 a.m. and 9:00 p.m. (Zhang et al., 1999). The leeward side of a street usually had CO concentrations about twice as high as that observed on the windward side owing to wind vortices that carried the pollutant to the leeward side and ascended to the leeward roof edge (Xie et al., 2003). As with NO_x, the old city core (Liwan, Yuexie, and Dongshan Districts) displayed a higher CO concentration because of dense transportation networks and closely space buildings.

9.2.4 Urban Land Use and Air Pollution Patterns

Guangzhou proper has experienced a series of drastic changes in administrative boundaries. These changes affect the computation of urban and built-up areas in different periods of time. The current city jurisdiction came into effect in 1988, governing eight urban districts and four adjacent counties. The 1960 and 1984 urban maps cover the urban core area only. The remote sensing–GIS analysis indicates that urban and built-up land had expanded by more than six times from 1960 through 1997. In 1960, the urban and built-up area was 64.2 km^2

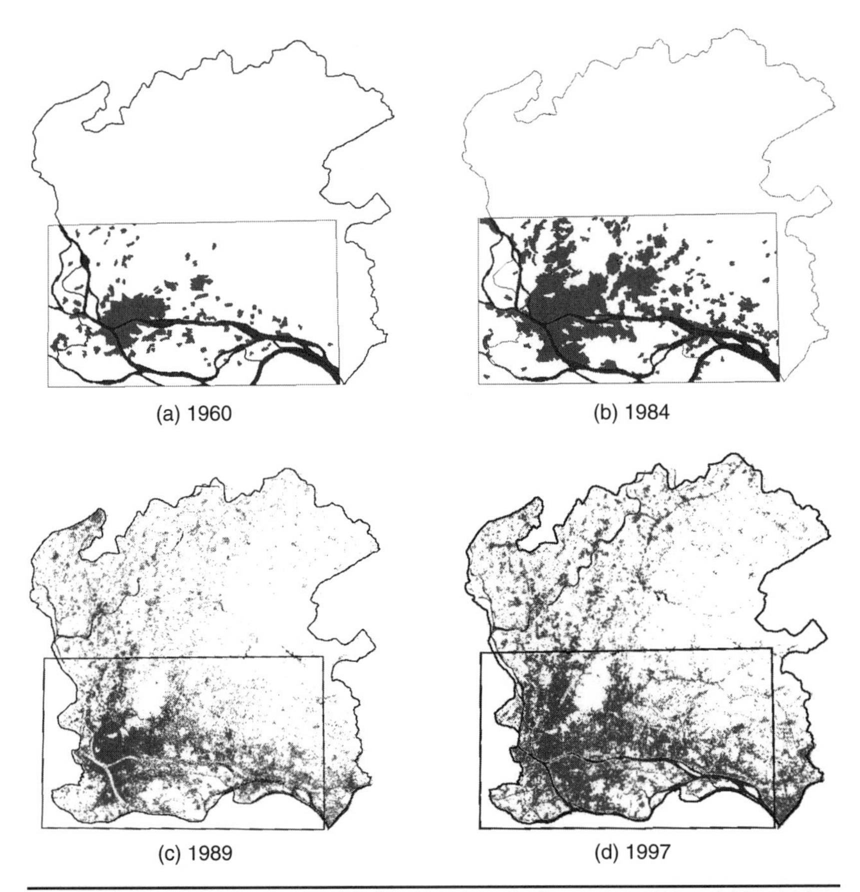

FIGURE 9.5 Changes in urban and built-up land use in Guangzhou, China, 1960–1997. (*Adapted from Weng et al., 2003.*)

(Fig. 9.5*a*), and in 1984, it was 159.6 km² (see Fig. 9.5*b*). During this 24-year period, urban land use (in the rectangular box) increased by 95.4 km², or 149 percent. The rectangular box defines the limit of the 1984 aerial photograph survey, which lies between 23°2′30″ and 23°13′40″N in latitude and 113°10′00″ and 113°34′00″E in longitude, covering the then urban area. Moreover, the LULC maps derived from the Landsat TM images show that the urban and built-up area was 194.8 km² in 1989 (see Fig. 9.5*c*) and 295.2 km² in 1997 (see Fig. 9.5*d*). Urban land use increased by 100.4 km², or 1.5 percent, in the 8-year period. Overlaying the urban land-use maps with the city district boundary and major roads reveals the areal extent and spatial occurrence of urban and built-up areas and the expansion trend.

Before the Communists took over in 1949, there was an old city core area (the biggest polygon in Fig. 9.5*a*) extending from the Haizhu Bridge to some narrow and crowded streets in the Yuexiu and Dongshan

Districts. Most of the new development took place in the suburbs as organized clusters for accommodating industries, warehouses, and external transportation facilities, aside from a few developments on the outskirts of the old city core (Xu, 1985). Huangpu was designated as the outgoing port of the city. New development was directed to the suburbs in order to contain the growth of the inner city. When a project required a large piece of land, city planners would intentionally locate it in a remote area that had sufficient less productive land, especially in the eastern and southern suburbs. The built-up area increased from 36 km^2 in 1949 to 56.2 km^2 in 1954 (Guo, 2001). In the late 1950s, there was a significant increase in industrial and residential land uses. Factories were built in selected areas of Haizhu, Fangcun, and Huangpu Districts, whereas residential developments spread out in the Tianhe District to house higher-education, research, and medical units.

During the 1960s and 1970s, urban development was sluggish owing to continuous political movement. New factories were built in the Haizhu District, and production and port facilities were expanded on the northeast riverbank of the Fungcun District. The city's port function started to shift largely to the Huangpu District, where heavy chemical and power industries had been started. The status of Guangzhou as the only sanctioned foreign trade center in the prereform China warranted new housing developments on the northern fringe of the Yuexiu District. In addition, a major railway terminus was developed immediately north of the old city core. These developments promoted the city's northern and northwestern expansion. By 1978, the total built-up areas reached 89 km^2.

The 1980s witnessed dramatic urban development. A triangular economic and technological development zone (9.6 km^2) was established in Huangpu District to attract foreign industrial investment. The Huangpu New Port finished building eight 20,000-tonnage deepwater berths for ships. This construction, in conjunction with other port facilities, residential and office buildings, hotels, schools, and recreational facilities, shaped a modern Huangpu District. Furthermore, the Baiyun International Airport was reconstructed and expanded. New bridges were constructed to link with the island of Ho Nam (*Henan*), laying the foundation for future southern expansion. Figure 9.5*c* shows that two urban development corridors were becoming visible in 1989, one expanding eastward to Huangpu and the other expanding northward along the expressways.

The urban and built-up areas grew even faster in the 1990s owing primarily to the development of commercial housing. Commercial housing was scattered initially out in the Tianhe District. However, recent development shifted to the northern and southern banks of the Zhujiang River. A modern business and residential zone of 6.6 km^2, the Zhujiang New Town (*Zhujiang Xin Cheng*), was developed along the northern waterfronts of the Zhujiang immediately east of the old city core. In terms of industrial land use, there had been substantial development in the Tianhe High-Tech Industrial Park, whereas heavy industries continued to build up in the Huangpu District. By 1997, the urban and built-up areas reached 295.2 km^2. Figure 9.5*d* shows that a west-east-running urban corridor following the northern shore of the waterway has taken shape. In addition, northward urban sprawl along both sides of Baiyun Hill is now a conspicuous feature of the city's land use.

How have urban growth and associated LULC changes altered the local air pollution pattern in Guangzhou? To answer this question, several GIS analyses were implemented using the following procedures: First, using the buffer function in GIS, three GIS data layers were constructed. The first buffer shows 10 buffer zones around the major roads in the city (the road buffer hereafter), each with a width of 1000 m. The second and third buffer layers were created to delineate 10 buffer zones around the geometric center of Liwan District (the Liwan buffer hereafter) and Yuanchun Township of Tianhe District (the Yuanchun buffer hereafter), each having a width of 1000 m (Fig. 9.6). As discussed previously, proximity to the major roads and/or pollution centers has been shown to have an important impact on local air pollution patterns. Second, the urban map of 1989 (Fig. 9.5*c*) was overlaid with each buffer layer one by one to calculate the amount of urban/built-up land in each buffer zone. The density of urban/built-up land then was calculated by dividing the amount of urban/built-up land by the total land area in each zone. The 1989 urban map was used because it matched best the year (1992) when the data were collected to create the pollution maps. Finally, each of the concentration maps (SO_2, NO_x, and dust; Figs. 9.4*a*, 9.4*b*, and 9.4*d*) were overlaid with each of the buffer maps so that the concentration in each buffer zone could be calculated. The concentration map of TSPs (Fig. 9.4*c*) was excluded in these GIS analyses because insufficient data were obtained to cover the whole city proper.

For each pollutant, three correlation analyses were conducted between the densities of urban/built-up land in each buffer zone and the mean values of pollutant concentrations. Table 9.2 shows the results of the correlation analyses. For SO_2, a correlation coefficient of 0.33 was obtained (significance level 0.05) with the road buffer as the overlay layer, whereas correlation coefficients of 0.65 and 0.76 were found, respectively, with the Liwan buffer and the Yuanchun buffer as the overlay layer. This finding suggests that the spatial distribution of SO_2 concentration was significantly related to the distance from the pollution centers but showed little relation to distance from the major road network. The results of the correlation analysis for NO_x were different from those for SO_2. A correlation coefficient of 0.69 was observed with the road buffer zones as statistical units, indicating that the mapped pattern of NO_x was closely associated with that of

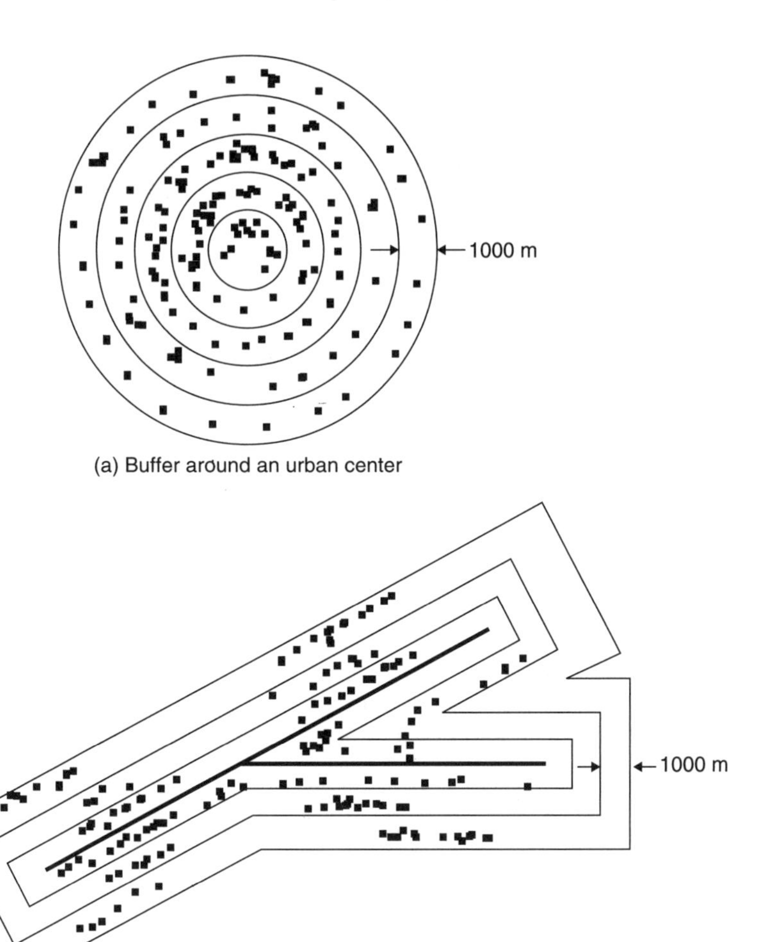

FIGURE 9.6 Diagram illustrating how the densities of urban/built-up land are calculated within the buffer zones using GIS.

Polutant	Densities of Urban/Built-up Use			LSTs	
	Road Buffer	Liwan Buffer	Yuanchun Buffer	1989 Map	1997 Map
SO_2	0.33	0.65	0.76	0.54	0.67
NO_x	0.69	0.39	0.42	0.68	0.72
Dust	0.53	0.64	0.68	0.36	0.44

TABLE 9.2 Correlation Coefficients between Pollutant Maps and Densities of Built-Up Use and LST Maps (Significant Level = 0.05)

sectoral urban expansion. A much weaker correlation was found with the Liwan buffer ($r = 0.39$) and the Yuanchun buffer ($r = 0.42$) as GIS overlays. These computations confirm our observation that NO_x was more a traffic-related rather than a built-up-related pollutant. Higher concentrations were observed along major radial transportation routes, such as the easterly development along the Guang-Shen Expressway and the Guang-Jiu Railway, the southerly development along the Guangzhou Da Road and Guang-Pan Highway, and the recent northwesterly and northeasterly development. Correlations between the densities of urban/built-up land and dust concentrations suggest that the pattern of dust pollution had a stronger linkage to pollution centers such as Liwan ($r = 0.64$) and Yuanchun ($r = 0.68$) than to the transportation network ($r = 0.53$). The concentrations of dust in the city conformed more to the pattern of concentric zonation and less to the pattern of sectoral radiation.

9.2.5 Urban Thermal Patterns and Air Pollution

A choropleth map (Fig. 9.7) was produced to show the spatial distribution of emissivity-corrected LSTs in 1989 and 1997. The statistics on LSTs from December 13, 1989 indicate that the lowest temperature was 16.98°C, the highest temperature was 25.23°C, and the mean was 21.17°C, with a standard deviation of 1.72. From Fig. 9.7 it becomes apparent that all the urban or built-up areas have a relatively high temperature. Some "hot spots" (the highest temperature class) can be clearly identified. In 1989, the most extensive hot spot was found in the western part of Haizhu District, an industrial region of the city. Another noticeable hot spot was detected in the southeast corner of the city proper, where the Guangzhou Economical and Technological Development Zone is located. There also were many smaller hot spots throughout the Tianhe District that are related to the sparsely distributed industries in that region. However, there were no extensive hot spots in the old urban areas such as Liwan, Yuexiu, and Dongshan Districts despite their high construction density. Apparently, commercial and residential areas are less effective in increasing LST. The lowest temperature class (16.98 to 19.45°C) appeared in the following three areas: (1) the eastern part of the Baiyun District around Maofeng Mountain, (2) Baiyun Hill, and (3) the southeastern part of the Haizhu District. These areas were substantially rural at one time and were covered mostly by forest. Both the northwestern Baiyun District and the western Fangcun District have a moderate temperature, ranging from 19.45 to 21.17°C, where cropland and dike-pond land prevail.

The spatial pattern of LSTs on August 29, 1997 is markedly different from that on December 13, 1989, as seen from Fig. 9.7. The difference reflects not only the differences in solar illumination, the state of vegetation, and atmospheric influences of the remotely sensed

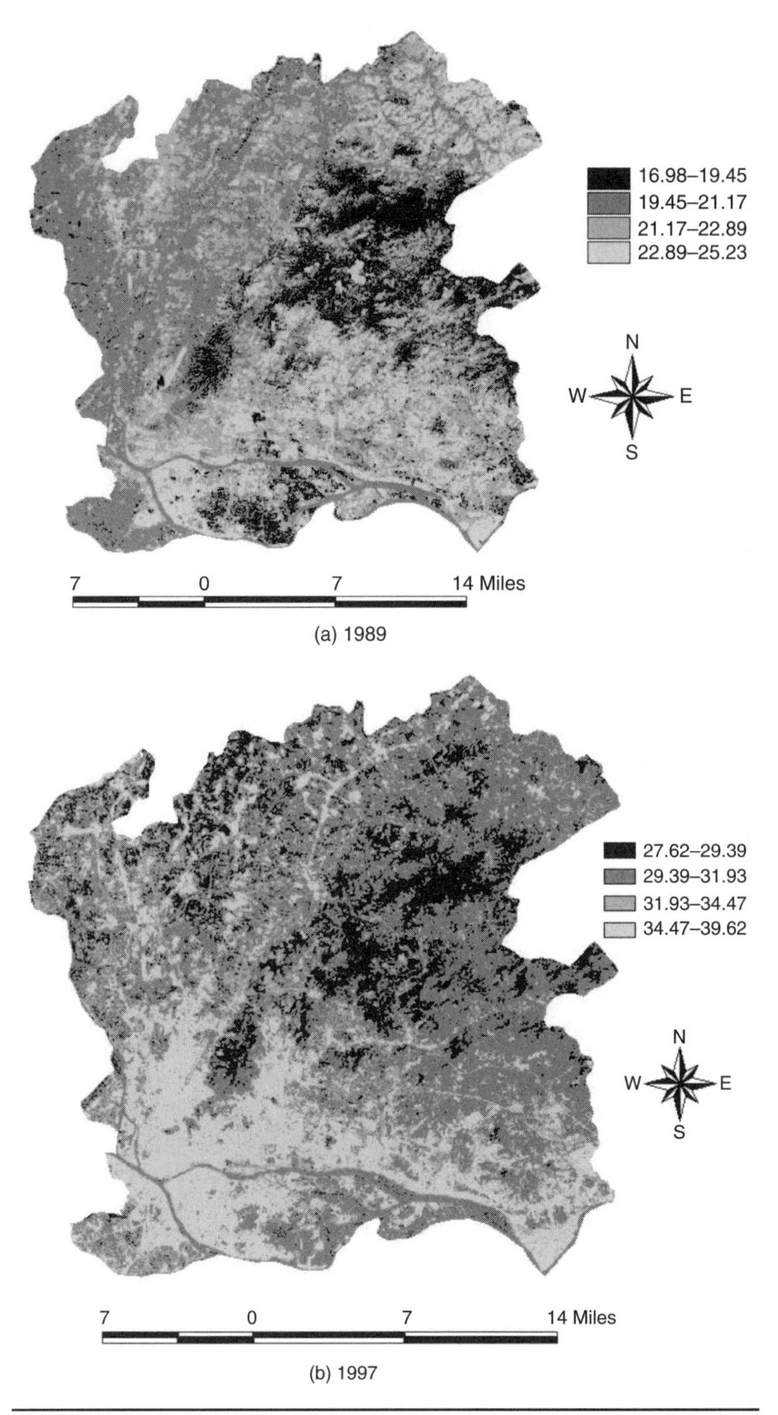

FIGURE 9.7 A choropleth map showing the geographic distribution of LSTs in 1989 and 1997. See also color insert.

Thematic Mapper dataset but also changes in LULC. The 1997 image was taken in the hottest month. The average temperature was 31.93°C, with a range between 27.62 and 39.62°C. The standard deviation also was larger (2.54) than that in 1989, indicating that the surfaces experienced a wider variation in LSTs. The urban and rural areas can be easily distinguished in Fig. 9.7. The urban areas showed a high temperature of over 34.47°C, whereas the rural settlements showed a minimal temperature of 31.93°C. A major hot spot expanded eastward from the urban core areas of Liwan, Yuexiu, and Dongshan Districts to Huangpu District, forming a high-temperature corridor. A major hot spot to the south of the Pearl River seemed to stretch out from the western Haizhu District to the Fangcun District. In addition, two large hot spots newly emerged, one centered at the Guangzhou Railroad Station and Baiyun International Airport and the other to the east of Baiyun Hill. Both areas had undergone a rapid urban sprawl since the 1990s. Numerous strip-shaped hot spots also were detectable along the northerly highways, such as the Guangzhou-Huaxian, Guangzhou-Huandong, and Guangzhou-Conghua highways.

The relationship between the LST maps and the three pollutant concentration maps (SO_2, NO_x, and dust) was investigated through a pixel-by-pixel correlation analysis after converting the concentration maps into raster format with a grid resolution of 30 m. The significance of each correlation coefficient was determined using a one-tail Student's *t* test. It was found that the concentrations of pollutants tended to positively correlate with LST values in both years but exhibited a stronger correlation with the 1997 (summer LST) map (see Table 9.2). The highest positive correlation was found for NO_x (0.72 with the 1997 LST map and 0.68 with the 1989 map). In both years, SO_2 exhibited a moderate correlation with LSTs (0.67 for 1997 and 0.54 for 1989). An even lower correlation was observed in the concentrations of dust in both years (0.44 in 1997 and 0.36 in 1989). This study has reconfirmed the existence of the relationship between pollutant measurements and LSTs, and the mechanisms of this correlation warrant further investigations. Previous studies suggested that the distinctive LST patterns are associated with the thermal characteristics of land-cover classes (Weng, 2003; Weng et al., 2004). Moreover, land-use zoning has been demonstrated to have a profound impact on the physical characteristics of urban land covers by imposing such restrictions as maximum building height and density and the extent of impervious surface and open space, land-use types, and activities. These restrictions would affect surface energy exchange, surface and subsurface hydrology, micro- to meso-scale weather and climate systems, and other environmental processes (Wilson et al., 2003). Apparently, knowledge about the interplay among LULC, air pollution, and thermal landscape should be integrated to assess the causes of the relationship between remote sensing-derived measurements and air pollution concentrations.

9.3 Summary

This study suggests that high-quality data and reliable information regarding LULC and LSTs can be derived with satellite remote sensing and GIS technologies and that the data so derived closely correspond to ambient air-quality measurements. The spatial patterns of air pollution are subject to the influence of many factors related to land-use activities, such as the division of functional districts, the distribution of land-use types, the occurrence of water bodies and parks, building and population densities, the layout of the transportation network, and air flushing rates. In the Guangzhou context, the impact of these factors varied with different pollutants, and the street canyon effect was particularly evident owing to Guangzhou's closely spaced high-rise buildings and low-wind environment. Because of the locations of industrial plants, high population density, clustering of the catering industry, and low air-flushing rates, two urban localities (namely the Liwan and Yuanchun Districts) became the pollution hubs for SO_2, dust, and other pollutants. The study demonstrates that GIS is effective in examining the spatial pattern of air pollution and its association with urban built-up density. Positive correlation between the concentrations of the pollutants probed and satellite-derived LST values, particularly with the summer LST map, indicates that both ambient air quality and LSTs were associated with land use.

9.4 Remote Sensing–GIS Integration for Studies of Urban Environments

The relationships and dynamics among urban environmental variables are apparently complex given the fact that they involve multiple variables and may vary with the biophysical and socioeconomic settings of a city under investigation. Future studies of these interplays are warranted, noting that it may occur on different spatial and temporal scales. The accuracy of pollutant maps is of particular significant in a type of analysis similar to the case study because they would influence the correlation analyses between the pollutants and urban built-up densities/LSTs. In this study, the maps were produced by the city's environmental monitoring authority based on data collected in the fixed and mobile observations. Contours were drawn by interpolation of point data that represented annual averages of pollutant measurements. The contours were digitized manually as vector GIS data layers. To examine the relationship between the pollutants and urban built-up densities, a vector GIS analysis was conducted after computing the densities of urban/built-up within buffer zones. On the other hand, in analyzing the relationship between the pollutants and surface temperatures, a raster GIS approach was applied by converting the pollutant maps into the raster format and conducting a pixel-by-pixel correlation analysis. Data conversion between the vector and raster formats tends to introduce error. This error will be augmented in the process of analysis, where two or more data layers are used. Error-propagation modeling and sensitivity analysis should be conducted in future research of a similar type.

References

Anderson, J. R., Hardy, E. E., Roach, J. T., and Witmer, R. E. 1976. A Land Use and Land Cover Classification Systems for Use with Remote Sensing Data, USGS Professional Paper, U.S. Geological Survey, Washington.

Artis, D. A., and Carnahan, W. H. 1982. Survey of emissivity variability in thermography of urban areas. *Remote Sensing of Environment* **12**, 313–329.

Balling, R. C., and Brazell, S. W. 1988. High resolution surface temperature patterns in a complex urban terrain. *Photogrammetric Engineering and Remote Sensing* **54,** 1289–1293.

Brivio, P. A., Gempvese, G., Massari, S., et al. 1995. Atmospheric pollution and satellite remotely sensed surface temperature in metropolitan areas. In *EARSeL Advances in Remote Sensing Pollution Monitoring and Geographical Information Systems,* pp. 40–46. Paris: EARSeL.

Byrne, G. F. 1979. Remotely sensed land cover temperature and soil water status: A brief review. *Remote Sensing of Environment* **8,** 291–305.

Carnahan, W. H., and Larson, R. C. 1990. An analysis of an urban heat sink. *Remote Sensing of Environment* **33,** 65–71.

Carlson, T. N., Gillies, R. R., and Perry, E. M. 1994. A method to make use of thermal infrared temperature and NDVI measurements to infer surface soil water content and fractional vegetation cover. *Remote Sensing Review* **9,** 161–173.

DeWitt, J., and Brennan, M. 2001. Taking the heat. *Imaging Notes* **16,** 20–23.

Gallo, K. P., and Owen, T. W. 1998. Assessment of urban heat island: A multi-sensor perspective for the Dallas–Ft. Worth, USA region. *Geocartography International* **13,** 35–41.

Gallo, K. P., McNab, A. L., Karl, T. R., et al. 1993. The use of NOAA AVHRR data for assessment of the urban heat island effect. *Journal of Applied Meteorology* **32,** 899–908.

Gibbons, D. E., and Wukelic, G. E. 1989. Application of Landsat Thematic Mapper data for coastal thermal plum analysis at Diablo Canyon. *Photogrammetric Engineering and Remote Sensing* **55,** 903–909.

Gillies, R. R., and Carlson, T. N. 1995. Thermal remote sensing of surface soil water content with partial vegetation cover for incorporation into climate models. *Journal of Applied Meteorology* **34,** 745–756.

Goward, S. N., Xue, Y., and Czajkowski, K. P. 2002. Evaluating land surface moisture conditions from the remotely sensed temperature/vegetation index measurements: An exploration with the simplified simple biosphere model. *Remote Sensing of Environment* **79,** 225–242.

Guangdong Statistical Bureau. 1999. *Guangdong Statistical Yearbook.* Beijing: China Statistics Press (in Chinese).

Guangzhou Bureau of City Planning. 1996. *Master Plan of Guangzhou City, 1991–2010 (Guangzhou Shi Chengshi Zongti Guihua).* Guangzhou, China: Guangzhou Bureau of City Planning (in Chinese).

Guangzhou City Local Chronicle Compilation Committee (GZCLCCC). 1995. *Gazetteer of Guangzhou,* Vol. 3. Guangzhou, China: Guangzhou Press (in Chinese).

Guangzhou Municipality Planning Committee. 1997. *Atlas of Guangzhou Natural Resources.* Guangzhou, China: Guangdong Province Atlas Publishing House (in Chinese).

Guangzhou Statistical Bureau. 1999. *Fifty Years in Guangzhou, 1949–1999.* Beijing: China Statistics Press (in Chinese).

Guo, H. 2001. *Land* use and land cover changes and environmental impacts in Guangzhou. Master thesis, Department of Geography, South China Normal University, Guangzhou, China (in Chinese).

Kidder, S. Q., and Wu, H. T. 1987. A multispectral study of the St. Louis area under snow-covered conditions using NOAA-7 AVHRR data. *Remote Sensing of Environment* **22,** 159–172.

Kim, H. H. 1992. Urban heat islands. *International Journal of Remote Sensing* **13,** 2319–2336.

Malaret, E., Bartolucci, L. A., Lozano, D. F., et al. 1985. Landsat-4 and Landsat-5 Thematic Mapper data quality analysis. *Photogrammetric Engineering and Remote Sensing* **51,** 1407–1416.

Markham, B. L., and Barker, J. K. 1985. Spectral characteristics of the Landsat Thematic Mapper sensors. *International Journal of Remote Sensing* **6,** 697–716.

Marsh, W. M., and Grossa, J. M., Jr. 2002. *Environmental Geography: Science, Land Use, and Earth Systems,* 2nd ed. New York: Wiley.

Nichol, J. E. 1994. A GIS-based approach to microclimate monitoring in Singapore's high-rise housing estates. *Photogrammetric Engineering and Remote Sensing* **60,** 1225–1232.

Nichol, J. E. 1996. High-resolution surface temperature patterns related to urban morphology in a tropical city: A satellite-based study. *Journal of Applied Meteorology* **35,** 135–146.

Oke, T. R. 1982. The energetic basis of the urban heat island. *Quarterly Journal of the Royal Meteorological Society* **108,** 1–24.

Poli, U., Pignatoro, F., Rocchi, V., and Bracco, L. 1994. Study of the heat island over the city of Rome from Landsat TM satellite in relation with urban air pollution. In *Remote Sensing: From Research to Operational Applications in the New Europe,* Edited by R. Vaughan, pp. 413–422. Berlin: Springer-Hungarica.

Qian, Z., Zhang, J., Wei, F., et al. 2001. Long-term ambient air pollution levels in four Chinese cities: Inter-city and intra-city concentration gradients for epidemiological studies. *Journal of Exposure Analysis and Environmental Epidemiology* **11,** 341–351.

Quattrochi, D. A., and Luvall, J. C. 1999. High Spatial Resolution Airborne Multispectral Thermal Infrared Data to Support Analysis and Modeling Tasks in the EOS IDS Project Atlanta, Global Hydrology and Climate Center, NASA, Huntsville, AL; available at *www.ghcc.msfc.nasa.gov/atlanta/* (last date accessed June 26, 2005).

Quattrochi, D. A., and Ridd, M. K. 1994. Measurement and analysis of thermal energy responses from discrete urban surfaces using remote sensing data. *International Journal of Remote Sensing* **15,** 1991–2022.

Roth, M., Oke, T. R., and Emery, W. J. 1989. Satellite derived urban heat islands from three coastal cities and the utilisation of such data in urban climatology. *International Journal of Remote Sensing* **10,** 1699–1720.

Streutker, D. R. 2002. A remote sensing study of the urban heat island of Houston, Texas. *International Journal of Remote Sensing* **23,** 2595–2608.

Streutker, D. R. 2003. Satellite-measured growth of the urban heat island of Houston, Texas. *Remote Sensing of Environment* **85,** 282–289.

Voogt, J. A., and Oke, T. R. 2003. Thermal remote sensing of urban climate. *Remote Sensing of Environment* **86,** 370–384.

Wang, S., Shao, M., and Zhang, Y. 2001. Influence of fuel quality on vehicular NO_x emissions. *Journal of Environmental Science* **13,** 265–271.

Ward, L., and Baleynaud, J.-M. 1999. Observing air quality over the city of Nates by means of Landsat thermal infrared data. *International Journal of Remote Sensing* **20,** 947–959.

Weng, Q. 2001. A remote sensing–GIS evaluation of urban expansion and its impact on surface temperature in the Zhujiang Delta, China. *International Journal of Remote Sensing* **22,** 1999–2014.

Weng, Q. 2003. Fractal analysis of satellite-detected urban heat island effect. *Photogrammetric Engineering and Remote Sensing* **69,** 555–566.

Weng, Q., and Yang, S. 2006. Urban air pollution patterns, land use, and thermal landscape: An examination of the linkage using GIS. *Environmental Monitoring and Assessment* **117,** 463–489.

Weng, Q., Qiao, L., Yang, S., and Guo, H. 2003. Guangzhou's growth and urban planning, 1960–1997: An analysis through remote sensing. *Asian Geographer* **22,** 77–92.

Weng, Q., Lu, D., and Schubring, J. 2004. Estimation of land surface temperature-vegetation abundance relationship for urban heat island studies. *Remote Sensing of Environment* **89,** 467–483.

Wilson, J. S., Clay, M., Martin, E., et al. 2003. Evaluating environmental influences of zoning in urban ecosystems with remote sensing. *Remote Sensing of Environment* **86,** 303–321.

Xie, M., and Chen, N. 2001. Changes in Guangzhou's air pollution and strategies for control them based on an analysis of changes in the sanitary district. Unpublished paper, Guangzhou Municipality Environmental Monitoring Central Station, Guangzhou, China.

Xie, S., Zhang, Y., Li, Q., and Tang, X. 2003. Spatial distribution of traffic-related pollution concentrations in street canyons. *Atmospheric Environment* **37,** 3213–3224.

Xu, J. 1990. *A Collection of Essays on Lingnan Historical Geography* (*Lingnan Lishi Dili Lunji*). Guangzhou, China: Zhongshan University Press (in Chinese).

Xu, X. 1985. Guangzhou: China's southern gateway. In *Chinese Cities: The Growth of the Metropolis since 1949,* edited by Victor F. S. Sit. Hong Kong: Oxford University Press.

Zhang, Y., Xie, S., Zeng, L., and Wang, H. 1999. Traffic emission and its impact on air quality in Guangzhou area. *Journal of Environmental Science* **11,** 355–360.

CHAPTER 10

Population Estimation

The large world population has produced great pressures on global resources, the environment, and sustainable development (Lo, 1986a; Sutton et al., 1997). The pressure from population increase often results in urban expansion at the expense of decreased nonurban lands such as agricultural land and forest. Timely and accurate population estimation and the spatial distribution of population and its dynamics become considerably significant in understanding the effects of population increase on social, economic, and environmental problems. Moreover, population information at different levels, such as national, regional, and local, is very important for many purposes, such as urban planning, resource management, service allocation, and so on. Conventional census methods of population estimation are time-consuming, costly, and difficult to update. Besides, the census interval is often too long for many types of applications; for example, the U.S. Census is conducted every 10 years. Owing to large migrations, population distributions can change quickly; thus Census data are frequently found to be obsolete. Therefore, it is necessary to develop suitable techniques for estimating population in an accurate and timely manner on different spatial scales.

Geographic research of population estimation started as early as in the 1930s. John K. Wright, a geographer working at the American Geographical Society, pioneered in population-estimation study by producing a map of the population distribution of Cape Cod, Massachusetts (Wright, 1936). Wright termed his method *dasymetric mapping,* in which the breaks in the population-distribution map were related to types of land use. With the advent of mature geographic information systems (GIS) technology, some applications suggest that Wright's seminal work can be applied to areal interpolation (Flowerdew and Green, 1992) may be suitable for statistically modeling a wide range of phenomena, including population. Following the idea of dasymetric mapping and implemented through use of the pycnophylactic interpolation method, the National Center for Geographic Information and Analysis created global raster images of population distribution for a set of 15,000 administrative units in the Global Demography Project (Tobler et al., 1995). Remote sensing techniques have been used for population estimation since the 1950s, when Porter (1956) estimated population in a settlement in Liberia by counting the number of huts on aerial photographs and by multiplying that number by mean occupants per hut derived from ground sampling survey. With advances in remote sensing and GIS technology, remotely sensed data have become an important resource in population estimation owing to their strengths in data coverage, reasonable accuracy, and low cost (Jensen and Cowen, 1999; Lo, 1995). Different methods have been developed to estimate population based on aerial photographs and satellite imagery (Harvey, 2002a, 2002b; Langford et al., 1991; Lo, 1986a, 1986b, 1995, 2001; Lo and Welch, 1977; Qiu et al., 2003; Sutton, 1997; Sutton et al., 1997, 2001; Watkins and Morrow-Jones, 1985; Weeks et al., 2000). Satellite-based variables have been combined recently with other geographic variables to produce a comparatively high-resolution (30 arc sec) population database for the entire globe (Dobson, 2003; Dobson et al., 2000, 2003). This chapter explores the potential of integration of Landsat Enhanced Thematic Mapping Plus (ETM+) data with U.S. Census data for estimation of population density at the level of block groups based on a case study in the city of Indianapolis, Indiana.

10.1 Approaches to Population Estimation with Remote Sensing–GIS Techniques

Population/housing estimation using remotely sensed data has attracted increasing interest since 1960s. Scientists have made great efforts to develop techniques suitable for population/housing estimation using remotely sensed data (Harvey, 2002a, 2002b; Langford et al., 1991; Lo, 1986a, 1986b, 1995, 2001; Lo and Welch, 1977; Sutton et al., 1997, 2001). Four categories of population-estimation methods were summarized (Table 10.1), including (1) measurement of built-up areas, (2) counts of dwelling units, (3) measurement of different land-use areas, and (4) spectral radiance of individual pixels.

10.1.1 Measurements of Built-Up Areas

This method was based on the allometric growth model, which Huxley (1932) described initially as the relationship between growth of part of an organ and the whole organism. Then Nordbeck (1965) and Tobler (1969) introduced this method into population estimation. The relationship between a built-up area of settlement and population size can be expressed as $r = aP^b$, where r is the radius of a circle of settlement, a is a coefficient, P is population size of that settlement,

Method	Data Sets Used	Advantages	Disadvantages	References
Measurements of built-up areas	Aerial photographs, Landsat images, and low-resolution DMSP images, radar images	Can estimate population at different scales, such as national and global	Difficult to accurately delineate urban or built-up areas	Lo, 2001; Lo and Welch, 1977; Sutton, 1997; Sutton et al., 1997, 2001; Wellar, 1969
Counts of dwelling units	High-resolution images such as aerial photographs	Can estimate population with high accuracy at a local level	Not suitable for large areas owing to efficiency considerations; difficult to distinguish high-rise apartment buildings from multistory office buildings	Hsu, 1971; Lindgren, 1971
Measurement in different land-use areas	Aerial photographs and medium-resolution images such as Landsat and SPOT	Can provide population estimation on different scales with reasonable accuracy	Accuracy depends on the classification results; the presence of multistoried housing affects accuracy	Langford et al., 1991; Lo, 1995
Digital image analysis	Landsat MSS, TM, ETM+, and SPOT	Analyze images and implement easily; suitable for local population estimation; model is simple and robust	Selection of remote sensing variables arbitrary; models difficult to transfer to other image scenes; reference data at pixel level not available; suitable techniques needed to disaggregate demographic data	Harvey, 2002a, 2002b; Lisaka and Hegedus, 1982; Li and Weng, 2005, 2006; Lo, 1995

TABLE 10.1 Summary of Population-Estimation Methods in the Literature

and b is an exponent. The population size can be computed by measuring the area of a settlement using photography or imagery. Nordbeck (1965) developed a model ($A = 0.0015P^{0.88}$) for U.S. cities, where A is the area measured in square miles. A similar research project was conducted in Houston and San Antonio, Texas, using Gemini XII photographs to measure the areas of 10 settlements (Wellar, 1969). The accuracy of population estimation was found to be higher for smaller settlements with populations of fewer than 10,000 people than for large settlements. Lo and Welch (1977) also used the allometric growth model to estimate population in Chinese cities. They modified the preceding equation to $P = aA^b$, where P is estimated population and A is built-up area of settlement. A correlation coefficient of 0.75 was obtained based on census data in 1953 and the built-up areas of 124 cities measured from maps, and a model $P = 74{,}696A^{0.7246}$ was developed. Afterwards, this model was applied to estimate population of 13 cities in 1972 through 1974 using Landsat Multispectral Scanner (MSS) image. It was found that population was underestimated when it was greater than 2.5 million in large cities.

Holz and colleagues (1973) developed a complicated regression model to explain the relationship between population and land area by taking roads and other urban areas into account:

$$P_i = a + b_1L_i + b_2P_j - b_3D_{ij} + b_4A_i \tag{10.1}$$

where P_i = the population of urban area i
L_i = the number of direct roads L between i and the other urban area
P_j = the population of the nearest large urban area j
D_{ij} = the highway distance between urban area i and the nearest larger urban area j
A_i = the observable occupied dwelling area of urban area i

L_i and A_i were extracted from large-scale photographs, and then stepwise linear regression was used. Two models were obtained for 40 urban centers in the Tennessee Valley in 1953 and 1963 with coefficients of correlation of 0.95 and 0.88, respectively. Ogrosky (1975) improved the regression model using infrared aerial photography at a scale of 1:135,000 and achieved a higher correlation (0.973) between population and the logarithm of an image area classified as urban in the Puget Sound Region.

In addition to aerial photograph and Landsat imagery, low-resolution nighttime images (2.7 km) from the Defense Meteorological Satellite Program (DMSP) also were used to map human settlements (Elvidge et al., 1995, 1997) and urban extent (Imhoff et al., 1997) and to estimate population nationally and globally (Lo, 2001; Sutton, 1997; Sutton et al., 1997, 2001; Welch and Zupko, 1980). For example, Welch and Zupko

(1980) used DMSP images acquired on February 15, 1975, to study quantitative relationships between nighttime lights and population and between nighttime lights and energy use in 35 cities in the United States. The mean volumes of illuminated urban-area domes of individual cities were found to be strongly correlated with energy consumption (correlation coefficient 0.89) and with population (correlation coefficient 0.96) using the model $r = aP^b$. Sutton and colleagues (1997) compared gridded vector population-density data derived from the 1992 U.S. Census block-group level and DMSP imagery of the continental United States. A strong correlation between DMSP nighttime imagery and human population density was found at a range of spatial scales, including aggregation to state and county levels. The areas of saturated clusters are strongly correlated with populations with a coefficient R^2 of 0.63. Lo (2001) used radiance-calibrated DMSPOLS nighttime lights data acquired between March 1996 and January and February 1997 to model population in China at the provincial, county, and city levels. The allometric growth models $P = aA^b$ were developed based on light areas and responding population. Meanwhile, linear regression models, $PD = a + bX$, also were established based on light intensity [digital number (DN) value per pixel, where PD is population density and X is light density. It was found that DMSP images could provide reasonably accurate predictions of urban population at provincial and city levels using the allometric growth model, where light volume was used as the independent variable. The linear regression model is most suitable for estimation of urban population density, where light density was used as independent variable. Sutton and colleagues (2001) also used DMSP OLS images to estimate the population of all cities of the world based on areal extension in the images employing the allometric growth model. They identified 22,920 urban clusters on DMSP OLS images. They measured the areal extents of those clusters by counting the number of contiguously lighted pixels in the image for each isolated urban area and used 1383 clusters to develop a regression model to estimate the urban population of every nation in the world. Based on information about percent of population in urban areas for every nation, the population of every nation was estimated. Then the estimated population of every nation was aggregated to a total global population. A total of 6.3 billion people was estimated in the world. The DMSP nighttime images provided an inexpensive means for mapping population size and spatial distribution of the human population.

10.1.2 Counts of Dwelling Units

This method was regarded as the most accurate remote sensing method (Forester, 1985; Haack et al., 1997; Holz, 1988; Jensen and Cowen, 1999; Lindgren, 1985; Lo; 1986a, 1986b, 1995). The method assumes that (1) the imagery used has sufficiently high spatial resolution to identify the types of individual buildings, (2) the average number of persons per dwelling unit is available, (3) the number of homeless and seasonal and migratory workers can be estimated, and (4) all dwelling units are occupied. Hsu (1971) used aerial photography at a scale of 1:5000 to estimate and map the population distribution of 1963 for Atlanta, Georgia. In this research, it was assumed that a dwelling unit was occupied by one household (in fact, this assumption may not be valid in urban environment). The number of persons per household was obtained from Census statistics. The number of dwelling units in a grid of 0.25 mi^2 was counted from aerial photography. The population density was computed using following formula:

$$\text{Population density} = (\text{persons per household} \times \text{number of dwelling units})/\text{grid-cell area}$$

Similar research was conducted using color infrared photography at medium scale (1:20,000) to improve the results of dwelling-unit estimation for Boston metropolitan area (Lindgren, 1971). A correct identification of 99.5 percent of the residential structures was achieved. The main difficulty using this approach is in distinguishing high-rise apartment buildings from multistory office buildings.

10.1.3 Measurement of Different Land-Use Areas

This method involves classification of remotely sensed images into different land-use categories and focuses on residential areas, which are further subdivided into different classes according to the cultural characteristics of the study area (Lo, 1986). Kraus and colleagues (1974) used this approach to estimate the population of four cities in California (Fresno, Bakersfield, Santa Barbara, and Salinas) with panchromatic photography, color infrared photography, and Census block data. Thompson (1975) refined this method by using base population and residential land-use changes from aerial photography. Langford and colleagues (1991) used a Landsat Thematic Mapper (TM) image covering 49 wards of Leicestershire, United Kingdom, to estimate population. The TM image first was classified into five land-use classes using principal components analysis (PCA) and supervised classification. Then the pixels of each category within each ward were counted in aid of ERDAS IMAGINE, and a correlation between population and land-cover pixel counts was computed. It was found that ward population had a relatively high positive correlation with the number of pixel in industry, commerce, dense residential, and ordinary residential categories, respectively, and had low negative correlations with those in areas of no population and agriculture. Webster (1996) developed models to estimate dwelling densities in the 47 suburbs of Harare, Zimbabwe, based on measures of tones (six TM bands), measures of texture (three measures derived from classification of pixels into urban and nonurban: urban pixel density, homogeneity, and entropy), and measures of context (distance from the city center) using SPOT and TM images. The R^2 values ranged from 0.69 to 0.81. Chen (2002) studied the relationship between areal census dwelling data and residential densities classified from a Landsat TM image covering 13 census collection districts (CD) in Hornsby Heights, Sydney, Australia. First, three residential density levels were identified using a combination of a texture statistic and six bands; then the correlations between areal census data and residential densities classified were tested. It was found that correlations between areal census dwelling data and areas of different residential densities were higher than those between areal census dwelling data and aggregated area of a whole residential area.

10.1.4 Spectral Radiance

Hsu (1973) proposed the potential use of Landsat MSS multispectral radiance data cell by cell (1 × 1 km) for population estimation through a multiple regression model using ground-truth data and low-altitude aerial photography. Iisaka and Hegedus (1982) studied population distribution in residential sections of suburban Tokyo, Japan, using MSS data. Of the grid cells (500 × 500 m), 88 were selected, representing a wide range of population-density values. Two multiple linear regression models were developed in which population was used as a dependent variable, and mean reflectance values of four MSS bands were calculated over the 10 × 10 pixel-grid squares were used as explanatory variables. Correlation coefficients of 0.77 and 0.899 for 1972 and 1979 were obtained, respectively. It was observed that correlation between building density and population density in the central business district (CBD) was very weak.

Lo (1995) used two approaches, including spectral radiance values of image pixels and counts of pixels in residential classes, to estimate population and dwelling-unit densities in 44 tertiary-planning units (TPUs) in Kowloon, Hong Kong, employing multispectral SPOT imagery. Five different regression models were developed for estimation of population and dwelling densities using 12 TPUs. In four cases, the models were linear, and the dependent variable was population or dwelling density. The independent variables were means of SPOT bands 1, 2, and 3; mean of SPOT band 3 alone; percentages of pixels classified as high- and low-density residential use in each TPU; and proportion of high-density residential-use pixels in each TPU. In the fifth case, the model was the allometric growth model, the dependent variable was population or dwelling counts, and the independent variables were the number of pixels in the high-density residential class. The models were validated by applying them to 44 TPUs. It was found that the allometric growth model was best on a macro scale. On a micro scale, estimation accuracy was not satisfactory owing to highly mixed land use and difficulty in distinguishing residential from non-residential use.

Harvey (2002a) refined the method used by Iisaka and Hegedus (1982) and Lo (1995) by introducing a number of standard spectral transformations (squares of 6 basic band means, 15 band-mean to band-mean cross-products, 15 pairwise band-to-band ratios, and 15 pairwise difference-to-sum ratios of the TM data) into regression models for population estimation in Ballarat, Sydney, Australia, using Landsat TM images. In this study, 132 collection districts (CDs) of Ballarat were used to develop the models. The dependent variable was population density of each CD or its logarithmic and square-root transformation. A number of models were established through stepwise regression analysis. The results showed that incorporation of spectral transformations and application of either the square-root or the logarithmic transformation to population density increased the correlation coefficient. Of these models, six were validated by being applied to a nearby culturally and demographically similar area, Geelong. Three of the most complex models produced median proportional errors for the population of individual CDs with a range of 17 to 21 percent. Median proportional errors for the total population of Ballarat were within 3 percent. When these models were applied to the Geelong area, the R^2 decreased and median proportional errors increased. Similar to other estimation methods, all these models overestimated population in low-density rural sections and underestimated them in high-density urban sections.

In another study, Harvey (2002b) used a new method based on individual TM pixels for Ballarat and Geelong, Australia. The TM image was first classified into residential and non-residential classes using a supervised maximum likelihood classifier. The initial ground census populations of CDs were assigned to each pixel uniformly by using the formula $P_i = P/n$, where n is the number of pixels seen as residential within a CD and P is the population in a CD. An expectation-maximization (EM) algorithm was used to iteratively regress P_i against the spectral indicators such as means of TM bands and band ratio, band-difference to band-sum ratio, hue transformation, and spatial standard deviations of hue and reestimate of pixel population. By comparing these models with previous models (Harvey, 2002a), the estimation accuracy based on pixels for extremes of population density, which usually were over- or underestimated, was much improved. In addition, the models based on pixels were more robust; that is, when applying these models to a second image, the estimation accuracy also increased.

10.2 Case Study: Population Estimation Using Landsat ETM+ Imagery

10.2.1 Study Area and Datasets

The city of Indianapolis, Indiana, has been chosen to implement this study. A detailed description and a map of the study area can be found in Sec. 3.2.1 of Chap. 3. A landsat 7 ETM+ image (row/path: 32/21) dated June 22, 2000, was used in this research. Atmospheric conditions were clear at the time of image acquisition, and the image was acquired through the U.S. Geological Survey's (USGS's) Earth Resource Observation Systems Data Center, which had corrected the radiometric and geometric distortions of the image to a quality level of 1G before delivery.

Population data at block group level were obtained from an ESRI Data and Maps CD, which was provided by the ESRI Company, based on a combination of topologically integrated geographic encoding and referencing (TIGER) files and 2000 U.S. Census population data. Because of different coordinate systems used for the Census data and the ETM+ image, the geographic coordinates of the Census data were converted to the Universal Transverse Mercator (UTM) to match with those of the ETM+ image.

10.2.2 Methods

Principal Component Analysis

Remotely sensed data, such as visible bands in Landsat TM/ETM+ images, are highly correlated between the adjacent spectral bands (Barnsley, 1999). Several techniques have been developed to transform highly correlated bands into an orthogonal subset. PCA is used most commonly. After performing PCA, the original correlated bands are transformed into independent principal components (PCs), of which the first PC contains the largest portion of data variance, and the second PC contains the second largest data variance, and so on. The higher-numbered PCs often appear noisy because they contain very little variance of information (Richards, 1994). In this study, six ETM+ multispectral bands (1 to 5 and 7) were used to perform PCA. The first three PCs were used in population-estimation analysis because they accounted for 99 percent of total variance.

Vegetation Indices

Many vegetation indices have been developed based on the fact that plants reflect less in visible red light but more in near-infrared radiation than nonvegetated surfaces (Bannari et al., 1995; Jensen, 2000). Thus vegetation indices can enhance or extract some specific features

Vegetation Index	Abbr.	Formula	References
Normalized difference vegetation index	NDVI	$\frac{NIR-RED}{NIR+RED}$	Rouse et al., 1974
Soil adjusted vegetation index	SAVI	$\frac{(1+L)(NIR-RED)}{NIR+RED+L}$, $L = 0.5$	Huete, 1988
Renormalized difference vegetation index	RDVI	$\frac{NIR-RED}{\sqrt{NIR+RED}}$	Roujean and Breon, 1995
Transformed NDVI	TNDVI	$\sqrt{NDVI+0.5}$	Deering et al., 1975
Simple vegetation index	SVI	NIR – RED	
Simple ratio	RVI	NIR/RED	Birth and McVey, 1968

Note: NIR = near-infrared wavelength, ETM+ band 4; RED = red wavelength, ETM+ band 3.

TABLE 10.2 Definition of Vegetation Indices Used

that single spectral bands cannot. In this research, six vegetation indices, namely, the normalized difference vegetation index (NDVI), the soil adjusted vegetation index (SAVI), the renormalized difference vegetation index (RDVI), the transformed NDVI (TNDVI), the simple vegetation index (SVI), and the simple ratio (RVI), were examined to use for population estimation (Table 10.2).

Fraction Images

Spectral mixture analysis (SMA) is regarded as a physically based image processing tool that supports repeatable and accurate extraction of quantitative subpixel information (Mustard and Sunshine, 1999; Roberts et al., 1998; Smith et al., 1990). It assumes that the spectrum measured by a sensor is a linear combination of the spectra of all components within the pixel (Adams et al., 1995; Roberts et al., 1998). Because of its effectiveness in handling spectral mixture problems, SMA has been used widely in estimation of vegetation cover (Asner and Lobell, 2000; McGwire et al., 2000; Small, 2001; Smith et al., 1990), in vegetation or land-cover classification and change detection (Adams et al., 1995; Aguiar et al., 1999; Cochrane and Souza, 1998; Lu et al., 2003; Roberts et al., 1998), and in urban studies (Phinn et al.,

2002; Rashed et al., 2001; Small, 2001; Wu and Murray, 2003). In this study, SMA was used to develop green-vegetation and impervious-surface fraction images. Endmembers were identified initially from the ETM+ image based on high-resolution aerial photographs. The shade endmember was identified from the areas of clear and deep water, whereas green vegetation was selected from the areas of dense grass and cover crops. Different types of impervious surfaces were selected, from building roofs to highway intersections. An unconstrained least-squares solution was used to decompose the six ETM+ bands (1 through 5 and 7) into three fraction images (e.g., vegetation, impervious surface, and shade). The fractions represent the areal proportions of the endmembers within a pixel. The shade fraction was not used owing to its irrelevance to the population distribution. A detailed description of this procedure can be found in Lu and Weng (2004).

Texture Images

Texture often refers to the pattern of intensity of variations in an image. Many texture measures have been developed (Haralick, 1979; Haralick et al., 1973; He and Wang, 1990) and used for land-cover classification (Gong and Howarth, 1992; Marceau et al., 1990; Narasimha Rao *et al.*, 2002; Shaban and Dikshit, 2001). A common texture measure, variance, has been shown to be useful in improving land-cover classification (Shaban and Dikshit, 2001). In this study, variance was developed and used to examine its relationship with population. Landsat ETM+ bands 3 and 7, which correlate strongly with urban features, were used for deriving texture images with window sizes of 3×3, 5×5, and 7×7.

Temperature

A surface-temperature image was extracted from the ETM+ thermal infrared band (band 6). The procedure to develop the surface temperature involves three steps: (1) converting the digital number of ETM+ band 6 into spectral radiance, (2) converting the spectral radiance to at-satellite brightness temperature, which is also called *blackbody temperature,* and (3) converting the blackbody temperature to land surface temperature. A detailed description of how the temperature image was developed can be found in Weng and colleagues (2004).

Model Development

Since Census data and ETM+ data have different formats and spatial resolutions, they need to be integrated. With the help of ERDAS IMAGINE, remotely sensed data were aggregated to block-group level. The mean values of selected remote sensing variables at the block-group level were computed. The variables include radiances of ETM+ bands, principal components, vegetation indices, green-vegetation and impervious surface fractions, temperatures, and texture indicators. All these data then were exported into SPSS software for correlation and regression analysis.

Twenty-five percent of the total block groups (658) in the city were randomly selected. A 2.5 standard deviation was used to identify the outliers. A total of 162 samples was used for developing models with a non-stratified sampling scheme. The population density in Indianapolis was calculated to range from 0 to 7253 persons/km^2, whereas most block groups had a population density that ranged from 400 to 3000 persons/km^2 (Fig. 10.1).

Previous research has indicated that extremely high or low population density is difficult to estimate using remotely sensed data

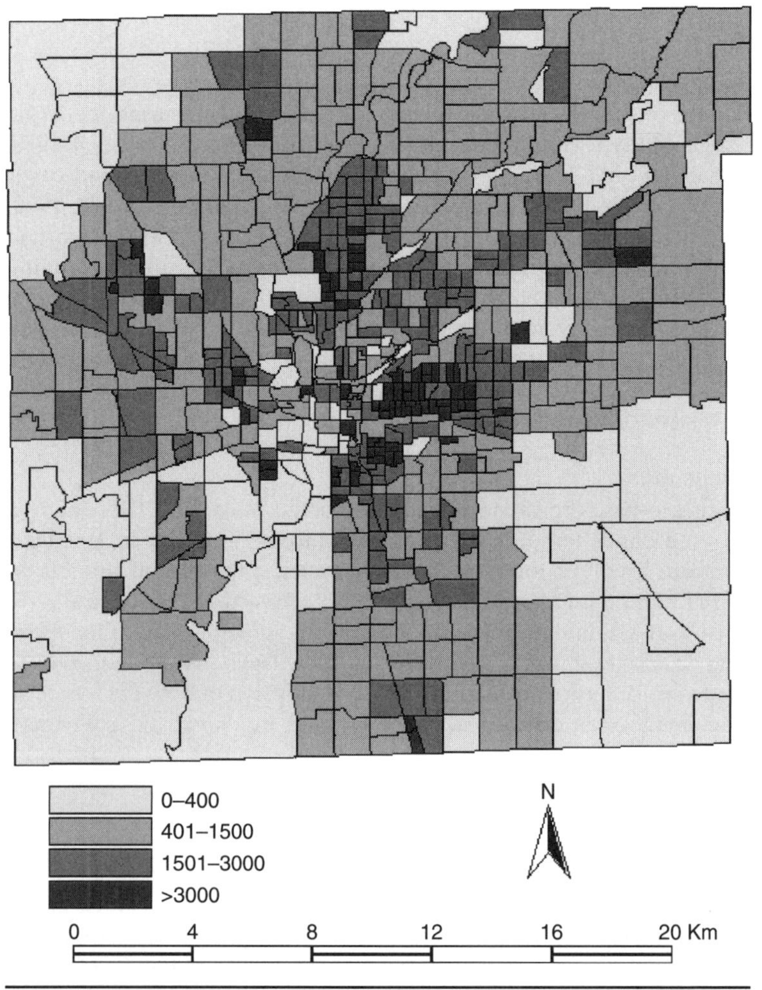

Figure 10.1 Population-density distribution by block groups in Indianapolis based on the 2000 Census.

Category	Samples	Min.	Max.	Mean	SD
Non-stratified	162* (175†) (658‡)	8	4479	1470.71	948.62
Low	77* (82‡)	1	393	208.94	123.11
Medium	114* (125†) (499‡)	402	2824	1417.31	676.97
High	70* (77‡)	3015	5189	3707.66	579.08

*Samples with outliers removed that finally were used for data analysis.
†Samples selected based on a random sampling technique.
‡Total number of block groups corresponding to population.

TABLE 10.3 Statistical Descriptions of Samples of Population Densities (persons/km²)

(Harvey, 2002a, 2000b; Lo, 1995); hence the population densities of the city were divided into three categories: low (fewer than 400 persons/km²), medium (401 to 3000 persons/km²), and high (more than 3000 persons/km²) based on the data distribution. All block groups in the low- and high-density categories were used for sampling owing to their limited number. For the medium-density category, samples were chosen using a random sampling technique. Table 10.3 summarizes the statistical characteristics of selected samples for different categories.

Pearson's correlation coefficients were computed between population densities and the remote sensing variables. Stepwise regression analysis was further applied to identify suitable variables for developing population-estimation models. The coefficient of determination (R^2) was used as an indicator to determine the robustness of a regression model. To improve model performance, various combinations of the remote sensing variables were explored, as well as the transformation of population densities (PD) into natural-logarithm (LPD) and square-root (SPD) forms.

Accuracy Assessment

Whenever a model is applied for prediction, there are always discrepancies between true and estimated values, and these are called *residuals.* It is necessary to validate whether the model fits training-set data, which is called *internal validation,* or to test its fitness with other data sets that are not used as training sets, which is called *external validation* (Harvey, 2002a). Relative and absolute error can be computed. For an individual case, the relative error can be expressed as

$$\text{RE} = (P_g - P_e)/P_g \times 100 \qquad (10.2)$$

where P_g and P_e are the reference and estimated values, respectively. The residual $(P_g - P_e)$ for individual cases may be negative or positive, so absolute values of the residuals are used to assess the overall performance of the model; that is,

$$\text{Overall relative error (RE)} = \frac{\sum_{k=1}^{n} |\text{RE}n|}{n} \tag{10.3}$$

$$\text{Overall absolute error (AE)} = \frac{\sum_{k=1}^{n} |P_g - P_e|}{n} \tag{10.4}$$

where n is the number of block groups used for accuracy assessment. The smaller the RE and AE, the better the models will be. A total of 483 unsampled block groups was used to assess the performance of models in the non-stratified sampling scheme. For the stratified sampling scheme, a total of 521 samples was used for accuracy assessment. A residual map was created based on the best estimation model for geographic analysis of predicted errors.

10.2.3 Result of Population Estimation Based on a Non-Stratified Sampling Method

Six groups of remote sensing variables were used to explore their relationship with population parameters, and their correlation coefficients are presented in Table 10.4.

Table 10.4 indicates that among the ETM+ spectral bands, band 4 was the most strongly correlated with population density; the transforms of population density into natural-logarithm or square-root forms did not improve the correlation coefficients of single ETM+ bands except for band 5. The principle components, especially PC2, improved the correlation with population parameters when compared with single ETM+ bands. All selected vegetation indices had a significant correlation with population density. The green-vegetation fraction had a better correlation with population density than the impervious-surface fraction. Selected textures, especially band 7 associated with a window size of 7 × 7, were strongly correlated with population density. Among all selected remote sensing variables, temperature was the most correlated variable with population density. Moreover, it was found that vegetation-related variables such as band 4, PC2, vegetation indices, and the green-vegetation fraction all had a negative correlation with population parameters. This is so because for a given area, more vegetation is often related to less built-up area and thus less population.

The strong correlations between population parameters and several remote sensing variables imply that a combination of temperature, textures, and spectral responses could be used to improve the population-estimation models. A series of estimation models was developed by performing stepwise regression analysis based on different

Variables		PD	SPD	LPD
Bands	B1	0.226*	0.160†	0.019
	B2	0.163†	0.096	–0.039
	B3	0.164†	0.096	–0.039
	B4	–0.255*	–0.209*	–0.108
	B5	–0.155†	–0.196†	–0.251*
	B7	0.068	0.003	–0.115
PCs	PC1	0.123	0.056	–0.073
	PC2	–0.319*	–0.283*	–0.190†
	PC3	–0.248*	–0.239*	–0.178†
VIs	NDVI	–0.244*	–0.182†	–0.052
	RDVI	–0.242*	–0.178†	–0.040
	SAVI	–0.245*	–0.182†	–0.053
	SVI	–0.221*	–0.156†	–0.023
	RVI	–0.385*	–0.337*	–0.206*
	TNDVI	–0.164†	–0.098	0.026
Frac.	GV	–0.231*	–0.171†	–0.045
	IMP	0.109	0.043	–0.082
Text.	B3_3 × 3	–0.196†	–0.223*	–0.267*
	B7_3 × 3	–0.295*	–0.326*	–0.347*
	B3_5 × 5	–0.280*	–0.317*	–0.360*
	B7_5 × 5	–0.368*	–0.406*	–0.427*
	B3_7 × 7	–0.322*	–0.364*	–0.407*
	B7_7 × 7	–0.402*	–0.444*	–0.463*
Temp.	TEMP	0.519*	0.513*	0.411*

Note: Bn = band *n*; PD = population density; SPD = square root of population density; LPD = natural logarithm of population density; PCs = principal components; VIs = vegetation indices; Frac. = fraction images; GV = green-vegetation fraction; IMP = impervious surface fraction; Text. = texture; Temp. = temperature.
*Correlation at 99 percent confidence level (two-tailed).
†Correlation at 95 percent confidence level (two-tailed).

Table 10.4 Relationships between Population Parameters and Remote Sensing Variables Based on Non-Stratified Samples

combinations of remote sensing variables. The predictors and R^2 of the regression models developed are presented in Table 10.5.

Table 10.5 indicates that any single group of remote sensing variables did not produce a satisfactory R^2 except for vegetation indices. Incorporation of vegetation-related variables or use of all variables provided better modeling results. The square-root form of population density improved the regression models, whereas the natural-logarithm form degraded the regression performance, with an exception in the textures. Table 10.6 summarizes the best-performing regression models and associated estimation errors.

Potential Variables	PD		SPD		LPD	
	Selected Var.	R^2	Selected Var.	R^2	Selected Var.	R^2
Bands	B4	0.065	B5, B1, B2	0.212	B1, B2, B5	0.160
PCs	PC2, PC3	0.159	PC2, PC3	0.134	PC2, PC3, PC1	0.107
VIs	RVI, TNDVI, SAVI, RDVI	0.622	TNDVI, SAVI, RDVI, RVI	0.645	TNDVI, SAVI, RDVI	0.548
Frac.	GV, IMP	0.079	GV, IMP	0.065		
Text.	B7_7 × 7, B7_3 × 3, B3_3 × 3	0.369	B7_7 × 7, B3_3 × 3, B3_5 × 5	0.465	B7_3 × 3, B7_5 × 5	0.448
Temp.	TEMP	0.269	TEMP	0.263	TEMP	0.169
VRV	RVI, TNDVI, SAVI, PC2, B4	0.768	RVI, TNDVI, SAVI, PC2, B4	0.797	RVI, TNDVI, SAVI, PC2, B4	0.678
B-temp.	Temp., B5	0.351	Temp., B7	0.376	Temp., B7	0.338
Mixture	B7_7 × 7, RVI, B2, TNDVI, SAVI, B5	0.785	TEMP, RVI, TNDVI, SAVI, B5, RDVI, SVI	0.828	TNDVI, SAVI, B5, TEMP, RVI	0.698

Note: VRV = vegetation-related variables, including band 4, PC2, VIs, and GV;
B-temp. = combination bands and temperature; Mixture = combination of all variables.

TABLE 10.5 Comparison of Regression Results for Population-Density Estimation Based on Non-Stratified Samples

Model		Variable	Regression Equation	R^2	RE	AE
PD	1	Mixture	$-83613.428 - 58.830 \times B7\_7 \times 7 + 5914.817 \times RVI + 117300.115 \times TNDVI - 65068.691 \times SAVI - 65.723 \times B5 + 64.369 \times B2$	0.785	204.3	505
	2	VRV	$-95394.477 + 6378.881 \times RVI + 132709.023 \times TNDVI - 73728.142 \times SAVI - 137.526 \times PC2 + 129.704 \times B4$	0.768	204.4	523
SPD	3	Mixture	$-1293.678 + 1.318 \times TEMP + 57.79 \times RVI + 1347.089 \times TNDVI - 789.683 \times SAVI - 1.124 \times B5 - 11.674 \times RDVI + 1.325 \times SVI$	0.828	123.1	439
	4	VRV	$-1226.463 + 72.752 \times RVI + 1754.789 \times TNDVI - 1.915 \times PC2 - 945.565 \times SAVI + 1.742 \times B4$	0.797	142.1	452

TABLE 10.6 Summary of Selected Estimation Models for Population-Density Estimation Based on Non-Stratified Samples

Overall, larger R^2 values resulted in fewer estimation errors. The regression models using a combination of spectral, texture, and temperature data provided the best estimation results. The R^2 value for the best model (model 3) reached 0.83, but the estimation errors were still high. Figure 10.2 shows population-density distribution estimated using this model. The overall relative errors were larger than 123 percent, and the overall absolute errors were greater than 439 persons/km^2 (the mean population density is 1470 persons/km^2). The extreme low and high population-density block groups were the main sources of error. Low-population-density block groups had a more severe impact on relative errors, whereas high-population-density block groups had more impact on absolute errors. These

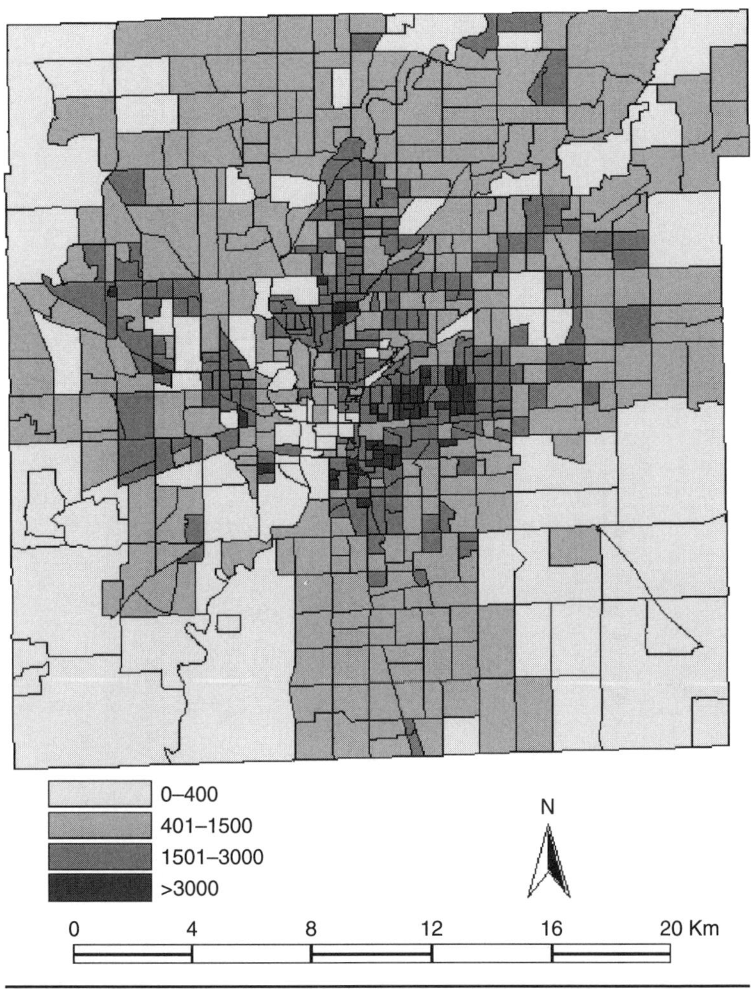

Figure 10.2 Population-density distribution estimated using the best regression model (model 3) based on non-stratified categories.

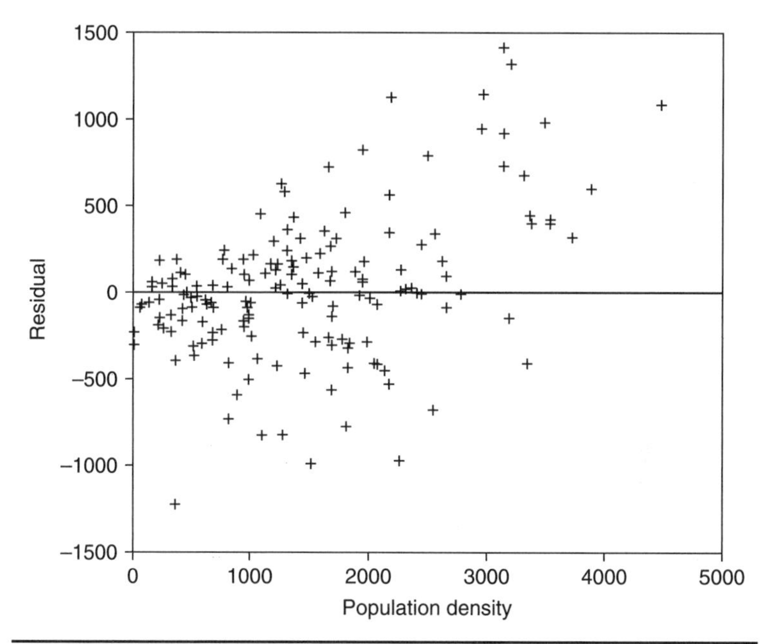

FIGURE 10.3 Residual distribution from model 4; negative indicates overestimated, and positive indicates underestimated.

impacts can be illustrated clearly in the scatter plot of the residuals. Figure 10.3 shows the residual distributions of the best model (model 3). It indicates that population in very low-density block groups was overestimated, whereas population in high-density area was greatly underestimated. The high estimation errors imply that no single model worked well for all levels of population density. In order to improve population-estimation results, separating the population density into subcategories such as low, medium, and high densities and developing models for each category become necessary.

10.2.4 Result of Population Estimation Based on Stratified Sampling Method

Table 10.7 shows correlation coefficients between population parameters and remote sensing variables in the low-, medium-, and high-population-density categories. It is clear that in the low-density category, correlations were not as strong as those in medium- and high-density categories. Similar to the non-stratified scheme, in the medium- and high-density categories, temperature had the strongest positive correlation with population, whereas vegetation-related variables had negative correlations with population. The low correlation between remote sensing variables and population in the low-density category implies that population estimation for these areas was more complicated, and the issue warrants further study.

Remote Sensing Variables		Low Density			Medium Density			High Density		
		PD	SPD	LPD	PD	SPD	LPD	PD	SPD	LPD
ETM	B1	–0.231†	–0.237†	–0.232†	0.398*	0.398*	0.39*	0.274†	0.269†	0.264†
	B2	–0.232†	–0.234†	–0.230†	0.340*	0.342*	0.338*	0.248†	0.243†	0.237†
	B3	–0.244†	–0.245†	–0.237†	0.349*	0.351*	0.346*	0.267†	0.263†	0.259†
	B4	0.223	0.231†	0.207	–0.354*	–0.335*	–0.304*	–0.371*	–0.371*	–0.371*
	B5	–0.164	–0.141	–0.132	–0.060	–0.047	–0.029	–0.085	–0.087	–0.089
	B7	–0.256†	–0.248†	–0.234†	0.243*	0.247*	0.246*	0.194	0.191	0.188
PCs	PC1	–0.249†	–0.247†	–0.237†	0.302*	0.304*	0.302*	0.231	0.227	0.223
	PC2	0.164	0.181	0.168	–0.391*	–0.373*	–0.347*	–0.379*	–0.378*	–0.377*
	PC3	–0.001	0.027	0.068	–0.269*	–0.291*	–0.316*	–0.019	–0.009	0.001
VIs	NDVI	0.253†	0.257†	0.237†	–0.388*	–0.376*	–0.354*	–0.346*	–0.344*	–0.342*
	RDVI	0.255†	0.257†	0.242†	–0.411*	–0.409*	–0.398*	–0.318*	–0.315*	–0.312*
	SAVI	0.253†	0.257†	0.237†	–0.388*	–0.377*	–0.354*	–0.347*	–0.345*	–0.342*
	SVI	0.252†	0.256†	0.241†	–0.392*	–0.384*	–0.365*	–0.340*	–0.337*	–0.335*
	RVI	0.211	0.210	0.191	–0.514*	–0.506*	–0.485*	–0.354*	–0.353*	–0.351*
	TNDVI	0.260†	0.266†	0.246†	–0.320*	–0.308*	–0.286*	–0.335*	–0.333*	–0.330*
Frac.	GV	0.254†	0.258†	0.238†	–0.388*	–0.376*	–0.353*	–0.357*	–0.356*	–0.355*
	IMP	–0.273†	–0.266†	–0.249†	0.291*	0.290*	0.283*	0.264†	0.262†	0.260†

Text.	B3_3 × 3	–0.119	–0.124	–0.129	–0.047	–0.025	0.001	–0.123	–0.125	–0.127
	B7_3 × 3	–0.149	–0.143	–0.133	–0.150	–0.138	–0.123	–0.130	–0.132	–0.133
	B3_5 × 5	–0.104	–0.111	–0.125	–0.133	–0.114	–0.090	–0.103	–0.105	–0.107
	B7_5 × 5	–0.142	–0.137	–0.132	–0.218†	–0.210†	–0.198†	–0.106	–0.108	–0.109
	B3_7 × 7	–0.093	–0.102	–0.122	–0.177	–0.159	–0.136	–0.080	–0.083	–0.085
	B7_7 × 7	–0.136	–0.132	–0.130	–0.251*	–0.244*	–0.234†	–0.076	–0.078	–0.080
Temp	TEMP	–0.234†	–0.231†	–0.212	0.622*	0.635*	0.637*	0.425*	0.423*	0.420*

*Correlation at 99 percent confidence level (two-tailed).
†Correlation at 95 percent confidence level (two-tailed).

TABLE 10.7 Relationships between Population Parameters and Remote Sensing Variables Based on Stratified Samples

The R^2 values for individual regression models are summarized in Table 10.8. In the low- and high-density categories, the highest R^2 were only 0.13 and 0.180, respectively. This indicates that the ETM+ data may not be suitable for population estimation in these categories. In the medium-density category, the combination of vegetation indices or vegetation-related variables and the incorporation of spectral response, textures, and temperature can provide good estimations, especially the latter, when R^2 reached as high as 0.87, 0.86, and 0.83 for different forms of dependent variables. Overall, the transforms of population density did not significantly improve the estimation in these categories.

Table 10.9 displays the four best models, selected based on R^2 and estimation errors. It shows that the results of population estimation for the low-density category using remote sensing variables were not satisfactory owing to high estimation errors. In the medium-density category, both models provided very good estimations using vegetation-related independent variables or using a combination of spectral, texture, and temperature variables. The R^2 values reached 0.83 and 0.86, respectively, with a relative error of less than 29 percent and an absolute error of less than 384 in both models (compared with a mean value of 1417). Figure 10.4 illustrates estimated population-density distribution using the best regression model (model 7) for this category. For the high-density category, the model using temperature as the only independent variable yielded the best estimation result. The relative error was only 11.4 percent, but absolute error reached 429 (compared with a mean value of 3707). Overall, the performance of the estimation models was much improved after stratification of population density into three categories. This finding implies that stratification based on population density is necessary for developing population-estimation models using remotely sensed data.

Figure 10.5 shows the distribution of residuals when a combination of models 5, 7, and 8 was applied to predict the population of Indianapolis in 2000. Most underestimations and overestimations were located in the central part of the city. For example, block groups 1, 2, 3, 4, and 5 (marked in Fig. 10.5) with very high population densities were greatly underestimated. These block groups usually had several multistory apartment buildings for residential use. On the other hand, most overestimated block groups were found in the downtown area, where commercial or service uses dominated. For example, the most overestimated block group, block group 3, had a population density of 678 persons/km^2, according to the Census data, but the estimated population density reached up to 1884 persons/km^2. The second most overestimated area was observed in block group 7, where university and residential uses shared the land.

Scale	Variable	PD		SPD		LPD	
		Selected Var.	R^2	Selected Var.	R^2	Selected Var.	R^2
Low	Bands	B7	0.066	B7	0.062	B3	0.056
	PCs	PC1	0.062	PC1	0.061	PC1	0.056
	VIs or VRV	TNDVI	0.068	TNDVI	0.071	TNDVI	0.061
	Frac.	IMP	0.075	IMP	0.071	IMP	0.062
	Temp.	TEMP	0.055	TEMP	0.053	TEMP	0.045
	Mixture	IMP, B3_7 × 7	0.130	IMP	0.071	IMP	0.062
Medium	Bands	B1, B2	0.290	B1, B2	0.283	B1, B2, B4	0.290
	PCs	PC2, PC3	0.277	PC2, PC3	0.277	PC2, PC3	0.269
	VIs	TNDVI, SAVI, RDVI, SVI	0.641	TNDVI, SAVI	0.611	TNDVI, SAVI	0.597
	Frac.	GV	0.151	GV	0.141	GV	0.124
	Text.	B7_7 × 7, B7_3 × 3	0.215	B7_7 × 7, B7_3 × 3	0.230	B3_3 × 3, B3_5 × 5	0.296
	Temp.	TEMP	0.387	TEMP	0.404	TEMP	0.407
	VRV	RVI, TNDVI, PC2, SAVI, RDVI, SVI	0.825	RVI, TNDVI, PC2, SAVI, RDVI, SVI, B4	0.829	TNDVI, PC2, RDVI, NDVI, SVI	0.794
	B-temp.	TEMP, B7	0.442	TEMP, B7	0.461	TEMP, B7	0.466
	Mixture	B3_7 × 7, RVI, TNDVI, SAVI, B5, PC1	0.869	TEMP, B3_7×7, RVI, TNDVI, SAVI, B5, PC1	0.863	TEMP, B7_7 × 7, RVI TNDVI, AVI, B5, RDVI	0.831

TABLE 10.8 Comparison of Regression Results for Different Population-Density Categories

Scale	Variable	PD		SPD		LPD	
		Selected Var.	R^2	Selected Var.	R^2	Selected Var.	R^2
High	Bands	B4	0.138	B4	0.138	B4	0.138
	PCs/VRV	PC2	0.144	PC2	0.143	PC2	0.142
	VIs	RVI	0.125	RVI	0.124	RVI	0.123
	Frac.	GV	0.128	GV	0.127	GV	0.126
	B-temp. mixture	TEMP	0.180	TEMP	0.179	TEMP	0.177

Table 10.8 Comparison of Regression Results for Different Population-Density Categories (*Continued*)

Model		Potential Var.	Dep. Var.	Regression Equation	R^2	RE	AE
Low	5	Mixture	PD	$296.190 - 9.136 \times IMP + 7.746 \times B3\_7 \times 7$	0.130	315.3	97
Medium	6	VRV	SPD	$-649.707 + 37.317 \times RVI + 1027.726 \times TNDVI - 1.579 \times PC2 - 538.330 \times SAVI + 0.608 \times B4 - 12.092 \times RDVI + 1.183 \times SVI$	0.829	28.9	357
	7	Mixture	SPD	$-966.765 + 0.684 \times TEMP - 0.632 \times B3\_7 \times 7 + 46.507 \times RVI + 1110.756 \times TNDVI - 581.426 \times SAVI - 1.003 \times B5 + 0.514 \times PC1$	0.863	28.4	384
High	8	B-temp./Mixture	SPD	$-565.332 + 2.047 \times TEMP$	0.179	11.4	429

Table 10.9 Best Estimation Models for Different Population-Density Categories Based on the Stratified Sampling Scheme

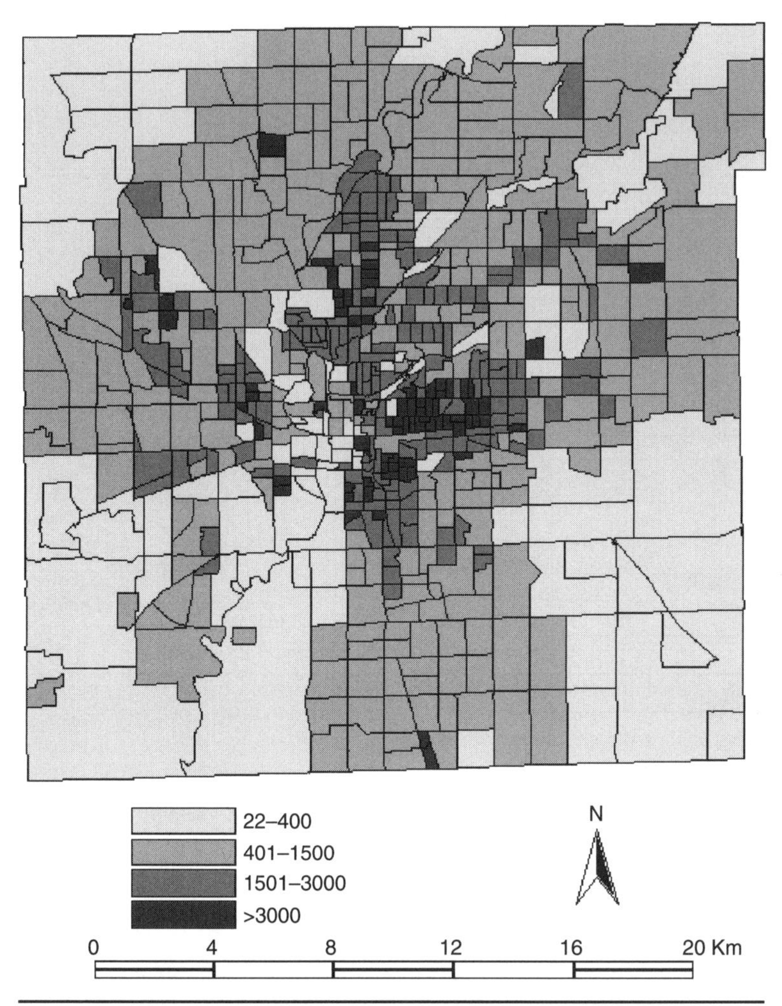

Figure 10.4 Estimated population-density distribution using the best regression model (model 7) based on stratified category.

Based on the models developed for estimating population densities, the population of individual block groups can be calculated, and the total population of the whole city can be summed up. The total population estimates were 832,792 with a relative error of 3.2 percent using the stratified scheme (i.e., combination of models 5, 7, and 8) and 789,756 with a relative error of 8.2 percent using the non-stratified scheme (i.e., model 3). It is concluded that remote sensing techniques can provide reasonably accurate estimation results for the total population and that dividing population density into different categories is more effective than conventional non-stratified methods.

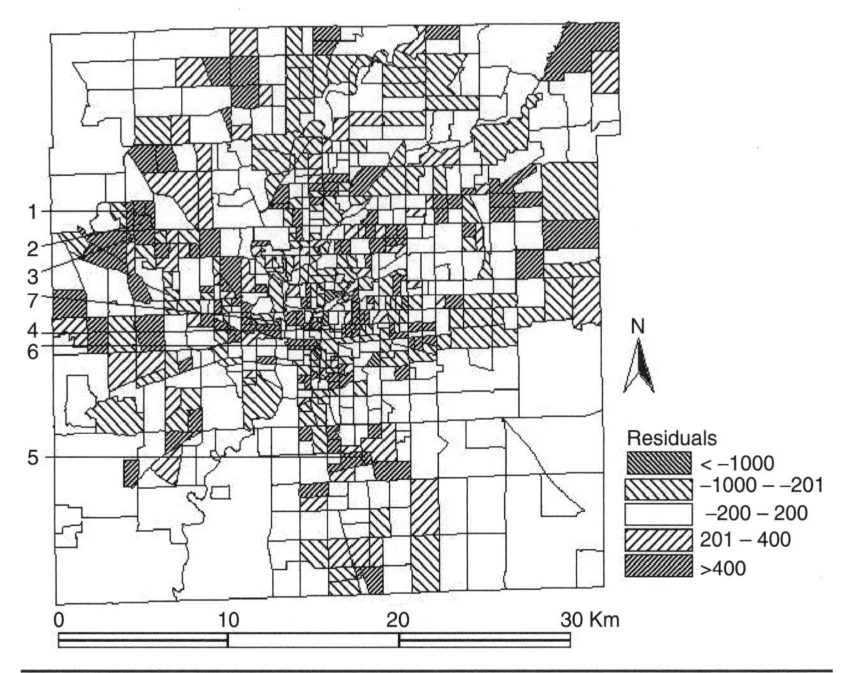

FIGURE 10.5 Residual map of estimated population densities based on the best model of the stratified sampling scheme. Negative value indicates overestimation, and positive value indicates underestimation.

10.3 Discussion

Using remote sensing techniques to estimate population density is still a challenging task in terms of theory and methodology (owing to remotely sensed data per se), the complexity of urban landscapes, and the complexity of population distribution. Remotely sensed data are associated with the characteristics of surface features but not directly related to population counts or population densities. For example, the areas with low population density may be located in industrial/commercial areas or in forest/agricultural areas, but the spectral characteristics of these landscapes are fundamentally different. The areas with high population density generally have a number of multistory apartment buildings. The distribution of these high-density areas varies greatly with urban spatial structure, which has identified three classic types: concentric zone, sector, and the multiple-nuclei city (Wheeler and Muller, 1981). In developed countries, substantial urban growth frequently occurs in suburban areas owing to the redistribution of population and/or decentralization of metropolitan urban functions, whereas in developing countries, urban expansion is more related to rapid population growth and industrialization. In Eastern cities, such as Hong Kong, Singapore, and Shanghai, high population density is often observed in the central part of the city, where high-rise residential buildings coalesce. Optical remote-sensing data, however, have not been linked directly to the vertical or internal features of such buildings for digital analysis of population estimation.

Remotely sensed data and census data are often collected with different formats and stored with different data structures. The values of any census variable are aggregated totals or mean values for the entire spatial extent of a census unit. In other words, census units are assumed to be homogeneous, no matter what types of variation in land use occur within them. A problem is created when unoccupied areas, such as water, airport, and forest, are given a population in the census. Remote sensing data have finer resolution than census data. When integrating these two types of data, a common method is to aggregate remotely sensed data to an appropriate census level. For example, this research aggregated various remote sensing variables at a block-group level. This aggregation has the potential to result in the same values for different block groups despite their differences in land-use and land-cover types, causing errors for population estimation.

Because of the complexity of population distribution, a single model is often difficult to fit all the data. Stratification of population density in this study has proved to be effective in improving estimation results. However, the method of stratifying population based on density may result in spatial discontinuity of the data, and it may be difficult to find suitable thresholds for stratification. The population categories identified here correspond, to some extent, to land-cover categories in urban residential areas in the National Land Cover Database (NLCD) (Vogelman et al., 1998, 2001). Low-intensity residential areas in the NLCD, mostly single-family housing units, relate to the medium-population-density category, whereas high-intensity residential areas, such as apartment complexes and row houses, relate to the high-population-density category. Moreover, given that various factors affect remotely sensed data quality, it is usually difficult to transfer the model developed in one site directly to other sites. Many factors need to be considered, including image acquisition date and time, the atmospheric condition when the image data were acquired, and the characteristics of the urban landscapes under investigation. Model transfer to nearby areas with similar socioeconomic conditions is favorable if these areas are within the same image scene or adjacent scenes in the same path of image acquisition.

10.4 Conclusions

Population-estimation models developed based on the integration of satellite imagery and census data have numerous applications. They can be used to provide information on intra-urban population distribution, which is essential in urban planning, natural-hazard risk assessment, disaster prevention and response, environmental impact assessment, transportation planning, economic decision making, and evaluation of quality of life. They also can be applied to validate urban growth models if the time-series image data become available.

This study demonstrates that Landsat ETM+ imagery could be used to provide reasonably accurate population-density estimation by combining various remote sensing-derived variables. Vegetation-related variables were especially effective. Remote sensing-based models were more suitable for estimation of population with medium-density than with low- and high-density areas. The stratification of population density into some categories and development of estimation models for individual categories improved model performance. Although population estimation using remotely sensed data is not straightforward, especially for low- and high- density regions, a major advantage of this approach is that it can provide a timely update of a population database and its spatial distribution, which is impossible by conventional census approaches.

More research is needed to improve population estimation through development of suitable models and use of multisource data, such as high spatial-resolution imagery and light detection and ranging (LiDAR) data, which are capable of identifying individual buildings and of measuring the heights of buildings. Comparative analyses of different methods are further suggested; for example, the method based on dwelling counts could be used to estimate population in low-density areas using high spatial-resolution remote sensing images. Further studies are warranted by incorporating Elvidge's radiance-calibrated nighttime lights imagery with original frequency data from satellite imagery because the former has been proven to be effective in estimating low- and high-population extremes (Elvidge et al., 1995, 1997). In addition, dynamic changes in population can be examined if multitemporal remote sensing and census data become available.

References

Adams, J. B., Sabol, D. E., Kapos, V., et al. 1995. Classification of multispectral images based on fractions of endmembers: Application to land cover change in the Brazilian Amazon. *Remote Sensing of Environment* **52,** 137–154.

Aguiar, A. P. D., Shimabukuro, Y. E., and Mascarenhas, N. D. A. 1999. Use of synthetic bands derived from mixing models in the multispectral classification of remote sensing images. *International Journal of Remote Sensing* **20,** 647–657.

Asner, G. P., and Lobell, D. B. 2000. A biogeophysical approach for automated SWIR unmixing of soils and vegetation. *Remote Sensing of Environment* **74,** 99–112.

Bannari, A., Morin, D., Bonn, F., and Huete, A. R. 1995. A review of vegetation indices. *Remote Sensing Reviews* **13,** 95–120.

Barnsley, M. J. 1999. Digital remote sensing data and their characteristics. In *Geographical Information Systems: Principles, Techniques, Applications, and Management,* 2nd ed., edited by P. Longley, M. Goodchild, D. J. Maguire, and D. W. Rhind, pp. 451–466. New York: Wiley.

Birth, G. S., and McVey, G. 1968. Measuring the color of growing turf with a reflectance spectrophotometer. *Agronomy Journal* **60,** 640–643.

Chen, K. 2002. An approach to linking remotely sensed data and areal census data. *International Journal of Remote Sensing* **23,** 37–48.

Cochrane, M. A. and Souza, C. M., Jr. 1998. Linear mixture model classification of burned forests in the eastern Amazon. *International Journal of Remote Sensing* **19,** 3433–3440.

Deering, D. W., Rouse, J. W., Haas, R. H., and Schell, J. A. 1975. Measuring forage production of grazing units from Landsat MSS data. In *Proceedings of Tenth International Symposium on Remote Sensing of Environment,* ERIM 2, Ann Arbor, MI, pp. 1169–1178.

Dobson, J. E. 2003. Estimating population at risk. In *Geographical Dimensions of Terrorism,* edited by S. L. Cutter, D. B. Richardson, and T. J. Wilbanks, pp. 161–167. New York: Routledge.

Dobson, J. E., Bright, E. A., Coleman, P. R., et al. 2000. LandScan: A global population database for estimating populations at risk. *Photogrammetric Engineering and Remote Sensing* **66,** 849–857.

Dobson, J. E., Bright, E. A., Coleman, P. R., and Bhaduri, B. L. 2003. LandScan2000: A new global population geography. In *Remotely-Sensed Cities,* edited by V. Mesev, pp. 267–279. London: Taylor & Francis.

Elvidge, C. D., Baugh, K. E., Kihn, E. A., et al. 1995. Mapping city lights with nighttime data from the DMSP operational linescan system. *Photogrammetric Engineering and Remote Sensing* **63,** 727–734.

Elvidge, C. D., Baugh, K. E., Hobson, V. R., Kihn, et al. 1997. Satellite inventory of human settlements using nocturnal radiation emissions: A contribution for the global toolchest. *Global Change Biology* **3,** 387–395.

Flowerdew, R., and Green, M. 1992. Developments in areal interpolation methods and GIS. *Annals of Regional Science* **26,** 67–78.

Forester, B. C. 1985. An examination of some problems and solutions in monitoring urban areas from satellite platforms. *International Journal of Remote Sensing* **6,** 39–151.

Gong, P., and Howarth, P. J. 1992. Frequency-based contextual classification and gray-level vector reduction for land-use identification. *Photogrammetric Engineering and Remote Sensing* **58,** 423–437.

Haack, B. N., Guptill, S. C., Holz, R. K., et al. 1997. Urban analysis and planning. In *Manual of Photographic Interpretation,* pp. 517–553. Bethesda, MD: American Society for Photogrammtry and Remote Sensing.

Haralick, R. M., Shanmugam, K., and Dinstein, I. 1973. Texture features for image classification. *IEEE Transactions on Systems, Man and Cybernetics* **3,** 610–621.

Haralick, R. M. 1979. Statistical and structural approaches to texture. *Proceedings of the IEEE* **67,** 786–804.

Harvey, J. T. 2002a. Estimation census district population from satellite imagery: Some approaches and limitations. *International Journal of Remote Sensing* **23,** 2071–2095.

Harvey, J. T. 2002b. Population estimation models based on individual TM pixels, *Photogrammetric Engineering and Remote Sensing.* **68,** 1181–1192.

He, D. C., and Wang, L. 1990. Texture unit, textural spectrum and texture analysis. *IEEE Transaction on Geoscience and Remote Sensing* **28,** 509–512.

Holz, R., Huff, D. L., and Mayfield, R. C. 1973. Urban spatial structure based on remote sensing imagery. *The Surveillant Science: Remote Sensing of Environment.* Edited by Holz, R. K., New Jersey, Palo Alto, 375–380.

Holz, R. 1988. Population estimation of Colonias in the Low Rio Grande Valley using remote sensing techniques. Paper presented at the Annual Meeting of the Association of American Geographers, Phoenix, Arizona.

Hsu, S. Y. 1971. Population estimation. *Photogrammetric Engineering* **37,** 449–454.

Hsu, S. Y. 1973. Population estimation from ERTS imagery: Methodology and evaluation. In *Proceedings of the American Society of Photogrammetry 39th Annual Meeting,* March 11-16, Washington, D. C., pp. 583–591.

Huete, A. R. 1988. A soil-adjusted vegetation index (SAVI). *Remote Sensing of Environment* **25,** 295–309.
Huxley, J. S., 1932. Problems of Relative Growth. Methuen, London.
Iisaka, J., and Hegedus, E. 1982. Population estimation from Landsat imagery. *Remote Sensing of Environment* **12,** 259–272.
Imhoff, M. L., Lowrence, W. T., Stutzer, D. C., and Elvidge, C. D. 1997. A technique for using composite DMSP/OLS city lights satellite data to map urban area. *Remote Sensing of Environment* **61,** 361–370.
Jensen, J. R. 2000. *Remote Sensing of the Environment: An Earth Resource Perspective.* Upper Saddle Rive, NJ: Prentice-Hall.
Jensen, J. R., and Cowen, D. C. 1999. Remote sensing of urban/suburban infrastructure and socio-economic attributes. *Photogrammetric Engineering and Remote Sensing* **65,** 611–622.
Kraus, S. P., Senger, L. W. and Ryerson, J. M., 1974, Estimating population from photographically determined residential land use types. *Remote sensing of environment,* **3,** 35–42.
Langford, M., Maguire, D. J., and Unwin, D. J. 1991. The areal interpolation problem: Estimating population using remote sensing in a GIS framework. In *Handing Geographical Information: Methodology and Potential Applications,* edited by L. Masser and M. Blakemore. New York: Longman and Wiley.
Li, G., and Weng, Q. 2005. Using Landsat ETM+ imagery to measure population density in Indianapolis, Indiana, USA. *Photogrammetric Engineering & Remote Sensing* **71,** 947–958.
Li, G. and Weng, Q. 2006. The integration of GIS and remote sensing for assessing urban quality of life: Model development and validation. In Weng, Q. and D. Quattrochi (eds.): *Urban Remote Sensing*. Boca Raton, FL: CRC/Taylor & Francis, pp. 311–336.
Lindgren, D. T. 1971. Dwelling unit estimation with color-IR photos. *Photogrammetric Engineering* **37,** 373–378.
Lindgren, D. T. 1985. *Land Use Planning and Remote Sensing.* Boston: Martinus Nijhoff.
Lo, C. P. 1986a. *Applied Remote Sensing.* New York: Longman.
Lo, C. P. 1986b. Accuracy of population estimation from medium-scale aerial photography. *Photogrammetric Engineering and Remote Sensing* **52,** 1859–1869.
Lo, C. P. 1995. Automated population and dwelling unit estimation from high resolution satellite images: A GIS approach. *International Journal of Remote Sensing* **16,** 17–34.
Lo, C. P. 2001. Modeling the population of China using DMSP operational linescan system nighttime data. *Photogrammetric Engineering and Remote Sensing* **67,** 1037–1047.
Lo, C. P., and Welch, R. 1977. Chinese urban population estimation. *Annals of the Association of American Geographers* **67,** 246–253.
Lu, D., and Weng, Q. 2004. Spectral mixture analysis of the urban landscape in Indianapolis with Landsat ETM+ imagery. *Photogrammetric Engineering & Remote Sensing* (in press).
Lu, D., Moran, E., and Batistella, M. 2003. Linear mixture model applied to Amazonian vegetation classification. *Remote Sensing of Environment* **87,** 456–469.
Marceau, D. J., Howarth, P. J., Dubois, J. M., and Gratton, D. J. 1990. Evaluation of the gray-level co-occurrence matrix method for land-cover classification using SPOT imagery. *IEEE Transactions on Geoscience and Remote Sensing* **28,** 513–519.
McGwire, K., Minor, T., and Fenstermaker, L. 2000. Hyperspectral mixture modeling for quantifying sparse vegetation cover in arid environments. *Remote Sensing of Environment* **72,** 360–374.
Mustard, J. F., and Sunshine, J. M. 1999. Spectral analysis for earth science: Investigations using remote sensing data. In *Remote Sensing for the Earth Sciences: Manual of Remote Sensing,* 3rd ed., Vol. 3, edited by A. N. Renc, pp. 251–307. New York: Wiley.
Narasimha Rao, P. V., Sesha Sai, M. V. R., Sreenivas, K., et al. 2002. Textural analysis of IRS-1D panchromatic data for land cover classification. *International Journal of Remote Sensing* **23,** 3327–3345.

Nordbeck, S., 1965. The law of allometric growth. Michigan Inter-University Community of mathematical Geographers, paper 7.
Ogrosky, C. E. 1975. Population estimation from satellite imagery. *Photogrammetric Engineering and Remote Sensing* **41,** 707–712.
Phinn, S., Stanford, M., Scarth, P., et al. 2002. Monitoring the composition of urban environments based on the vegetation-impervious surface-soil (VIS) model by subpixel analysis techniques. *International Journal of Remote Sensing* **23,** 4131–4153.
Porter, P. W. 1956. Population distribution and land use in Liberia. Ph.D. dissertation, London School of Economics and Political Science, London.
Qiu, F., Woller, K. L., and Briggs, R. 2003. Modeling urban population growth from remotely sensed imagery and TIGER GIS road data. *Photogrammetric Engineering and Remote Sensing* **69,** 1031–1042.
Rashed, T., Weeks, J. R., Gadalla, M. S., and Hill, A. G. 2001. Revealing the anatomy of cities through spectral mixture analysis of multisepctral satellite imagery: A case study of the Greater Cairo region, Egypt. *Geocarto International* **16,** 5–15.
Richards, J. A. 1994. *Remote Sensing Digital Image Analysis: An Introduction.* Berlin: Springer-Verlag.
Roberts, D. A., Batista, G. T., Pereira, J. L. G., et al. 1998. Change identification using multitemporal spectral mixture analysis: Applications in eastern Amazonia. In *Remote Sensing Change Detection: Environmental Monitoring Methods and Applications,* edited by R. S. Lunetta and C. D. Elvidge, pp. 137–161. Ann Arbor, MI: Ann Arbor Press.
Roujean, J. L., and Breon, F. M. 1995. Estimating PAR absorbed by vegetation from bidirectional reflectance measurements. *Remote Sensing of Environment* **51,** 375–384.
Rouse, J. W., Haas, R. H., Schell, J. A., and Deering, D. W. 1974. Monitoring vegetation systems in the Great Plains with ERTS. In *Proceedings of Third Earth Resources Technology Satellite-1 Symposium,* Greenbelt, MD, December 1973, NASA SP-351, pp. 310–317.
Shaban, M. A., and Dikshit, O. 2001. Improvement of classification in urban areas by the use of textural features: The case study of Lucknow city, Uttar Pradesh. *International Journal of Remote Sensing* **22,** 565–593.
Small, C. 2001. Estimation of urban vegetation abundance by spectral mixture analysis. *International Journal of Remote Sensing* **22,** 1305–1334.
Smith, M. O., Ustin, S. L., Adams, J. B., and Gillespie, A. R. 1990. Vegetation in deserts: I. A regional measure of abundance from multispectral images. *Remote Sensing of Environment* **31,** 1–26.
Sutton, P. 1997. Modeling population density with nighttime satellite imagery and GIS. *Computers, Environment, and Urban Systems* **21,** 227–244
Sutton, P., Roberts, D., Elvidge, C. D., and Meij, H. 1997. A comparison of nighttime satellite imagery and population density for the continental United States. *Photogrammetric Engineering and Remote Sensing* **63,** 1303–1313.
Sutton, P., Roberts, D., Elvidge, C. D., and Baugh, K. 2001. Census from heaven: An estimate of the global human population using night-time satellite imagery. *International Journal of Remote Sensing* **22,** 3061–3076.
Thomson, D., 1975, Small area population estimation using land use data derived from high altitude aircraft photography, Proceedings of the American Society of Photogrammetry (Fall Convention), 673–696.
Tobler, W., 1969, Satellite confirmation of settlement size coefficients. Area, 30–40.
Tobler, W. R., Deichmann, U., Gottsegen, J., and Maloy, K. 1995. *The Global Demography Project.* Technical Report No. 95-6, National Center for Geographic Information and Analysis, UCSB, Santa Barbara, CA.
Vogelman, J. E., Sohl, T. L., and Howard, S. M. 1998. Regional characterization of land cover using multiple sources of date. *Photogrammetric Engineering and Remote Sensing* **64,** 45–57.
Vogelman, J. E., Howard, S. M., Yang, L., et al. 2001. Completion of the 1990s national land cover data set for the conterminous United States from Landsat Thematic Mapper data and ancillary data sources. *Photogrammetric Engineering and Remote Sensing* **67,** 650–662.

Watkins, J. F., and Morrow-Jones, H. A. 1985. Small area population estimates using aerial photography. *Photogrammetric Engineering and Remote Sensing* **51,** 1933–1935.

Webster, C. J. 1996. Population and dwelling unit estimates from space. *Third World Planning Review* **18,** 155–176.

Weeks, J. R., Gadalla, M. S., Rashed, T., et al. 2000. Spatial variability in fertility in Menoufia, Egypt, assessed through the application of remote sensing and GIS technologies. *Environment and Planning A* **32,** 695–714.

Welch, R., and Zupko, S. 1980. Urbanized area energy utilization patterns from DMSP data. *Photogrammetric Engineering and Remote Sensing* **46,** 1107–1121.

Wellar, B. S. 1969. The role of space photography in urban and transportation data series. In *Proceedings of the Sixth International Symposium on Remote sensing of Environment*, October 15, 1969, Ann Arbor, Michigan, Vol. II, pp. 831–854.

Weng, Q., Lu, D., and Schubring, J. 2004. Estimation of land surface temperature-vegetation abundance relationship for urban heat island studies. *Remote Sensing of Environment* **89,** 467–483.

Wheeler, J. O., and Muller, P. O. 1981. *Economic Geography,* pp. 133–137. New York: Wiley.

Wright, J. K. 1936. A method of mapping densities of population with Cape Cod as an example. *Geographical Review* **26,** 103–110.

Wu, C., and Murray, A. T. 2003. Estimating impervious surface distribution by spectral mixture analysis. *Remote Sensing of Environment* **84,** 493–505.

CHAPTER 11

Quality of Life Assessment

Spatial variation and social stratification in quality of life (QOL) persists throughout all postindustrial cities owing to the uneven development among nations, regions, and cities. The study of quality of urban life has drawn increasing interest from a number of disciplines, such as planning, geography, sociology, economics, psychology, political science, behavioral medicine, marketing, and management (Kirby, 1999; Foo, 2001), and is becoming an important tool for policy evaluation, rating of places, and urban planning and management. QOL is a great research topic and relates many different aspects of the lives of human beings. There has been a great increase in amount of time, effort, and resources being concentrated on QOL studies. Over 200 communities in the United States and over 589 in the world have conducted QOL indicator projects (Barsell and Maser, 2004). Studies have been adopted by governments and public agencies to assess and compare changes in QOL within and between communities, cities, regions, and countries.

Although much exploration for the assessment of QOL has been conducted (Cornwell, 2004; Schyns and Boelhouwer, 2004), a universally acceptable definition of and approach to QOL assessment is lacked. Most previous work on urban QOL assessment used only socioeconomic variables from census data. Remote sensing-derived variables have been used for research related to socioeconomic conditions through the integration of ancillary data and the use of modeling. Geographic Information Systems (GIS) provide a powerful tool for data integration of physical and socioeconomic parameters for the development of advanced models. This chapter explores the integration of remote sensing and GIS for modeling QOL assessment with a case study of Indianapolis, Indiana. Although some explorations have been conducted previously using remote sensing and GIS techniques (Lo and Faber, 1997; Weber and Hirsch, 1992), numerous problems still remain to be solved in terms of how to develop a synthetic QOL index and how to better couple remote sensing and socioeconomic data, which are different in data model, format, and structure.

11.1 Assessing Quality of Life

11.1.1 Concept of QOL

There is no certainty as to the origin of term *quality of life* (QOL). American economists Ordway (1953) and Osborn (1954) are among the earliest people to use this term to address their concern about the ecological dangers of unlimited economic growth. The concept of QOL has been changing with the time and has become more complex. At the beginning, QOL was defined as a good standard of living, which is a measure of economic well-being. In 1960, the concept of QOL expanded to include education, health, and well-being (Fallowfield, 1990). In a speech, President Lyndon B. Johnson stated, "Goals cannot be measured by the size of our bank account; they can only be measured in the quality of lives that our people lead." This statement brought to the forefront the idea that QOL is not necessarily a simple function of material wealth. Increasing realizations of the importance of other aspects, such as environmental, social, psychological, and political health, inspired researchers to develop a more comprehensive system to evaluate the overall health of nations, regions, and locals and the well-being of citizens.

At present, there is no universally acceptable definition of QOL, and there is a great deal of ambiguity and controversy over the concept of QOL, its elements, and its indicators. Various concepts concerning urban environmental quality and QOL can be found in the literature. Kamp and colleagues (2003) reviewed the definitions of QOL and other main concepts related to urban environmental quality and human well-being, such as livability, sustainability, and environment quality. There is neither a comprehensive conceptual framework concerning urban QOL and human well-being nor any agreed-on indicator system to evaluate physical, spatial, and social aspects of urban quality because different disciplines address the concept from the perspective of their own research interests and objectives. For example, public health studies probably focus on communities, whereas medical specialists focus on individual patients; sociological researchers may choose to focus on the structure and content of groups (such as ages and races), communities, and societies; psychological researchers may look at any individual-based characteristics such as well-being, mental health, and the like; economists may direct their researchers to focus on economic development; and environmentalists perhaps focus on the impact of the physical environment on QOL. The fact that most definitions of QOL are similar implies that QOL is physical and psychological satisfaction of individuals or societies with their living conditions, including health, social, economic, and environmental factors.

Although no consensus on the definition of the QOL has been reached, this has not prevented QOL to be an academic discipline in its own right since 1970s, when a peer-reviewed scientific journal,

Social Indicators Research, was established (Scottish Executive Social Research, 2005). Since then, a great number of studies related to QOL and well-being issues have been released. Many community organizations at the national, provincial, municipal, and neighborhood levels have conducted QOL research. These studies have provided much information for governments and public agencies on which to formulate strategies and policies to improve QOL. Moreover, an association called the International Society for Quality of Life Studies (ISQOLS) specifically serves as a forum for all academic and professional researchers interested in the field of QOL; ISQOLS coordinates and stimulates interdisciplinary research on QOL within policy, behavioral, social, medical, and environmental sciences.

Indicators can be defined as measurable or observable variables and parameters that indicate the status of a particular system or phenomena. Over many years, researchers working to construct a conceptual framework to improve QOL have attempted to identify comprehensive and distinct categories representing major factors that affect QOL. Friedman (1997) proposed two approaches to identify QOL indicators: societal and personal indicators of QOL. Moreover, QOL has been assessed on different geographic scales ranging from individual to global level, such as census block, neighborhood, community, city, state, country, and the like.

11.1.2 QOL Domains and Models

QOL is a multidimensional construct encompassing many different dimensions, of which economic, social/cultural, and physical are the core domains. Each domain consists of many factors affecting the well-being of individuals and societies. Many researchers tried to define or classify those factors (Campbell et al., 1976; Flanagan, 1978, 1982). For example, Flanagan (1978, 1982) identified 15 factors defining QOL and grouped them into 5 major categories: (1) physical and material well-being; (2) relationships with others, including spouse, children, family members, and friends; (3) civic, community, and social activities; (4) personal development and achievement; and (5) leisure. Some models are highly theoretical, whereas others are strictly empirical, depending on different disciplines. As with the definitions of QOL, there is little consensus about which models should be employed.

Lawrence (2001) developed a holistic framework of human ecology that incorporates anthropologic, biologic, epidemiologic, psychological, and sociological perspectives. Two approaches—subjective and objective—were used to evaluate the roles of individuals, social groups, and institutions. From an ecological point of view, the human ecosystem is related to other ecosystems. Camagni and colleagues (1997) developed a model to approach sustainability and livability. The physical, social, and economic aspects form the materials of society. These three domains have been used by Shafer and colleagues (2000)

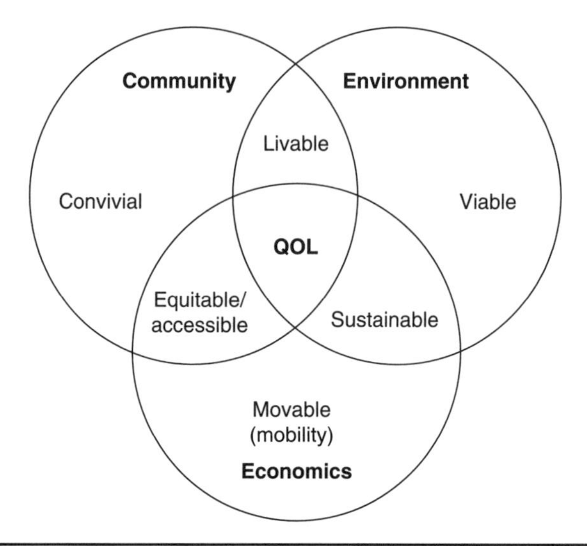

FIGURE 11.1 Shafer's model showing the factors contributing to community QOL from a human ecological perspective. (*Adapted from Shafer et al., 2000.*)

to conceptualize QOL (Fig. 11.1). They consider the quality of place (livability) to be the result of the interaction between the physical and social domains, and sustainability is the result of the interaction between the physical and economic domains. The interaction among three domains is alternatively defined as sustainability and QOL. The model implies that social, spatial, environmental, economic, and land-use planning cannot be done in isolation owing to the relationships between them.

11.1.3 Application of Remote Sensing and GIS in QOL Studies

Although many QOL studies do consider physical environmental impact, most of the data were obtained from field measurement and sampling (e.g., air quality and water quality are sampled at observation stations). As such, they could not provide detailed spatial patterns. Remote sensing data provide a great resource for extracting environmental variables for QOL analysis. Incorporating urban biophysical variables derived from remote sensing data with socioeconomic variables extracted from census data to assess QOL was pioneered by Green (1957), who employed aerial photographs to extract physical data, including housing density, the number of single-family homes, land uses adjacent to/within residential areas, and distance of residential areas to the central business district, and then combined those data with socioeconomic data, such as education, crime rate, and rental rates, to rank each residential area of Birmingham, Alabama, in terms "residential desirability." Later on, Green and Monier (1957) used the same method in other cities of the United States. Poverty of cities as an aspect of QOL also has been studied based on housing density and other indicators derived from aerial photograph by Mumbower and Donoghue (1967) and Metivier and McCoy (1971).

Advances in digital remote sensing and GIS technology have made QOL research more efficient to conduct based on the integration of remote sensing imagery and census socioeconomic data. For example, Weber and Hirsch (1992) developed urban life indices by combining remotely sensed Le Systeme Pour l'Observation de la Terre (SPOT) data with conventional census data, such as population and housing data, for Stasbourg, France. Strong correlations between census and remotely sensed data were found, mostly with housing-related data. Three urban QOL indices were developed based on the mixed data. They were interpreted as housing index, attractivity index, and repulsion index. However, each of these three indices describes only one aspect of QOL and cannot give a whole picture of QOL for a specific unit. Lo and Faber (1997) created QOL maps for the Athens (Clarke County), Georgia by integrating environmental factors including land use/land cover (LULC), surface temperature, and vegetation index [normalized difference vegetation index (NDVI)] derived from Landsat Thematic Mapper (TM) and census variables such as population density, per-capita income, median home value, and percentage of college graduates using principal component analysis (PCA) and GIS overlay methods, respectively. However, the first principal component, which the authors interpreted as the "greenness of the environment," explains only 54.2 percent of total variance; therefore, the measure of QOL based on the first principal component was not comprehensive because it did not incorporate the second principal component, that is, "personal traits," or higher orders of components. On the other hand, the GIS overlay method, which sums up ranked data layers (variables), was not able to remove redundant information existing in the highly correlated datasets.

11.2 Case Study: QOL Assessment in Indianapolis with Integration of Remote Sensing and GIS

11.2.1 Study Area and Datasets

Indianapolis, Indiana, was chosen to implement this study. A detailed description and a map of the study area can be found in Sec. 3.2.1 in Chapter 3. There are two primary data sources: U.S. Census 2000 and Landsat Enhanced Thematic Mapping Plus (ETM+). The Census 2000 data from the U.S. Census Bureau used in this study include tabular data stored in Summary Files 1 and 3, which contain information about population, housing, income, and education, and spatial data, called *topologically integrated geographic encoding and referencing* (TIGER) data, which contain data representing the positions and boundaries of legal and statistical entities. These two types of data are linked by Census geographic entity codes. The U.S. Census has a hierarchical structure composed of 10 basic levels: United States, region, division, state, county, county subdivision, place, Census tract, block group, and block. The block-group level was selected in this study. A Landsat 7 ETM+ image (row/path: 32/21) dated on June 22, 2000, was used. Atmospheric conditions were clear at the time of image acquisition, and the image was acquired through the U.S. Geological Survey (USGS) Earth Resource Observation Systems Data Center, which had corrected the radiometric and geometric distortions of the images to a quality level of 1G before delivery. The Census data and satellite image were coregistered to Universal Transverse Mercator (UTM) system before the integration.

11.2.2 Extraction of Socioeconomic Variables from Census Data

Selection of socioeconomic variables is based on the commonly used variables in previous studies (Lo and Faber, 1997; Smith, 1973; Weber and Hirsch, 1992). These variables include population density, housing density, median family income, median household income, per-capita income, median house value, median number of rooms, percentage of college above graduates, unemployment rate, and percentage of families under the poverty level. Initially, a total of 26 variables was extracted from Census 2000 Summary Files 1 and 3. A series of processes was performed to obtain the variables selected. A TIGER shape file of block group was downloaded from the Internet. The socioeconomic variables then were integrated with the TIGER shape file using geographic entity codes as attributes of the shape file.

11.2.3 Extraction of Environmental Variables

Previous studies show that vegetation greenness and urban land use within given districts are important indicators of QOL, with high greenness and low percentage of urban use being of higher quality. Greenness relates to vegetation and can be measured using vegetation indices such as the NDVI. However, NDVI values are affected by many other external factors, such as view angle, soil background, seasons, and differences in row direction and spacing in agricultural fields; therefore, it does not measure the amount of vegetation well (Weng et al., 2004). Urban land uses, such as transportation, commercial, and industrial uses, may be described as impervious surface, although impervious surface is not limited to urban uses. Impervious surface also may include some features in residential areas, such as buildings and sidewalks. Vegetation abundance and impervious surface are more accurate representations of urban morphologic composition. They can be obtained by using the technique of spectral mixture analysis (SMA).

In this study, three endmembers were identified initially from the ETM+ image based on high-resolution aerial photographs. The shade endmember was identified from the areas of clear and deep water, whereas green vegetation was selected from the areas of dense grass and cover crops. Different types of impervious surfaces were selected from building roofs and highway intersections. The radiances of these initial endmembers were compared with those of the endmembers selected from the scatterplot of Thematic Mapper (TM) 3 and TM 4 and the scatterplot of TM 4 and TM 5. The endmembers whose curves are similar but are located at the vertices of the scatter plot were finally used. A constrained least-squares solution was used to decompose the six ETM+ bands (1 to 5 and 7) into three fraction images (i.e., vegetation, impervious surface, and shade).

Temperature is an important factor affecting human comfort. High surface temperature is seen to be undesirable by most people; therefore, it can be used as an indicator of environmental quality (Lo and Faber, 1997; Nichol and Wong, 2005). Urban heat islands are a common phenomenon in the cities in which the urban area shows a higher temperature than the rural area. The thermal infrared band of ETM+ provides the source to extract surface temperatures. The procedure to extract land surface temperatures involves three steps: (1) converting the digital numbers of Landsat ETM+ band 6 into spectral radiance, (2) converting the spectral radiance to at-satellite brightness temperature, which is also called *blackbody temperature,* and (3) converting the blackbody temperature to land surface temperature. A detailed description of the procedures for extracting temperature images from Landsat ETM+ imagery can be found in Weng and colleagues (2004).

Since Census data and ETM+ data have different formats and spatial resolutions, they need to be integrated. With the help of the GIS function in ERDAS IMAGINE, remote sensing data were aggregated at the block-group level, and the mean values of green vegetation, impervious surface, and temperature were calculated for each block group. All these data then were exported into SPSS software for further analysis.

11.2.4 Statistical Analysis and Development of a QOL Index

Factor analysis is a statistical technique used to determine the number of underlying dimensions contained in a set of observed variables. The underlying dimensions are referred to as *factors*. These factors explain most of the variability among a large number of observed variables. In factor analysis, the first factor explains most of the variance in the data, and each successive factor explains less of the variance

(Tabachnick and Fidell, 1996). The number of factors to be selected depends on the percentage of variance explained by each factor. There are different factor-extraction methods. This study employed principal component analysis (PCA). Factors whose eigenvalues were greater than 1 were extracted (Kaiser, 1960).

Each factor can be viewed as one aspect of QOL. Therefore, factor scores can be used as a single index indicating the aspect with which the factor associates. A synthetic QOL index is a composite of different aspects. It is computed by the following equation:

$$\mathrm{QOL} = \sum_{1}^{n} F_i W_i \tag{11.1}$$

where n = the number of factors selected
F_i = the factor i score
W_i = the percentage of variance factor i explains

QOL maps were created to show the geographic patterns of QOL.

Ideally, either single or synthetic QOL scores developed based on factor analyses should be related to real QOL, and further, the approach can be validated. However, there were no such data available. Therefore, in this study, QOL scores created from factors were related to original indicators by developing regression models. For a single QOL score, predictors were those that had large loadings on the corresponding factor; for a synthetic QOL score, predictors included variables that had the highest correlation with the corresponding factors. These models can be applied to predict QOL in further studies.

11.2.5 Geographic Patterns of Environmental and Socioeconomic Variables

The distribution of per-capita income by block groups in Marion County shows that the highest per-capita incomes were found in the north, northeast, and northwest portions of the county, whereas the lowest per-capita incomes were found in the center of the county. Three environmental variables, including green-vegetation fraction, impervious surface fraction, and shade fraction, were further extracted from the Landsat image using SMA. The green-vegetation fraction showed that the highest values of green vegetation were observed in forest, grassland, and cropland areas, whereas the lowest values were found in the urban and water areas. In contrast, the highest values of the impervious surface fraction were found in the urban area, whereas the lowest values were found in forest, grassland, and water areas. The temperature image derived from ETM+ band 6 indicated that high surface temperatures were found in the urban area, especially downtown, whereas low temperatures were found in vegetated areas and water bodies. These remote sensing variables then were aggregated at the block-group level, and their mean values for each block group were calculated.

Pearson's correlation coefficient was computed to give a preliminary analysis of the relationships among all variables. Table 11.1 displays the correlation matrix. Green vegetation had a significant positive relationship with all income variables ($r = 0.336$ to 0.467), median house value ($r = 0.340$), median number of rooms ($r = 0.490$), and education level ($r = 0.301$) and a negative relationship with density variables ($r = -0.226$ and -0.265), temperature ($r = -0.772$), impervious surface ($r = -0.871$), percentage of poverty ($r = -0.421$), and unemployment rate ($r = -0.284$). Percentage of college graduates had a very high correlation with income variables and house characteristics, which indicates that well-educated people make more money and live well. The relationships between impervious surface and temperature and other variables were in contrast to vegetation. Because high correlations existed among these variables, it is necessary to reduce the data dimension and redundancy.

11.2.6 Factor Analysis Results

As a general guide in interpreting factor analysis results, the suitability of data for factor analysis was first checked based on Kaiser-Meyer-Olkin (KMO) and Bartlett test values. Only when KMO was greater than 0.5 and the significance level of the Bartlett test was less than 0.1 were the data acceptable for factor analysis. The second step was to validate the variables based on communality of variables. Small values indicate that variables do not fit well with the factor solution and should be dropped from the analysis. Initially, all 13 variables were input for processing. The KMO (0.847) and Bartlett tests (significant level 0.000) indicated that the data were suitable for factor analysis. However, there were three variables, namely, median number of rooms, unemployment rate, and percentage of families under the poverty level, with low communality values. These three variables were dropped from the analysis. Therefore, 10 variables were finally entered into the factor analysis. Based on the rule that the minimum eigenvalue should not be less than 1, three factors were extracted from the factor analysis. For the purpose of easy interpretation, the factor solution was rotated using varimax rotation (Table 11.2). The first factor (factor 1) explained about 40.67 percent of the total variance, the second factor (factor 2) accounted for 24.69 percent, and the third factor (factor 3) explained 21.86 percent. Together, the first three factors explained more than 87.2 percent of the variance.

Interpreting factor loadings is the key in factor analysis. Factor loadings are measurements of relationships between variables and factors. Generally speaking, only variables with loadings greater than 0.32 should be considered (Tabachnick and Fidell, 1996). Comrey and Lee (1992) suggested a range of values to interpret the strength of the

	PD	HD	GV	IMP	T	MHI	MFI	PCI	POV	PCG	UNEMP	MHV
HD	0.917*											
GV	–0.226*	–0.265*										
IMP	0.065*	0.085	–0.871*									
T	0.510*	0.506*	–0.722*	0.652*								
MHI	–0.297*	–0.328*	0.467*	–0.521*	–0.536*							
MFI	–0.273*	–0.264*	0.419*	–0.508*	–0.491*	0.926*						
PCI	–0.270*	–0.194*	0.336*	–0.482*	–0.453*	0.808*	0.856*					
POV	0.344*	0.357*	–0.421*	0.370*	0.427*	–0.623*	–0.622*	–0.524*				
PCG	–0.262*	–0.181*	0.301*	–0.426*	–0.399*	0.700*	0.746*	0.818*	–0.437*			
UNEMP	0.235*	0.188*	–0.284*	0.265*	0.273*	–0.436*	–0.465*	–0.435*	0.561*	–0.459*		
MHV	–0.210*	–0.160*	0.340*	–0.451*	–0.402*	0.720*	0.740*	0.791*	–0.372*	0.725*	–0.343*	
MR	–0.092†	–0.186*	0.490*	–0.522*	–0.386*	0.695*	0.604*	0.458*	–0.367*	0.384*	–0.313*	0.479*

Note: PD = population density; HD = housing density; GV = green vegetation; IMP = impervious surface; T = temperature; MFI = median household income; MFI = median family income; PCI = per-capita income; POV = percentage of families under poverty level; PCG = percentage of college or above graduates; UNEMP = unemployment rate; MHV = median house value; MR = median number of rooms.

*Correlation at the 99 percent confidence level (two-tailed).

†Correlation at the 95 percent confidence level (two-tailed).

Table 11.1 Correlation Matrix of Variables

Indicator	Communality	
	13 Variables	10 Variables
Population density	0.933	0.947
Housing density	0.939	0.949
Green vegetation	0.920	0.932
Impervious surface	0.914	0.931
Temperature	0.781	0.816
Median household income	0.854	0.837
Median family income	0.879	0.874
Per-capita income	0.850	0.887
Percentage of college graduates	0.758	0.787
Median house value	0.710	0.762
Median number of rooms	0.496	
Percentage of families under poverty	0.515	
Unemployment rate	0.349	

TABLE 11.2 Communality for the 13 Variables and 10 Variables

relationships between variables and factors. Loadings of 0.71 and higher are considered excellent, 0.63 is very good, 0.55 is good, 0.45 is fair, and 0.32 is poor. Table 11.3 presents factor loadings on each variable. Factor 1 has strong positive loadings (>0.8) on five variables, including median household income, median family income, per-capita income, median house value, and percentage of college or above graduates. Apparently, factor 1 is associated with material welfare. The higher the score on factor 1, the better is the QOL in economic respects. Factor 2 has a high positive loading on green vegetation (0.94) and negative loadings on impervious surface (–0.904) and surface temperature (–0.716). Factor 2 is clearly related to environmental conditions. The higher the score on factor 2, the better is the environment quality. Factor 3 shows high positive factor loadings on population density and housing density and thus is related to crowdedness. The higher the score on factor 3, the smaller is the space in which people live.

The factor scores can be used as indices to represent the QOL in different dimensions. The distribution of each factor was mapped in Figs. 11.2, 11.3, and 11.4, respectively. Factor 1, the economic sector of QOL, has a similar distribution pattern as per-capita income because

	Factor 1	Factor 2	Factor 3
Population density	–0.178	–0.085	0.953
House density	–0.116	–0.132	0.958
Green vegetation	0.159	0.940	–0.153
Impervious surface	–0.328	–0.905	–0.061
Temperature	–0.283	–0.716	0.472
Median household income	0.835	0.295	–0.230
Median family income	0.885	0.244	–0.176
Per-capita income	0.918	0.168	–0.129
Percentage of college graduates	0.871	0.152	–0.070
Median house value	0.853	0.174	–0.069
Initial eigenvalues	5.520	1.770	1.430
Percent of variance	40.67	24.69	21.56
Cumulative percent	40.67	65.36	87.21

Table 11.3 Rotated Factor Loading Matrix

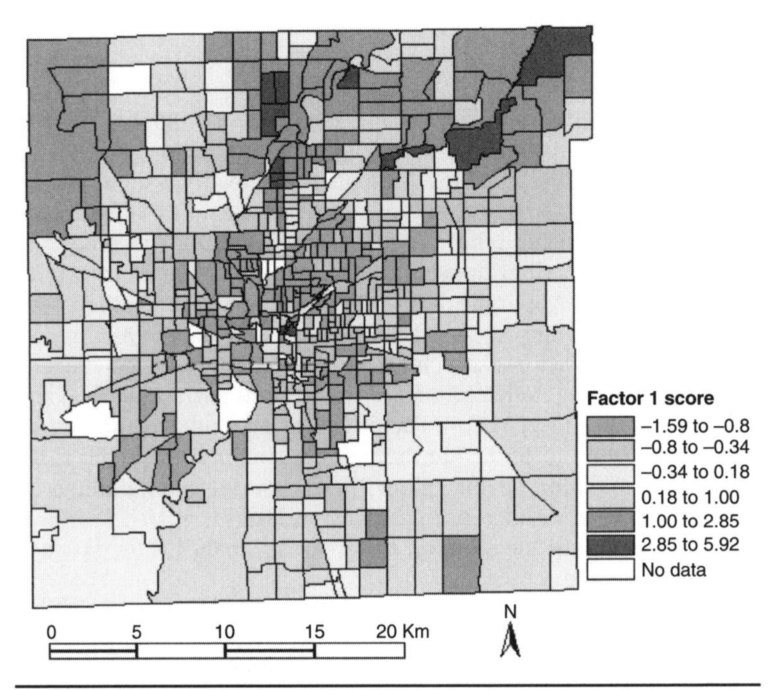

Figure 11.2 The first factor score—economic index. (*Adapted from Li and Weng, 2007.*) See also color insert.

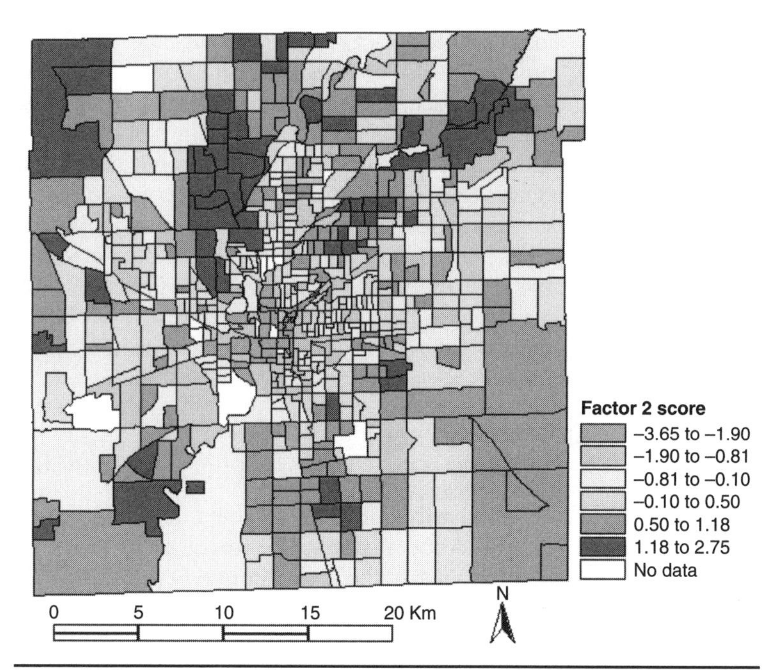

FIGURE 11.3 The second factor score—environmental index. (*Adapted from Li and Weng, 2007.*) See also color insert.

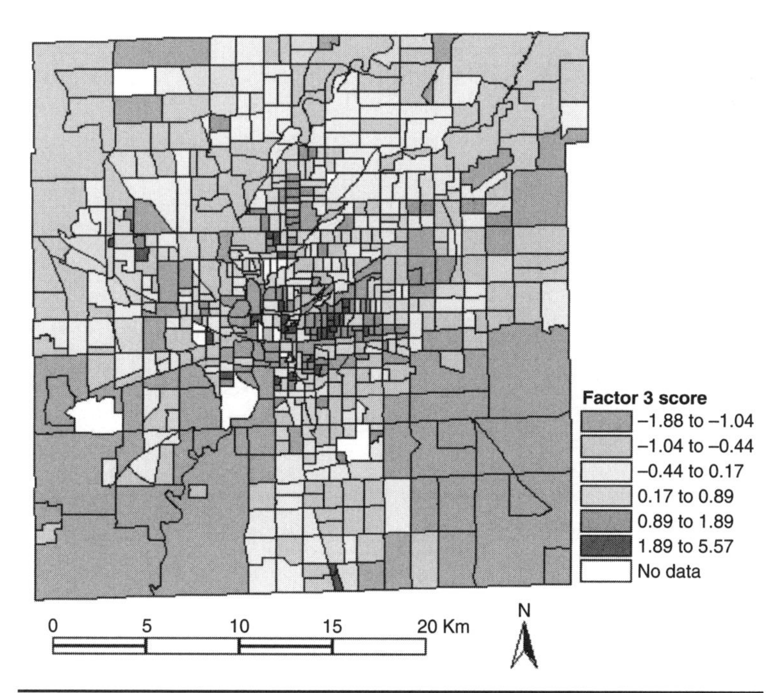

FIGURE 11.4 The third factor score—crowdedness. (*Adapted from Li and Weng, 2007.*) See also color insert.

it has the largest loading on the per-capita income variable. Similarly, factor 2, the environmental sector, has a distribution pattern similar to that of green vegetation. Factor 3, which represents crowdedness, has a similar distribution with housing density. It is noted that there were some non-residential block groups lacking data because these block groups missed at least one type of socioeconomic variable.

Development of a synthetic QOL index involved combination of the three factors that represent different aspects of QOL. Factors 1 and 2 have a positive contribution to QOL, whereas factor 3 has a negative correlation with QOL. The aggregate score for each block group then was obtained by adding weighted factor scores of the three factors using the following equation:

$$\text{QOL} = (40.666 \times \text{factor } 1 + 24.689 \times \text{factor } 2 - 21.859 \times \text{factor } 3)/100 \tag{11.2}$$

Figure 11.5 shows the distribution of QOL scores. The QOL scores ranged from –1.15 to 2.84. About 5 percent of the block groups had scores greater than 0.9, and most of them were found in the surrounding areas of the county, especially to the north. These block groups

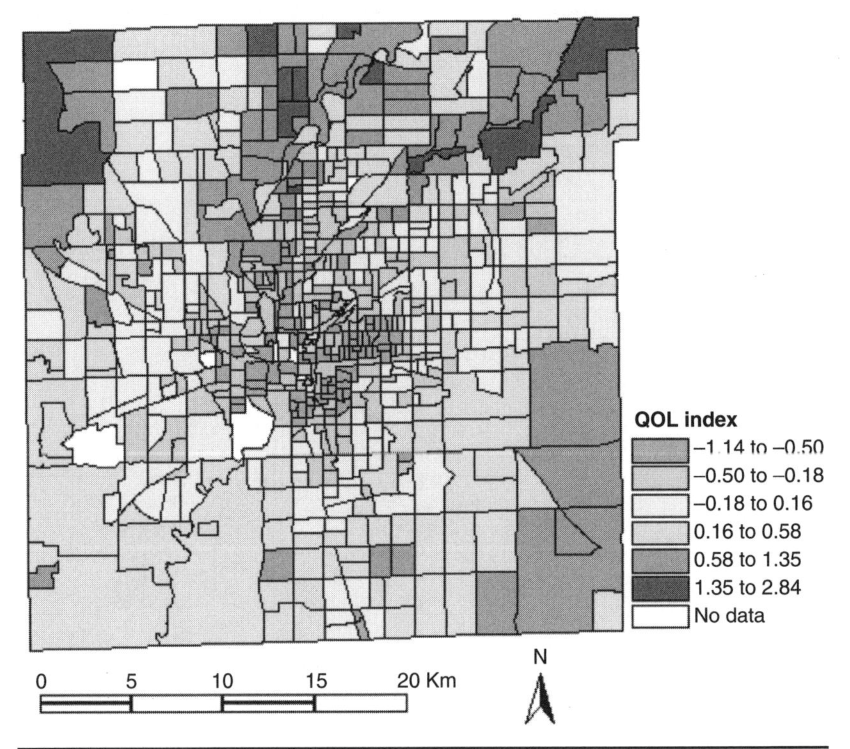

Figure 11.5 Synthetic quality of life index. (*Adapted from Li and Weng, 2007.*) See also color insert.

were characterized by low population density, large green-vegetation coverage, low temperature, less impervious surface, and high family income. Block groups with scores ranging from –1.15 to –0.3 accounted for 30 percent. Most of them were found in the city center, which was characterized by less green vegetation, high population density, and low per-capita income.

11.2.7 Result of Regression Analysis

Once the QOL indices were created based on factor analysis, regression analysis can be applied to relate QOL index values to environmental and socioeconomic variables. For a specific aspect of QOL, factor scores were regressed against the variables that had high loadings. Since factor 1 had high correlations with income, home value, and percentage of college or above graduates, these variables were used as predictive variables in the regression model for the economic aspect of QOL. Factor 2 had high correlation with green vegetation, impervious surface, and surface temperature, so these variables were employed in developing an environmental QOL model. Population- and housing-density variables were used to develop a crowdedness index model. Three variables, namely, per-capita income, green vegetation, and housing density, which had the highest loading on the corresponding factor, were used in developing a synthetic QOL model. Table 11.4 presents the best models selected based on R^2 and ease of implementation. All regression models produced a high value of R^2, especially for the synthetic model, in which R^2 reached 0.94.

Model	Predictors	Coefficients	R^2
Economic QOL	Constant	–1.698	0.92
	Per-capita income	4.388×10^{-5}	
	Median house value	5.093×10^{-6}	
	Percentage of college or above graduates	1.537×10^{-2}	
Environmental QOL	Constant	–1.143	0.91
	Green vegetation	7.244×10^{-2}	
	Impervious surface	5.871×10^{-2}	
Crowdedness	Constant	–1.282	0.92
	Housing density	1.720×10^{-3}	
Synthetic QOL	Constant	–1.178	0.94
	House density	-2.756×10^{-4}	
	Green vegetation	2.007×10^{-2}	
	Per capita income	3.372×10^{-5}	

TABLE 11.4 Selected QOL Estimation Models

11.3 Discussion and Conclusions

This chapter has presented a methodology to develop measures for the QOL in Indianapolis, Indiana, based on the integration of remote sensing imagery and Census data. Correlation analysis explored the relationship between environmental and socioeconomic characteristics and found that green vegetation had a strong positive correlation with income, house value, and education level and a negative relationship with temperature, impervious surface, and population/housing density. Factor analysis provided an effective way to reduce data dimensions and redundancy. Three factors were derived from 10 original variables, representing the economic, environmental, and demographic dimensions of the QOL, respectively. Regression analysis allowed for prediction of QOL based on environmental and socioeconomic variables. An important issue encountered was how to integrate different indicators into a synthetic index. There is currently no compelling theory for combining different indicators into one index (Schyns and Boelhouwer, 2004). Because of the lack of available criteria for weighing the indicators, this study applied a rather pragmatic solution: the factors as component indicators and the percentage of variance that a factor explains as associated weights.

This research also has demonstrated that GIS can provide an effective platform for integrating different data models from different data sources, such as remote sensing and census socioeconomic data, and for creating a comprehensive database to assess QOL. This would help urban managers and policymakers in formulating the strategies of urban development plans. However, several issues raised in the integration of disparate data should be kept in mind. Remote sensing and census data are collected for "different purpose, at different scales, and with different underlying assumptions about the nature of the geographic features" (Huang and Yasuoka, 2000). Remote sensing data are digital records of spectral information about ground features with raster format and often exhibit continuous spatial variation. Census socioeconomic data usually relate to administrative units such as blocks, block groups, tracts, counties, and states and tend to be more discrete in nature with sharp discontinuities between adjacent areas. More often, socioeconomic data are integrated into vector GIS as the attributes of its spatial units for various mapping and spatial analysis purposes. Integration of remote sensing and census socioeconomic data involves the conversion between data models. In this study, remote sensing data were aggregated to census block groups with raster-to-vector conversion, where values were assumed to be uniform throughout block groups. This would lead to loss of spatial information existing in remote sensing data. In addition, the census has different scales (levels), and integration of remote sensing data with different scales of census data would produce the so-called modifiable area unit problem. Therefore, finding desirable aggregation units is important to reduce the loss of spatial information from remote sensing data. Another method of data integration is through vector-to-raster conversion by rasterization or surface interpolation to produce a raster layer for each socioeconomic variable. More research is needed on disaggregating census data into individual pixels to match remote sensing data for the purpose of data integration.

QOL is a great research topic and concerns many different aspects of human life. There has been a great increase in amount of time, effort, and resources concentrated on QOL studies. However, there are still vast differences of opinions on the indicators and ingredients of QOL. Ideally, QOL research needs to incorporate every dimension of QOL and combine objective and subjective measurements together. In order to conduct such huge project, coalitions of different organization, such as nongovernmental groups and local government, are necessary. Owing to difficulties in collecting all the data related to QOL, especially for detailed geographic units such as census block groups, this study was conducted only from socioeconomic and environmental perspectives and explored the spatial variations of QOL, which may help city planners to understand problems and to find solutions to issues faced by their communities.

References

Andrew, K. 1999. Quality of life in cities. *Cities* **16,** 221–222.

Barsell, K., and Maser, E. 2004. Taking indicators to the next level: Truckee Meadows tomorrow launches quality of life compacts. In *Community Quality of Life Indicators,* Social Indicators Research Series 22, edited by M. J. Sirgy, D. R. Rahtz, and D.-J. Lee, pp. 53–74. Boston, MA: Kluwer Academic Publishers.

Camagni, R., Capello, R., Nijkamp, P. 1997. Towards sustainable city policy: an economy–environment technology nexus. *Ecology and Economy,* **24,** 103–118.

Campbell, A., Converse, P.E., and Rodgers, W. L., 1976. The Quality of American Life: Perceptions, Evaluation, and Satisfactions. New York: Russell Sage Foundation.

Comrey, A. L., and Lee, H. B. 1992. *A first Course in Factor Analysis,* 2nd ed. Hillsdale, NJ: Erlbuum.

Cornwell, T. L. 2004. Vital Signs: quality of life indicators for Virginia's technology corridor. In *Community Quality of Life Indicators Best Cases, Social Indicators Research Series 22,* edited by Joseph Sirgy, Don Rahtz, and Dong-Jin Lee, Kluwer academic publishers, Dordrecht/boston/London: Kluwer academic publishers, pp. 1–27.

Fallowfield, L. 1990. The Quality of Life: The Missing Measurement in Health Care. London: Souvenir Press.

Flanagan, J. C., 1978. A research approach to improving our quality of life. *American Psychologist,* **33,** 138–147.

Flanagan, J. C., 1982. Measurement of quality of life: current state of the art. *Archives of Physical Medicine and Rehabilitation,* **63,** 56–59.

Friedman, M., 1997. *Improving the quality of life: A holistic scientific strategy*. Westport, Connecticut, London, pp. 19–57.

Foo, T. S. 2001. Quality of life in cities. *Cities* **18,** 1–2.

Green, N. E. 1957. Aerial photographic interpretation and the social structure of the city. *Photogrammetric Engineering* **23,** 89–96.

Green, N. E., and Monier, R. B., 1957. Aerial photographic interpretation and the human geography of the city. *Professional Geographers* **9,** 2–5.

Huang, T., and Yasuoka, Y. 2000. Integration and application of socio-economic and environmental data within GIS for development study in Thailand; available at *www.gisdevelopment.net/aars/acrs/2000/ts7/gdi004pf.htm.* Last accessed on June 15, 2008.

Kaiser, H. 1960. The application of electronic computers to factor analysis. *Education and Psychology* **52,** 621–652.

Kamp, I. V., Leidelmeijer, K., Marsman, G., and Hollander, A. U. 2003. Urban environmental quality and human well-being towards a conceptual framework and demarcation of concepts: A literature study. *Landscape and Urban Planning* **65,** 5–18.

Kirby, A., 1999, Quality of life in cities. *Cities* **16,** 221–222.

Lawrence, R. J., 2001. Human Ecology. In *Our Fragile World: Challenges and Opportunities for Sustainable Development,* edited by M. K. Tolba, **1,** 675–693.

Li, G., and Weng, Q. 2007. Measuring the quality of life in city of Indianapolis by integration of remote sensing and census data. *International Journal of Remote Sensing* **28,** 249–267.

Lo, C. P., and Faber, B. J. 1997. Integration of Landsat Thematic Mapper and census data for quality of life assessment. *Remote Sensing of Environment* **62,** 143–157.

Metivier, E. D., and McCoy, R. M. 1971. Mapping urban poverty housing from aerial photographs. In *Proceedings of the Seventh International Symposium on Remote Sensing of Environment,* Ann Arbor, MI, pp. 1563–1569.

Mumbower, L. E., and Donoghue, J. 1967. Urban poverty study. *Photogrammetric Engineering* **33,** 610–618.

Nichol, J., and Wong, M. S. 2005. Modeling urban environmental quality in a tropical city. *Landscape and Urban Planning* **73,** 49–58.

Ordway, S. 1953. *Resources and the American Dream: Including a Theory of the Limit of Growth.* New York: Ronald Press.

Osborn, F. 1954. *The Limits of the Earth.* Boston, MA: Little, Brown.

Schyns, P., and Boelhouwer, J. 2004. The state of city Amsterdam monitor: Measuring quality of life in Amsterdam. In *Community Quality of Life Indicators, Social Indicators Research Series 22,* edited by M. J., Sirgy, D. R., Rahtz, and D.-J. Lee, pp. 133–152. Boston: Kluwer Academic Publishers.

Scottish Executive Social Research, 2005. Well-Being and Quality of Life: Measuring the Benefits of Culture and Sport: A Literature Review and Thinkpiece, p7.

Shafer, C.S., Koo Lee, B., and Turner, S. 2000. A tale of three greenway trails: user perceptions related to quality of life. *Landscape and Urban Planning,* **49,** 163–178.

Smith, D. M. 1973. *The Geography of Social Well-Being in the United States.* New York: McGraw-Hill.

Tabachnick, B., and Fidell, L. 1996. *Using Multivariate Statistics,* 3rd ed. New York: Harper Collins College.

Weber, C., and Hirsch, J. 1992. Some urban measurements from SPOT data: Urban life quality indices. *International Journal of Remote Sensing* **13,** 3251–3261.

Weng, Q., Lu, D., and Schubring, J. 2004. Estimation of land surface temperature–vegetation abundance relationship for urban heat island studies. *Remote Sensing of Environment* **89,** 467–483.

CHAPTER 12

Urban and Regional Development

The interface between socioeconomic development, urban growth, and environmental change has not been well understood, especially in developing countries. In the Zhujiang Delta, China, accelerated economic growth since 1978 has brought about a dramatic change in land use and land cover (LULC) patterns and created severe environmental consequences. Rapid industrialization and urbanization have resulted in the loss of a significant amount of agricultural land. This chapter develops an integrated approach to remote sensing and GIS to investigate the relationships between socioeconomic drivers and urban growth patterns. Measurement of LULC change in the Zhujiang Delta in the period of 1989 to 1997 is first presented via an integrated approach of remote sensing, geographical information systems (GIS), and field survey. Subsequently, these satellite-detected LULC changes are linked to socioeconomic development indicators by correlation and regression analysis. Finally, the spatial pattern of urban LULC change is measured and analyzed using the concept of distance decay and Theil's entropy index.

12.1 Regional LULC Change

Knowledge of LULC change is significant to a range of issues and themes in geography and central to the study of environmental change and human-environment interactions. It is also important for planning and management practices. Because of its implications for global environmental change and sustainable development, the International Geosphere-Biosphere Program (IGBP) and the Human Dimensions of Global Environmental Change Program (HDP) have recently jointly addressed the issue of LULC. The results of this joint effort are a series of important reports (Turner et al., 1995).

Globally, the most important land uses, spatially and economically, are related to urbanization, cultivation, forestry, recreation, and conservation. These and other land uses have altered the properties of the earth's land surface by transforming the biophysical states of land cover. A range of human driving forces, including social, economic, political, and cultural variables, determines the patterns of land use. Changes in land use can have strong impacts on local, regional, and even global environments, including climate, atmospheric chemistry, hydrology, water quality, soil condition, and ecological complexity. Environmental changes, in turn, may have feedback effects on land covers, land uses, and human driving forces. This series of interactions forms one core of research in human-environment interactions.

This section first discusses the concepts of LULC and their relationships, followed by a brief examination of the interplay between land-use change and land-cover change. Finally, this section will look into the forces that drive LULC change.

12.1.1 Definitions of Land Use and Land Cover

Land use can be defined as the human use of land. Settlement, cultivation, rangeland, and recreation are examples. Land use involves both the *manner* in which the biophysical attributes of the land are manipulated and the *purpose* for which the land is used (Turner et al., 1995). *Biophysical manipulation* refers to the specific ways in which human beings treat vegetation, soil, and water for their purpose, such as the cut-burn-hoe-weed sequence in many slash-and-burn agricultural systems; the use of fertilizers, pesticides, and irrigation for mechanized cultivation on arid lands; and so on. On the other hand, purposes of land use underlie the manipulation, such as forestry, parks, livestock herding, suburbia, and farmlands. Because land use has been a concern of many disciplines, to give it a perfect definition is notoriously difficult (Campell, 1983).

Land cover can be defined as the biophysical state of the earth's surface and immediate subsurface, including biota, soil, topography, surface and ground water, and human structures (Turner et al., 1995). In other words, it describes both natural and human-made coverings of the earth's surface. Examples of land covers include forest, grassland, cropland, wetland, urban structures, and so forth.

Land use, as just defined, is abstract and not always directly observable; in contrast, land cover is concrete and therefore subject to direct observation (Campell, 1983). Another distinction between land use and land cover lies in the emphasis of economic function in the concept of land use. The distinctions become more important as the scale of a study becomes larger and the level of detail becomes finer (Campell, 1983).

12.1.2 Dynamics of Land Use and Land Cover and Their Interplay

The relationship between land use and land cover is not always direct and obvious. A single class of land cover may support multiple uses. Forest, a land cover dominated by woody species, may be exploited for uses as varied as recreation, timber production, and wildlife conservation (Turner et al., 1993). Moreover, a land-cover type may be used for contrasting activities during different seasons of the year. Some farmland might be employed alternatively as cropland and as pasture at different times in an agricultural calendar (Campell, 1983). On the the other hand, a single land use may involve the maintenance of several distinct land covers in that certain farming systems combine cultivated land, wood lots, improved pasture, and settlements (Meyer and Turner, 1994).

Land use and land cover are linked by the proximate sources of change, the human actions that directly alter the physical environment (Meyer and Turner, 1994; Turner et al., 1995). It is through proximate sources that human goals of land use are translated into changed physical states of land cover (Meyer and Turner, 1994). Examples of proximate sources include cutting and burning, fertilizer application, species transfer, plowing, irrigation, drainage, livestock pasturing, pasture improving, and so on. Further complication occurs when this linkage is studied across different spatial and temporal scales (Turner et al., 1995).

Generally speaking, land-use changes can be subsumed under three main types: (1) urbanization, the conversion of other types of land to uses associated with the growth of populations and economies; (2) agriculture, including primarily the conversion of forests and wetlands to agriculture and losses of cropland owing to urbanization and other factors; and (3) forestry, the conversion of forests to other uses (deforestation) (Jacobson and Price, 1991). Changes in land use are frequent causes of land-cover change. Through the course of human history, most land-cover change has been driven by human use instead of natural change. In this sense, land-cover change is an immediate response to land-use changes (Turner et al., 1993). The proximate sources produce alterations in the properties of the earth's surface in a form ranging from minor modification of the existing cover to complete conversion to a new cover type. However, the use-to-cover relationship is not unidirectional. Land-cover changes cause environmental changes, which, in turn, may have feedback effects on land cover, land uses, and human driving forces that shape the direction and intensity of land use (Fig. 12.1). In this chapter, land-cover change is used as equivalent to

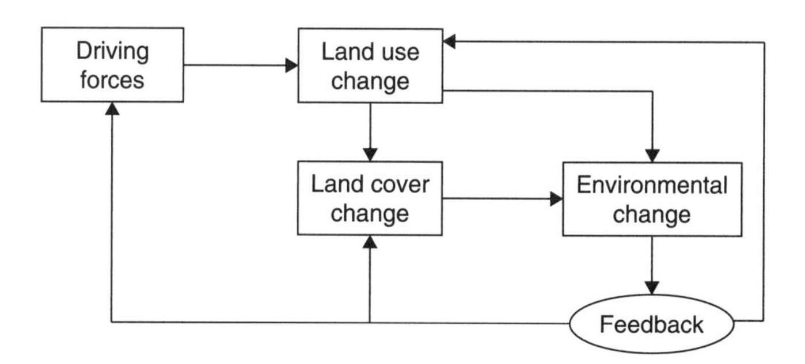

FIGURE 12.1 The interplay between land use and land cover change.

land-cover conversion for the purpose of gearing to the type of information that remote sensing technology can yield.

12.1.3 Driving Forces in LULC Change

Land use is a product of interactions between the biophysical forces and the human driving forces (Fig. 12.2). However, through much of the course of human history, land has been a basic element of production and consumption. Land use is changed mostly by human use and not by natural processes. A major effort therefore is oriented toward developing a profound knowledge of the interactions between the human driving forces and land use. The possible human forces driving land-use change can be grouped into five categories, as suggested by the literature as important explanatory variables (Meyer and Turner, 1992; Stern et al., 1992). They are population, level of affluence, technology, political and economic institutions, and cultural attitudes and values.

Population

Each person in a population makes some demand on the environment and the social system for the essentials of life, including food, water, clothing, shelter, and so on (Stern et al., 1992). On a global scale, population growth has been positively associated with the expansion of agricultural and urban land, land intensification, and deforestation. However, in some areas of the world, for example, in western Europe, afforestation is found to accompany the further increase of population density (Turner et al., 1996). This controversy indicates the importance of further studies between population and land-use change.

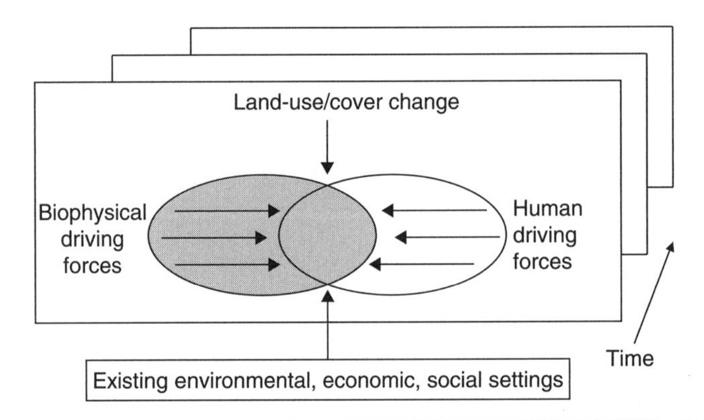

FIGURE 12.2 Conceptual model of human-environment interactions in LULC change.

Level of Affluence

Economic activity has long been a major source of land-use change. Today, economic activity is so extensive that it produces environmental change on a global level (Stern et al., 1992) through changes in land use and land cover. High levels of affluence and economic development are often associated with higher resource demands, although they could be reduced by advanced technologies (Turner et al., 1993). On the other hand, poverty usually has an affinity with environmental degradation, as has been observed widely in developing countries. The amount of land and resources needed for a given amount of economic growth depends, among other things, on the patterns of goods and services produced, the population and resources base for agricultural development, the forms of national political organization, and development policies.

Technology

The development of technology changes the usefulness and societal demand for various natural resources (Turner et al., 1996), as well as the manner, scope, and intensity of resource use. Different kinds of technology produce different environmental impacts from the same process (e.g., fossil fuel and nuclear energy production have different influences) (Stern et al., 1992). Two particular examples are transportation infrastructure and energy, which have important implications for an underdeveloped society and inevitably alter its land uses.

Political-Economic Institutions

It is reasonable that the political and economic institutions that control the exchange of goods and services, as well as structure the decisions, of large human groups have a strong influence on land uses and their changes. These institutions include economic and governmental institutions at all levels of aggregation (Stern et al., 1992), to which land uses must respond. Political and economic institutions can affect land use along many causal pathways: markets, governments, and the international political economy. Markets are always imperfect, and the impact of economic activity on the environment depends on which imperfect-market method of environment management is being used (Stern et al., 1992). Governmental policies can have significant impacts on land use and therefore on land cover and the environment. The international political economy, with its global division of labor and wealth, can promote environmental abuses, particularly in the third world (Stern et al., 1992).

Cultural Attitudes and Values

Cultural attitudes and values are related to material possessions, and the relationship between humanity and nature is often seen as lying at the root of environmental degradation (Stern et al., 1992). This relationship is clearly reflected in land-use activities. The effects of cultural attitudes and values on land use and the environment are of long term and should be observed from a historical perspective and using a comparative method. However, within single lifetimes, attitudes and values also may have significant influence on resource-using behavior, even when social and economic variables are held constant (Stern et al., 1992).

12.2 Case Study: Urban Growth and Socioeconomic Development in the Zhujiang Delta, China

A detailed description of the study area can be found in Sec. 8.2 in Chapter 8. The objective of this chapter is to examine the relationship between urban growth and driving forces in the Zhujiang Delta by employing an integrated approach of remote sensing, GIS, and spatial modeling. The case study will answer the following questions:

1. How has land-use/cover changed in the Zhujiang Delta between 1989 and 1997 under the influences of rapid industrialization and urbanization?
2. What are the underlying causes or driving forces behind those LULC changes?

12.2.1 Urban Growth Analysis

LULC patterns for 1989 and 1997 were mapped using Landsat Thematic Mapper (TM) data (dates: December 13, 1989 and August 29, 1997). Seven categories were classified, including (1) urban or built-up land, (2) barren land, (3) cropland, (4) horticulture farms, (5) dike-pond land, (6) forest land, and (7) water. A supervised classification with the maximum likelihood algorithm was conducted to classify the Landsat images using bands 2 (green), 3 (red), and 4 (near infrared). The accuracy of the classification was verified by field checking or comparing with existing LULC maps that have been field checked.

Post classification change detection was conducted by producing a change matrix. Quantitative areal data on the overall changes, as well as gains and losses in each category, were compiled. In order to analyze urban LULC change, an image of urban and built-up land was extracted from each LULC map. The two extracted images then were overlaid and recoded to obtain an urban land-change (expansion) image. This image was further overlaid with several geographic reference maps to analyze the patterns of urban growth, including a map of administrative boundaries, major roads, and urban centers.

12.2.2 Driving Forces Analysis

Land use is changed more by human use than by natural processes. This is especially true in the Zhujiang Delta, where human uses have influenced every inch of the land. The possible human forces driving LULC change can be grouped into five categories: population size, level of affluence, technology, political and economic institutions, and cultural attitudes and values. Accordingly, the following variables were selected for modeling (based on the availability of data): (1) population density, (2) per-capita gross value of industrial output (GVIO), (3) per-capita gross value of agricultural output (GVAO), (4) urban population percentage, (5) per-capita value of gross domestic production, and (6) per-capita grain production. No independent variables were selected from the category of technology because it is difficult to find a surrogate for it. None were from the categories of political and economic institutions, as well as cultural attitudes and values, because there is little possibility of conducting an investigation of the statistical relationships of these three sets of driving forces with the LULC change. Furthermore, the delta is a small region, has a highly similar (if not identical) cultural background, and is under the same political regime.

The relationship between urban growth and the independent variables was first investigated by correlation analysis because it can provide a general support for the explanatory variables under evaluation. However, the conceptually and empirically interrelated nature of the variables suggests that the analysis should be conducted in a multivariate framework. Using urban growth rate in each county/city as a dependent variable and the preceding six independent variables, a multiple linear regression model was applied to analyze the driving forces of urban growth.

12.2.3 Urban LULC Modeling

In an attempt to account for the spatial pattern of land use in a logical, consistent way, economic geographers and regional scientists have long been exploring the field. The earliest work by Von Thunen delineated the concentric zonation pattern of agricultural land uses around a city. His analysis has been extended in recent decades from the countryside into the heart of a town. Regional scientists suggest that a pattern of concentric zonation of land use also can be found within a city because of land users' consideration of maximum accessibility. If the assumptions employed in these analyses are correct, a city should be surrounded by concentric zones of different types of agriculture and forest lands. The characteristics of urban encroachment over agricultural and other natural uses around an expanding city should be sufficiently consistent in reference to the distance from the center of the city. To what extent do the cities in the Zhujiang Delta and their environs conform to the patterns of concentric zonation, and what is the relationship between urban expansion and accessibility? The techniques of GIS in conjunction with spatial modeling can facilitate these analyses.

Many methods have been developed for measuring the spatial pattern, the degree of spatial concentration and dispersion, of a geographic phenomenon or process. The well-known measures include variance, standard deviation, mean deviation, Gini coefficient, Hoover's concentration index, and Theil's entropy index. Each has some advantages and disadvantages (Mulligan, 1991). The measure employed in this study is a version of the Theil index of inequality derived from the notion of entropy in information theory (Theil, 1967). The idea behind information entropy is that unlikely events receive more weight than those that conform to expectations (Nissan and Carter, 1994). This measure is expressed as

$$H(y) = \sum_{i=1}^{N} y_i \log(1/y_i) \tag{12.1}$$

where y_i is the share of a geographic variable in the ith region in a total of N regions, and it is required to be nonnegative and to add to 1:

$$\sum_{i=1}^{N} y_i = 1 \qquad y_i \geq 0 i = 1,\ldots,N \tag{12.2}$$

The entropy index is always between 0 and 1. If one share is 1 and all others are 0, then $H(y) = 0$ (the minimum entropy value), and we have the maximum degree of concentration. If all shares are equal and hence equal to $1/N$, we have $H(y) = \log N$, which is the maximum entropy value and also the minimum degree of concentration. Obviously, the larger the entropy index, the lower is the degree of concentration. Therefore, this index can be regarded as an inverse measure of concentration, that is, a measure of dispersion.

In this study, the entropy index is applied to measure the spatial pattern of urban expansion in the Zhujiang Delta during the period of 1989 to 1997. Using the buffer function in GIS, two buffer images were generated showing the proximity to the city centers and the major roads. Ten buffer zones were created around a city center with a width of 1000 m, and 10 buffer zones were created around a major road with a width of 500 m. Local conditions have been taken into account in selecting these buffer widths. Each of the two buffer images was overlaid with the urban expansion image obtained by digital image processing to calculate the amount of urban expansion in each zone. The density of urban expansion then was calculated by dividing the amount of urban expansion by the total amount of land in each buffer zone. These values of density are the shares for computing the entropy index for each city and are used to construct distance-decay functions of urban expansion.

12.2.4 Urban Growth in the Zhujiang Delta, 1989–1997

The remote sensing–GIS analysis indicates that urban or built-up land has expanded by 47.68 percent (65,690 ha) in the Zhujiang Delta

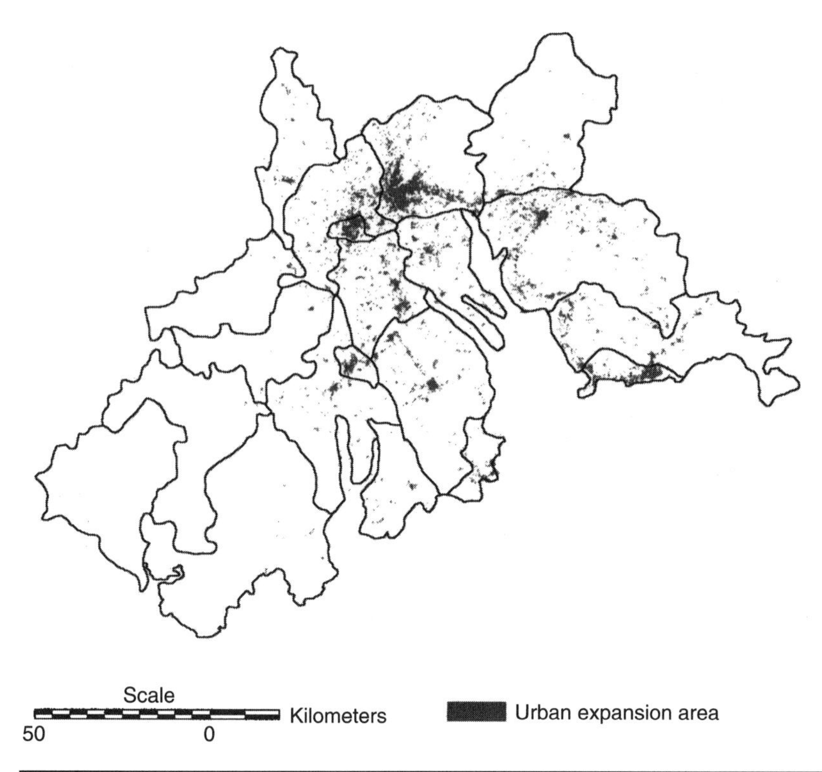

FIGURE 12.3 Urban expansion detected by Landsat imagery in the Zhujiang Delta, 1989–1997. (*Adapted from Weng, 2001.*)

during the period from 1989 to 1997. Overlaying the 1989 and 1997 LULC maps reveals that most of the increase in urban or built-up land comes from cropland (37.92 percent) and horticulture farms (16.05 percent). Figure 12.3 shows the areal extent and spatial occurrence of the urban expansion. The overlay of this map with the city/county boundaries reveals the spatial occurrence of urban expansion within administrative regions. In absolute terms, the greatest urban expansion occurred in Dongguan (23478.90 ha), Baoan (14941.08 ha), Nanhai (8004.1 ha), and Zhuhai (5869.71 ha). However, in percentage terms, the largest increase in urban or built-up land occurred in Zhuhai (1100.00 percent), followed by Shenzhen (306.65 percent), Baoan (233.33 percent), and Dongguan (125.71 percent). Massive urban sprawl in these areas can be ascribed to rural urbanization, which is a common phenomenon in postreform China. In contrast, the old cities, such as Guangzhou and Foshan, do not show a rapid increase in urban or built-up land because they have no land on which to expand further (because they already expanded fully in the past) and because of the concentration of urban enterprises in the city proper. Shenzhen and Zhuhai were designated as special economic zones at the same time, but the pace of urbanization in the two cities is quite different. Urban development in Shenzhen was mostly complete in the 1980s, whereas Zhuhai's urban expansion appears primarily during the period of 1989 to 1997 (5869.71 ha).

Spatial patterns of urban expansion can be examined by investigating the pattern and rate of change of urban LULC over space and time. Because proximity to a certain object, such as a city center or a road, has important implications in urban land development, urban expansion processes often show an intimate relationship with distance from these geographic objects. The density of urban expansion in the Zhujiang Delta between 1989 and 1997 is plotted against the average distance from a city's geometric center in Fig. 12.4. The figure shows that as the distance increases away from a city center, the density of urban expansion increases first to a peak and then slowly decreases. The break line has an average distance of 3500 m from a city center. A further inquiry into the density distribution curve reveals that the majority of urban expansion (72.85 percent) occurs in the zone from 2500 to 5500 m from a city center. Using a best-fit technique, a mathematical relationship between the density of urban expansion Y and the distance from a city center X can be established:

$$Y = -0.00575 + 0.0000885X \qquad X \leq 3500\text{ m}$$
$$Y = 18.975676e^{-0.00088X} \qquad X > 3500\text{ m} \qquad (12.3)$$

Urban expansion processes in the Zhujiang Delta during the period from 1989 to 1997 are further examined by plotting a distance decay

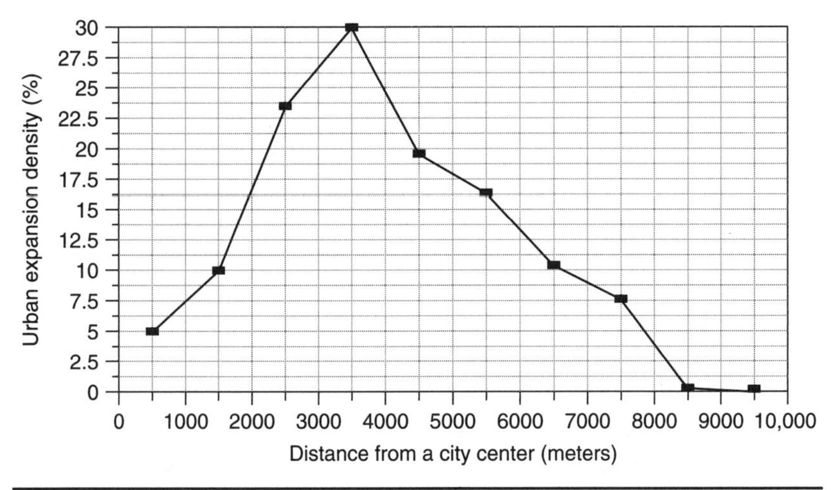

FIGURE 12.4 Urban expansion density as a function of distance from the geometric center of a city in the Zhujiang Delta. (*Adapted from Weng and Lo, 2001.*)

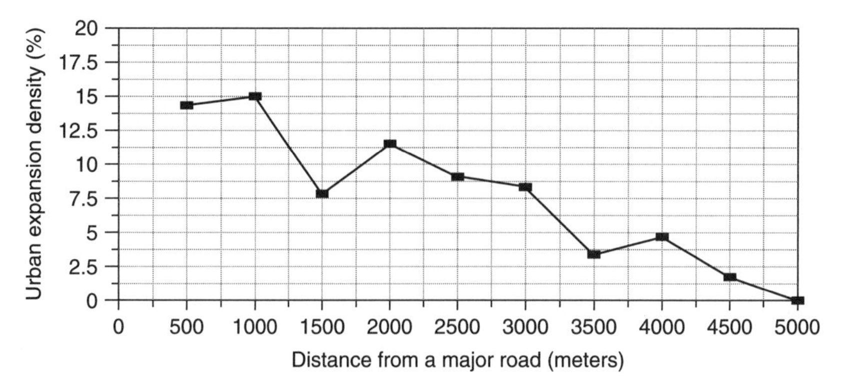

FIGURE 12.5 A distance decay curve of urban expansion in the Zhujiang Delta. (*Adapted from Weng and Lo, 2001.*)

curve from a major road and establishing a mathematical equation. Figure 12.5 indicates that the density of urban expansion decreases as the distance increases away from a major road. Most urban expansion (66 percent) can be observed within a distance of 2000 m from a major road. This rapid urban expansion pattern is vividly illustrated along the superhighway from Guangzhou to Hong Kong, as seen in Fig. 12.3, where Hong Kong investors seek sites for constructing factories and housing. The relationship between the density of urban expansion and the distance from a major road can be expressed mathematically as

$$Y = 0.2237267e^{-0.00046X} \tag{12.4}$$

12.2.5 Urban Growth and Socioeconomic Development

The relationship between urban growth and social and economic factors was explored initially by correlation analysis. Table 12.1 shows the results of correlation analysis between urban expansion rate and all possible explanatory variables. It was found that urban expansion rate is closely related to per-capita gross value of industrial and agricultural output in 1997 ($r = +0.6367$), per-capita gross value of industrial output in 1997 ($r = +0.6365$), and urban population percentage in 1997 ($r = +0.5024$). These values of multiple r's are higher than those in 1989. The increases in correlation coefficient indicate that the urban expansion process in the Zhujiang Delta has become more and more related to industrial development and urban population growth. The correlation between urban land development and urban population percentage suggests that there probably was an improvement in people's living space and per-capita share of urban infrastructure over the study period. In addition, a weak correlation was observed between urban land development and all agricultural variables.

Variable	Correlation Coefficient for 1989 Data*	Correlation Coefficient for 1997 Data*
Per-capita gross value of industrial and agricultural output	0.4173	0.6367
Per-capita gross value of agricultural output	0.0436	0.3910
Per-capita gross value of industrial output	0.4103	0.6365
Per-capita gross domestic production	0.4978	0.4082
Per-capita grain production	0.3656	0.3208
Population density	0.2524	0.16250
Urban population percentage	0.3723	0.5024

*Correlation coefficients are significant at the 0.05 level.

TABLE 12.1 Summary of Correlation Analysis between Urban Expansion Rate and Socioeconomic Variables by City/County

The mapped pattern of the urban expansion rate also shows a positive correlation with that of urban population increase ($r = +0.7507$), the rate of gross value of industrial and agricultural output ($r = +0.5024$), and the rate of grain production ($r = +0.5162$) (Table 12.2). These dynamic relationships suggest that urban population and overall economic performance in the region are the two most critical factors contributing to the urban dispersal. In other words, spatially dispersed urban land is more or less concentrated in the socioeconomically advanced region of the delta.

Correlation analysis can only provide a general support for the explanatory variables under evaluation, whereas the conceptually and empirically interrelated nature of these variables suggests that the analysis of the driving forces of LULC change should be conducted in a multivariate framework. The independent variables identified above may not be the most important determinants of LULC change because empirical and theoretical associations do not necessarily work under certain circumstances. Multiple regression models have the ability to identify the most significant factors and assess the relative importance of different variables (Wang, 1986). The dynamic behavior of land use can be explained by a few driving forces, as found in many ecological systems (Holling, 1992). Along this line, two multiple regression models have been developed to derive possible driving forces behind the LULC change. The first model consists of

Socioeconomic Variables	Urban Expansion Rate 1989–1997*	Cropland Loss Rate 1989–1997*
The rate of gross value of industrial and agricultural output	0.5024	0.0601
The rate of gross value of agricultural output	0.1773	0.0350
The rate of gross value of industrial output	0.4104	0.0272
The rate of gross domestic production	0.2547	0.3684
The rate of grain production	0.5162	0.1889
Population-density increment	0.1000	0.5447
The rate of urban population growth	0.7507	0.3118

*Correlation coefficients are significant at the 0.05 level.

Table 12.2 Summary of Correlation Analysis between the Rates of LULC Change and the Change Rates of Socioeconomic Variables by City/County

regression of the urban expansion rate against the possible driving forces, and the second model consists of regression of the decline rate of cropland against the forces. The modeling results are

$$\text{UREXP\_RATE} = 100.925 + 588.464\text{UBPOP\_RATE} - 0.08659\text{PCGVIAO97} + 0.08676\text{PCGVIO97} \quad (12.5)$$

This multiple regression is significant at the 0.05 level. The multiple coefficient of determination R^2 is 0.671, indicating that the predictors (UBPOP_RATE, PCGVIAO97, and PCGVIO97) account for 67.1 percent of variance in the urban expansion rate. The adjusted squared multiple R reduces this proportion to 0.562, a level expected when using this model in a new sample from the same population. The standardized multiple regression coefficient betas are 8.426 for PCGVIO97, –8.312 for PCGVIAO97, and 0.799 for UBPOP_RATE, respectively. Thus the order of importance in which socioeconomic factors drive urban expansion is PCGVIO97, PCGVIAO97, and UBPOP_RATE. This result suggests that the urban land dispersal in the delta is more related to industrialization than to urban population growth and that industrial development is the most likely cause of the urban land development.

12.2.6 Major Types of Urban Expansion

The computation of the entropy index indicates that the cities in the Zhujiang Delta have an average entropy of 0.64 from a city center, implying a relatively high degree of urban dispersion. However,

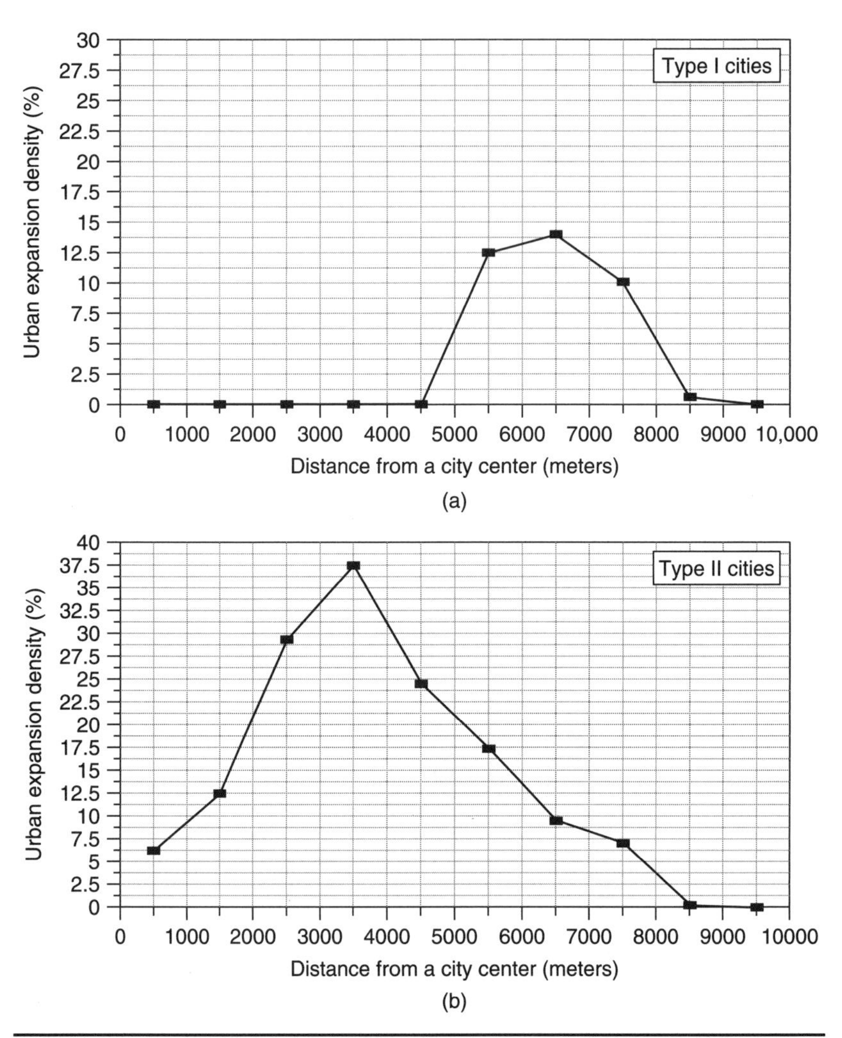

Figure 12.6 Two types of urban expansion patterns in the Zhujiang Delta.

significant differences in the entropy-index value persist among the cities examined. Two major types of urban expansion can be identified: concentrated and dispersed. The difference in the urban expansion pattern is further illustrated by a distance decay curve (Fig. 12.6). Guangzhou and Jiangmen belong to the first type, with an average value of the entropy index of 0.42. Most of the urban expansion in these cities in the study period is far away from the city centers. Guangzhou's urban development occurs primarily 6500 m outside the city center, whereas Jiangmen's urban development occurs 5500 m away from its center. The second type of urban expansion pattern has an average value in the entropy index of 0.69, including Zengcheng, Nanhai, Dongguan, Shunde, Xinhui, Zhongshan, Shenzhen, and Zhuhai. Urban development in these cities is found to be more scattered and spreads over to the suburban and surrounding rural areas from satellite images. The urban expansion zone for these cities lies between 2500 and 8500 m from city centers. These differences in urban expansion pattern reflect, to a large extent, the differences in the history of urban development. Guangzhou and Jiangmen have long been designed to function as pure urban centers, supported primarily by secondary and tertiary production. Most of the land near the urban centers was filled before 1978 when China initiated the economic reform policy. Recent urban development in these cities has had to seek spare land in the suburban areas, which are far from the city center. In contrast, cities that exhibited a more dispersed development pattern during the period of 1989 to 1997 were county-seat towns or even small towns before 1978. Urban development in these cities therefore has had much more freedom and is subject to the influence of the economic reform policies. In the Zhujiang Delta, the highly dispersed urban development pattern can be attributed to rural urbanization, the influence of Hong Kong and foreign investment, and the lack of urban development planning.

12.2.7 Summary

The use of Landsat TM data to detect LULC changes generally has been a success. The digital image classification, coupled with GIS, has demonstrated an ability to provide comprehensive information on the direction, nature, rate, and location of LULC changes as a result of rapid industrialization and urbanization. Given that digital LULC data are available, a further examination of the urban dispersal patterns or changes for other individual categories can be pursued. It has been demonstrated that the combined use of GIS and spatial modeling is sufficiently powerful to discern urban expansion patterns. The spatial process of urban expansion in the delta has shown an intimate relationship with the distance from major roads and from the geometric center of a city.

Satellite-derived LULC data can be linked with socioeconomic data to study the driving forces of LULC change or other environmental problems. Correlation and regression analyses are the two useful integrated approaches. The results of the analysis from this chapter suggest that urban land development in the delta is closely related to industrial development and urban population growth.

12.3 Discussion: Integration of Remote Sensing and GIS for Urban Growth Analysis

Methodologically, this chapter has focused on the development of an integrated approach of remote sensing, GIS, and spatial analysis for urban growth studies. It has been demonstrated that the integrated approach is sufficiently powerful and is needed for evaluating urban growth and its relationship with its driving forces.

The technique of remote sensing has been used to derive LULC maps by means of classification of satellite images. GIS then allows for detection of changes between the two years and extraction of information on changes in urban cover. These raster-based digital data have been further combined with vector-based socioeconomic data to investigate the driving forces of urban growth. Correlation and regression analyses are the two approaches for integration. Finally, the spatial patterns of urban expansion are examined by integrated use of remote sensing, GIS, and spatial analysis techniques. GIS are used to compute the value of satellite-derived urban expansion density in each buffer zone, and these data, in turn, become the input to subsequent spatial analyses, including the derivation of distance decay curves.

Technically, this chapter has used an integration strategy called *loose coupling* (Goodchild et al., 1992), in which remote sensing, GIS, and spatial analysis interplay by exchanging data files. This study demonstrates that this integration algorithm allows full use of existing analytical techniques and produces satisfactory results without use of any advanced computer programming language. However, this approach has limitations. For example, it may be error-prone in transferring data from one system to another, and is also time-consuming.

Another knotty problem in the integrated approach is related to the data model. Remote sensing systems collect data in raster mode, whereas most GIS datasets are built in vector format. The combined use of remotely sensed data with cartographic data generated in GIS requires going through three stages: rasterization/vectorization, registration, and overlay. A great many problems are produced in the format-conversion process that requires a lot of time and effort to sort out. This study has approached the raster-vector overlay problem by means of creation of raster masks. However, the best solution for this kind of problem can be achieved only when a closer integration comes into being, which will enable us to query raster pixels within vector polygons and to combine image statistics with socioeconomic data within polygons.

References

Campell, J. B. 1983. *Mapping the Land: Aerial Imagery for Land Use Information.* Washington, D. C.: Association of American Geographers.

Goodchild, M., Haining, R., Wise, S., et al. 1992. Integrating GIS and spatial data analysis: Problems and possibilities. *International Journal of Geographical Information Systems* **6,** 407–423.

Holling, C. S. 1992. Cross-scale morphology, geometry, and dynamics of ecosystems. *Ecological Monographs* **62,** 447–502.

Jacobson, H. K., and M. F. Price. 1991. A Framework for Research on the Human Dimensions of Global Environmental Change. Geneva: *International Social Science Council Human Dimensions of Global Environmental Change Program (Report No. 1).*

Meyer, W. B., and B. L. II., Turner. 1992. Human population growth and global land-use/land-cover change. *Annual Review of Ecology and Systematics* **23,** 39–61.

Meyer, W. B. and B. L. II, Turner. 1994. *Changes in Land Use and Land Cover: A Global Perspective.* Cambridge: Cambridge University Press.

Mulligan, G. F. 1991. Equality measures and facility location. *Papers in Regional Science,* 70(4), 345–365.

Nissan, E., and G. Carter. 1994. Measuring interstate and interregional income inequality in the United States. *Economic Development Quarterly* 8(4): 364–372.

Stern, P., Young, O. R., and D. Druckman. 1992. *Global Environmental Change: Understanding the Human Dimensions.* Washington, DC: National Academy Press.

Theil, H. 1967. *Economics and Information Theory.* Amsterdam: North-Holland.

Turner, B. L. II, Ross, R. H. and D. L. Skole. 1993. Relating Land Use and Global Land-cover Change: a Proposal for and IGBP-HDP Core Project. Stockholm: *International Geosphere-Bioshere Program and the Human Dimensions of Global Environmental Change Programme (IGBP Report No. 24 and HDP Report No. 5).*

Turner, B. L. II., Skole, D., Sanderson, S., et al. 1995. Land-Use and Land-Cover Change: Science and Research Plan. Stockhdm and Geneva: *International Geosphere-Bioshere Program and the Human Dimensions of Global Environmental Change Programme (IGBP Report No. 35 and HDP Report No. 7).*

Turner, B. L. II., Meyer, W. B., and D. L., Skole. 1996. Global land-use/land-cover change: toward an integrated programme of study. In *Issues in Global Change Research: Problems, Data and Programmes,* ed. L. A. Kosinshi, pp. 99–108. Geneva: *International Social Science Council Human Dimensions of Global Environmental Change Program (Report No. 6).*

Wang, X. 1986. *Multivariate Statistical Analysis of Geological Data.* Beijing: Science Press.

Weng, Q. 2001. A remote sensing–GIS evaluation of urban expansion and its impact on surface temperature in the Zhujiang Delta, China. *International Journal of Remote Sensing* **22,** 1999–2014.

Weng, Q., and Lo, C. P. 2001. Spatial analysis of urban growth impacts on vegetative greenness with Landsat TM data. *Geocarto International* **16,** 17–25.

CHAPTER 13

Public Health Applications

West Nile virus (WNV) is a mosquito-borne disease. It was first discovered in the West Nile district of Uganda in 1937. According to the documentation from the Centers for Disease Control and Prevention (CDC), WNV has been found in Africa, the Middle East, Europe, Oceania, West and Central Asia, and North America. It appeared in the New York metropolitan area in 1999. The spread of WNV has shown unique spatial patterns in different regions (Hodge and O'Connell, 2005; Komar, 2003; Pierson, 2005; Rainham, 2005; Ruiz et al., 2007). Environmental determinants, such as the presence of suitable habitats, temperatures, and climates, play important roles in WNV dissemination in North America (Ruiz et al., 2004; William et al., 2006).

Remote sensing and geographic information systems (GIS) techniques have been applied extensively in public and environmental health studies (Hay and Lennon, 1999; Lian et al., 2007; Ruiz et al., 2004, 2007; Sannier et al., 1998; Wang et al., 1999; Weng et al., 2004). Previous research includes examination of diverse epidemiologic issues, such as parasitic diseases and schistosomiasis, through the use of remote sensing and GIS as exclusive sources of information for studying epidemics. The accessibility of multitemporal satellite imagery effectively supports the study of epidemiology (Herbreteau et al., 2007). This chapter develops a multitemporal analysis of the relationship between environmental variables and WNV dissemination using an integration of remote sensing, GIS, and statistical techniques. The specific research objectives are to identify the spatial patterns of WNV outbreaks in different years and seasons in the city of Indianapolis, Indiana, and to examine the temporal dynamics of the relationships between WNV dissemination and environmental variables. Indianapolis is a typical Midwest city lying on a flat plain and has a temperate climate without pronounced wet or dry seasons. Therefore, this case study not only can offer valuable information for public health prevention and mosquito control in the study area but also can provide testimony relating to the spread of WNV in the other regions of the Midwest and beyond.

13.1 WNV Dissemination and Environmental Characteristics

WNV is a seasonal epidemic in North America that normally erupts in the summer and continues into the fall, presenting a threat to environmental health. Its natural cycle is bird-mosquito-bird and mammal. Mosquitoes, mainly the genus *Culex,* in particular the species *C. pipiens,* become infected when they feed on infected birds. Infected mosquitoes then spread WNV to birds and mammals such as humans when they bite. Fatal encephalitis in humans and horses is the most serious manifestation of WNV infection, as well as mortality in some birds.

Field and laboratory records of entomologic and ecologic observations have been used to examine how natural environmental constraints, such as water sources and climatic parameters, contribute to the transmission of WNV (Dohm and Turell, 2001; Dohm et al., 2002; Gingrich et al., 2006). *Culex* species appear to prefer certain land use and land cover (LULC) types (e.g., wetlands and specific grasslands) over certain others (e.g., exposed dry soils). Mosquitoes in canopy sites are believed to possess more infections than those in subterranean sites and on the ground (Anderson et al., 2006). Wetlands and stormwater ponds, especially those that are heavily shaded, provide an ideal environment for mosquito settlement. Ponds with plenty of sunshine and a shortage of vegetation are believed to be a poor environment for the growth of mosquitoes (Gingrich et al., 2006). WNV dissemination is found to be significantly related to average summer temperatures from 2002 to 2004 in the United States (Reisen et al., 2006). Doham and Turell (2001) found that the infection rates of WNV in mosquitoes were lower at cooler temperatures than when those vectors were maintained at warmer temperatures. The infection rates started to increase after 1 day of incubation at 26°C. WNV dissemination began more rapidly in mosquitoes settled at higher temperatures than in mosquitoes maintained at cooler temperatures (Dohm et al., 2002). Gingrich and colleagues (2006) detected a bimodal seasonal distribution of mosquitoes with peaks in early and late summer in Delaware in 2004.

Remote sensing and GIS have been applied in WNV studies in different regions with the application of statistical analysis (Cooke et al., 2006; Griffith, 2005; Lian et al., 2007; Liu and Weng, 2009; Liu et al., 2008; Ruiz et al., 2004, 2007). Ruiz and colleagues (2004) found that some environmental and social factors contributed to WNV dissemination in Chicago in 2002 by using GIS technologies and multi-step discriminant analysis. Those factors included distance to a WNV-positive dead bird specimen, age of housing, mosquito abatement, presence of vegetation, geologic factors, age, income, and race of the human population. Multiple mapping techniques were compared for WNV dissemination in the continental United States (Griffith, 2005). The results indicated that each mapping technique emphasized certain WNV risk factor(s) owing to the differences in modeling assumptions, statistical treatment, and error determination. No particular model performed better than all others. Cooke and colleagues (2006) estimated WNV risk in the state of Mississippi based on human and bird cases recorded in years 2002 and 2003 with the creation of avian GIS models. The results indicated that high road density, low stream density, and gentle slopes contributed to the dissemination of WNV in Mississippi. GIS and spatial-time statistics were applied for a risk analysis of the 2002 equine WNV epidemic in northeastern Texas (Lian et al., 2007). A total of nine nonrandom spatial-temporal equine case aggregations and five high-risk areas were detected in the study area. Ruiz and colleagues (2007) further examined the association of WNV infection and landscapes in Chicago and Detroit using GIS and statistical analysis. Their results showed that a higher WNV case rate occurred in the inner suburbs, where housing ages were around 48 to 68 years old and vegetation cover and population density were moderate.

13.2 Case Study: WNV Dissemination in Indianapolis, 2002–2007

13.2.1 Data Collection and Preprocessing

A detailed description of the study area can be found in Sec. 3.2.1 in Chapter 3. WNV epidemic records include the locations of positive sites in the city of Indianapolis, the number of positive mosquito pools in each of those sites, the number of mosquitoes in each pool, and the dates of the sampling records. The time period is monthly from April to October in 2002 through 2007. The information was provided by the Indiana State Department of Health. Since this study was to process a multitemporal analysis, all positive sites in each year were geocoded and visualized based on their physical locations using ArcGIS. Sites with positive WNV records in July through October from 2002 through 2007 also were geocoded, respectively, to identify monthly variations. WNV records in April through June were not used for studying the monthly variations owing to much fewer epidemic cases.

Environmental factors such as the presence of vegetation, mosquito abatement, age of housing, and stream density were believed to have contributed to the dissemination of WNV (Cooke et al., 2006; Lian et al., 2007; Ruiz et al., 2004). Some of these factors were used in this study, and other factors were added to the analysis owing to data accessibility and local environmental conditions. Land-use and land-cover (LULC) information was derived from three Advanced Spaceborne Thermal Emission and Reflection Radiometer (ASTER) images acquired on June 16, 2001, April 5, 2004, and October 13, 2006, respectively. An unsupervised image classification method was chosen to classify each image into six categories: urban, agriculture, forest, grasslands, water, and barren lands. Overall accuracy of image classification reached 88.33, 92.0, and 89.0 percent individually. The resulting LULC maps are shown in Fig. 7.1*b* (for the image of June 16, 2001) and Fig. 7.1*c* (for the image of April 5, 2004). Figure 13.1 shows the classified map for the image of October 13, 2006. It can be seen in the image that urban, forest, and grassland were dominant habitats, where urban fell in the central part of the city and forest was located mainly in the north mixed with grassland. Agriculture mainly lay in the southeastern and southwestern parts of the city. There were two major reservoirs, the Geist Reservoir and the Eagle Creek Reservoir, located in northern part. The White River runs north-south through the city in the center.

Other environmental data were collected from the Indiana Geological Survey Web site (http://igs.indiana.edu/). These data included estimated percentages of impervious surfaces in Indiana in 2001, digital elevation model with 1.5-m resolution, hydrogeologic terrains and settings, national wetland inventory data, pipe locations in the National Pollutant Discharge Elimination System (NPDES), facilities in NPDES, and industrial waste sites in Indiana. The outlines

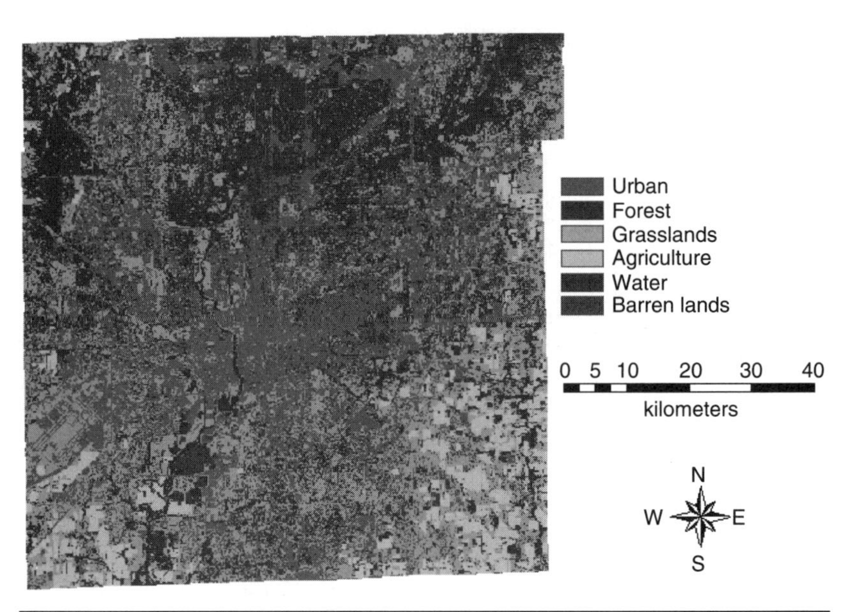

FIGURE 13.1 LULC map of Indianapolis, Indiana, on October 13, 2006. See also color insert.

of the U.S. Census block groups also were downloaded from the same Web site. There were a 658 Census block groups in this study area. This study was based on an assumption that selected environmental variables in a certain Census block groups contributed the most to WNV dissemination in that block group. As a result, all factors were summarized and analyzed at the Census block-group level. A centroid of each block group was calculated by using ArcGIS. The shortest distances from each centroid to the pollutants and industrial waste sites were calculated. The total length of streams and the area of wetlands in each block group also were computed in ArcGIS. Human population density in each block group was selected to examine the possible influence of human behavior on the spread of WNV. Table 13.1 lists all the variables used in the study.

Variable	Description	Mean for All 658 Block Groups in Indianapolis
Land use land cover (LULC)	Area percentage of each LULC category	Varied by seasons and years
Impervious surface	Average percentage of impervious surface	38.8 percent
Elevation	Elevation variation	12.9 m (42.3 ft)
Slope	Mean slope	1.6 degrees
Stream	Total stream length and stream density	466.7 m (1531.2 ft)
Wetland	Total size of wetlands	51,437.4 m^2 (553,667.6 ft^2)
Pollutant	Distance from centroid of each block group to the closest pollutant in National Pollutant Discharge Elimination System (NPDES)	3201.8 m (10,504.6 ft)
Pipe	Distance from centroid of each block group to the closest pipe in NPDES	7831.3 m (25,693.2 ft)
Waste industry	Distance from centroid of each block group to the closest waste industry site	804.9 m (2640.7 ft)
Population density	Human population density	1587/km^2

Note: Study unit: U.S. Census block groups.

Table 13.1 Environmental Variables Selected for the Study in Indiana, 2001

13.2.2 Plotting Epidemic Curves

WNV was first identified in Indiana in 2001, and its transmission was enhanced during the summer of 2002. *C. pipiens* was the main vector mosquito in the state. Six cumulative epidemic curves were created to show the peaks and temporal trends of mosquito WNV outbreaks in Indianapolis in years 2002 through 2007. Epidemic patterns could be identified based on monthly and annual comparisons. The spatial outbreaks of WNV were tracked to indicate the movement of WNV dissemination in the last 6 years.

13.2.3 Risk Area Estimation

It was significant to identify the risk areas of WNV mosquito outbreaks so that special care could be taken. A retrospective space-time permutation model in SaTScan software was selected to identify nonrandom WNV clusters in years 2002 through 2007. This space-time model was based on a null hypothesis that there was not any spatial, temporal, or spatio-temporal clusters in the study area. The p value for the most likely cluster would be larger than 0.05 with a 95-percent probability of change. Monte Carlo hypothesis testing was used to calculate the test statistic for all possible clusters and 999 random replications. If a possible cluster were among the 5 percent highest, then the significance level of the test would be 0.05 (Dwass, 1957). The centroids and radii of the most likely clusters with p values of less than 0.01 were recorded in a DBF (.dbf file format, a major feature of dBase) table and then visualized by using ArcGIS.

A *K*-means cluster analysis was developed to identify the high-risk areas for each month of July, August, September, and October. In order to increase the sampling size in the statistical analysis, the records collected in the same month but from different years were combined. Three variables were selected from each location: the coordinates of the location, the number of positive mosquito pools in each month, and the total number of mosquitoes in those pools. The spatial variations of high-risk areas in different months were identified based on the results of cluster analysis.

13.2.4 Discriminant Analysis

Some environmental variables are considered to have more influence than others on the spread of WNV. In order to identify which environmental factors played significant roles in WNV dissemination in the study area, a stepwise Wilks' lambda discriminant analysis was applied to compare the block groups with WNV cases with those without any cases relative to the set of environmental factors in Table 13.1. There were 658 block groups used in this study. The analysis was done by year.

13.2.5 Results

Seasonal Outbreaks of WNV

Figure 13.2 presents epidemic curves of the WNV mosquito from 2002 through 2007. In the figure, the *y* axis presents mosquito counts of WNV-positive tests. It becomes clear that year 2002 had positive mosquito records from July to October; year 2003, from June to October; year 2004, from May to October; and year 2005, from April to October; whereas the records started from June to September in 2006 and from May to August in 2007. The epidemic curve reaches a peak in August in each year of 2002 through 2006 and in July in 2007. The variations in the curves are consistent with the ecological observation that WNV erupted in the summer and continued into the fall. We can conclude that the outbreak of WNV was in August or July in Indianapolis in last 6 years. As for the spatial patterns of WNV outbreaks, WNV dissemination always started from central longitudinal corridor and spread out to the west and east. This observation indicates that some environmental conditions may have significantly affected the spatial patterns of the outbreaks. The results of discriminant analysis below provide important clues to explain this observation.

Risk Areas

Spatial-temporal WNV clusters were detected in each year of 2002 through 2007 using a space-time permutation model in the SaTScan program. Figure 13.3 shows that 2002 had six spatio-temporal clusters,

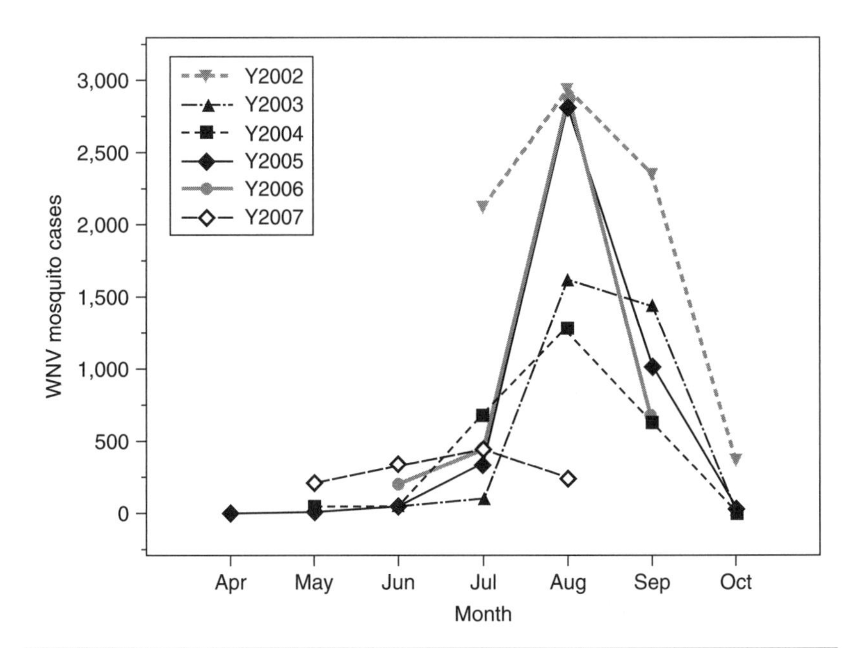

FIGURE 13.2 Epidemic curves of mosquito-borne WNV in 2002 through 2007.

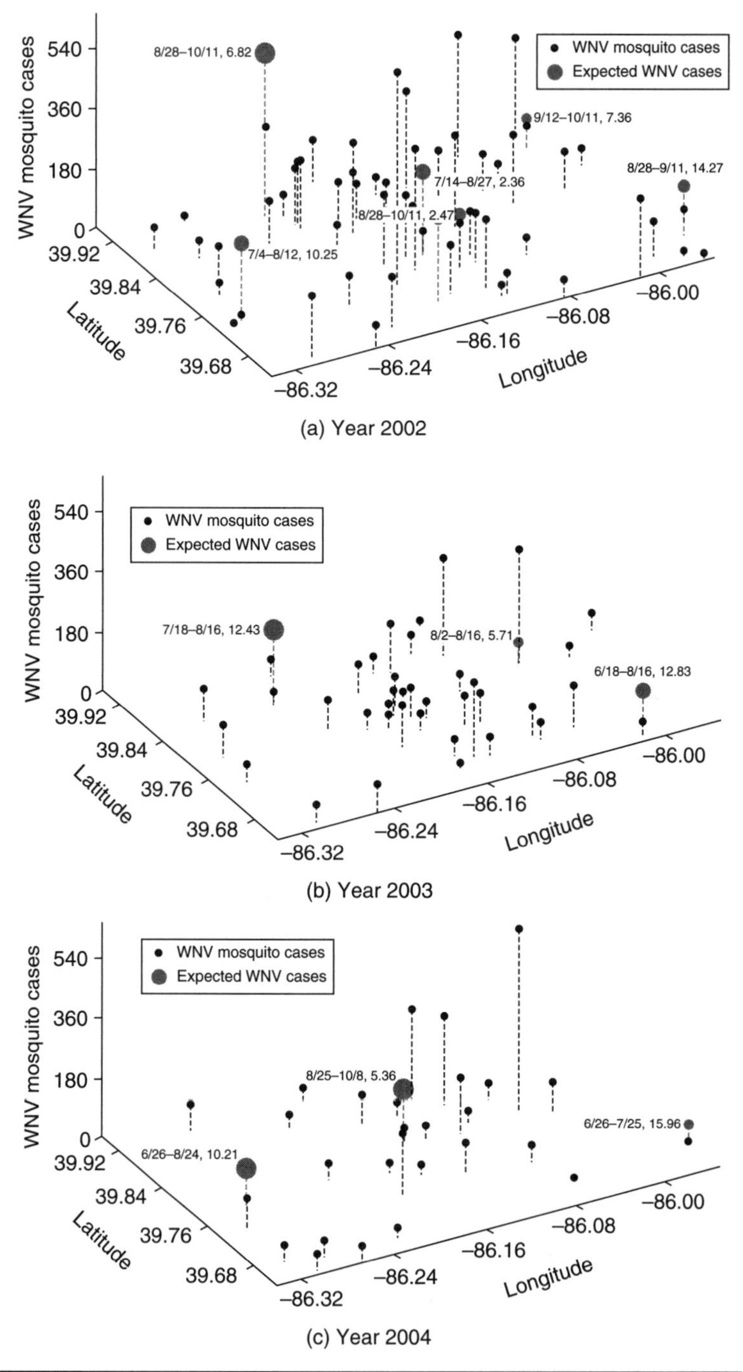

FIGURE 13.3 WNV mosquito cases and their spatial-temporal clusters in the years 2002 through 2007.

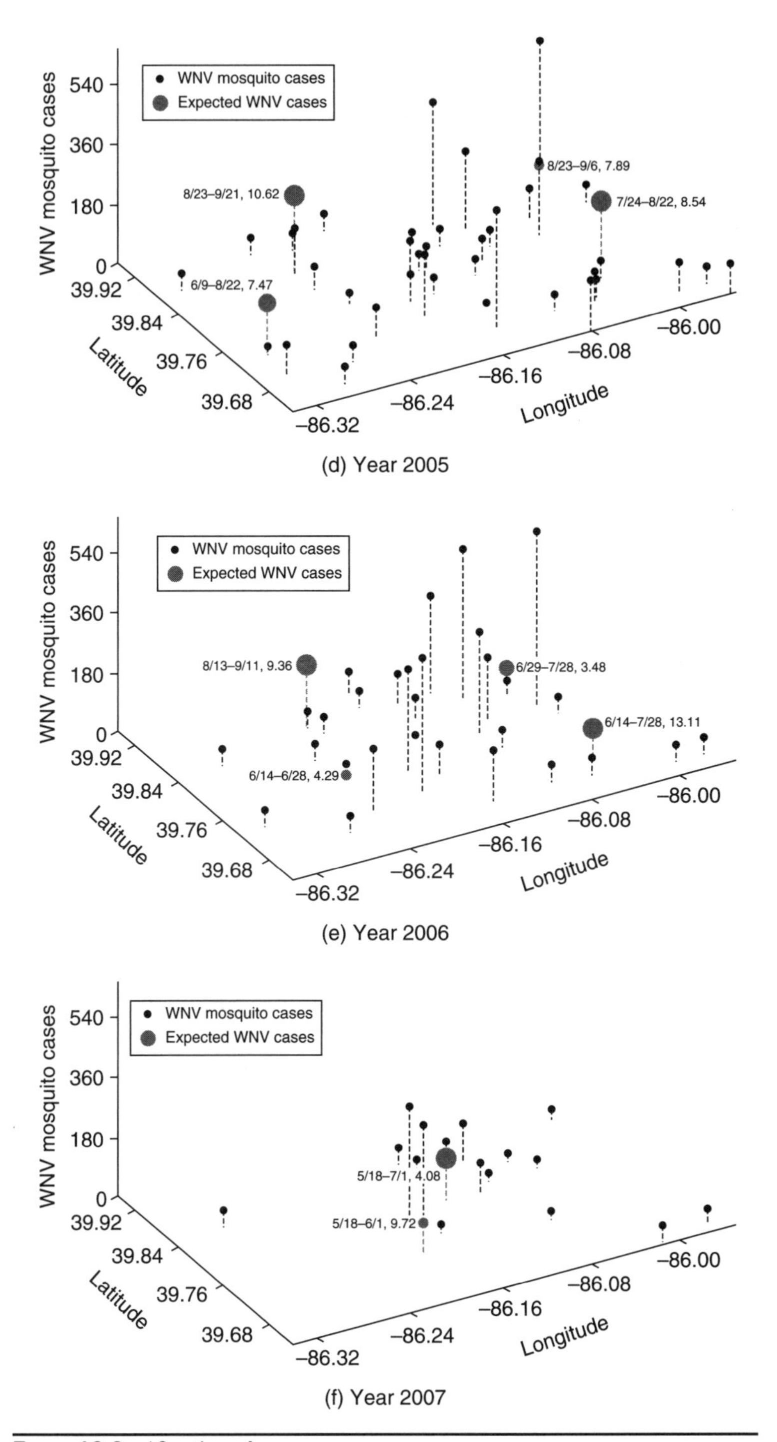

(d) Year 2005

(e) Year 2006

(f) Year 2007

Figure 13.3 (*Continued*)

whereas both 2003 and 2004 had three clusters. There were four clusters in each of 2005 and 2006 and two clusters in 2007. Each cluster had a certain time period of outbreak and individual radius. The p value for rejecting the null hypothesis of no clustering is 0.001. Although these spatial-temporal clusters appeared randomly in space, multiple clusters can be found on the southeast corner of the study area in 2002 through 2006 with large radii (more than 8 km) and a time period ranging from June to September. Two clusters from two different years (2002 and 2004) even shared the same centroid (39°69′N, 85°95′W) but with different radii (14.27 and 15.96 km) and periods of outbreak (8/28 through 9/11 and 6/26 through 7/25, respectively). The southeastern city was dominated by agriculture and grassland. A high concentration of clusters with large radii indicates that agriculture and grassland provided favorable and extensive habitats for mosquito breeding during the summer and early fall. Three clusters in 2002, 2004, and 2005 were found to share the same centroid on the vegetation land in the central west of the city. This finding supports the conclusion that mosquitos were attracted by grassland in the study area.

The result of K-means cluster analysis on individual months of July, August, September, and October shows that different months had different high-risk areas. There were two clusters in Julys, four clusters in Augusts, four clusters in Septembers, and two clusters in Octobers in the study period. Figure 13.4 shows the high-risk areas in the four months and their radii, which conform to the explosive transmission of WNV in Indianapolis during the summer. A cluster can be found in the southeastern corner of the city in each month of August and September. This finding supports the result of spatio-temporal analysis in which five WNV clusters were observed in the same area in 2002 through 2006. As noted in the preceding paragraph, this area was dominated by crops (mainly corn and soybean) and grass. In Indiana, corn usually is planted in April and May, and it becomes mature in September and October, whereas soybeans usually are planted in May and June and their leaves are shed mainly in September and October. Dense corn and soybean fields in August and September provide suitable temperature and sufficient moisture for mosquito breeding, which might not happen in July when small leaves are still growing, and in October, when crops are being harvested.

Two clusters appeared in the northwestern part of the city in August and September. The August cluster was close to the Eagle Creek Reservoir, and the September one included the reservoir. According to the records of a U.S. Geological Survey (USGS) streamflow gauging station, mean stream flow in the Eagle Creek watershed was highest in March and lowest in September. Wetlands were expected to appear when the water level decreased from July to August and September, which provided favorable habitats for mosquito

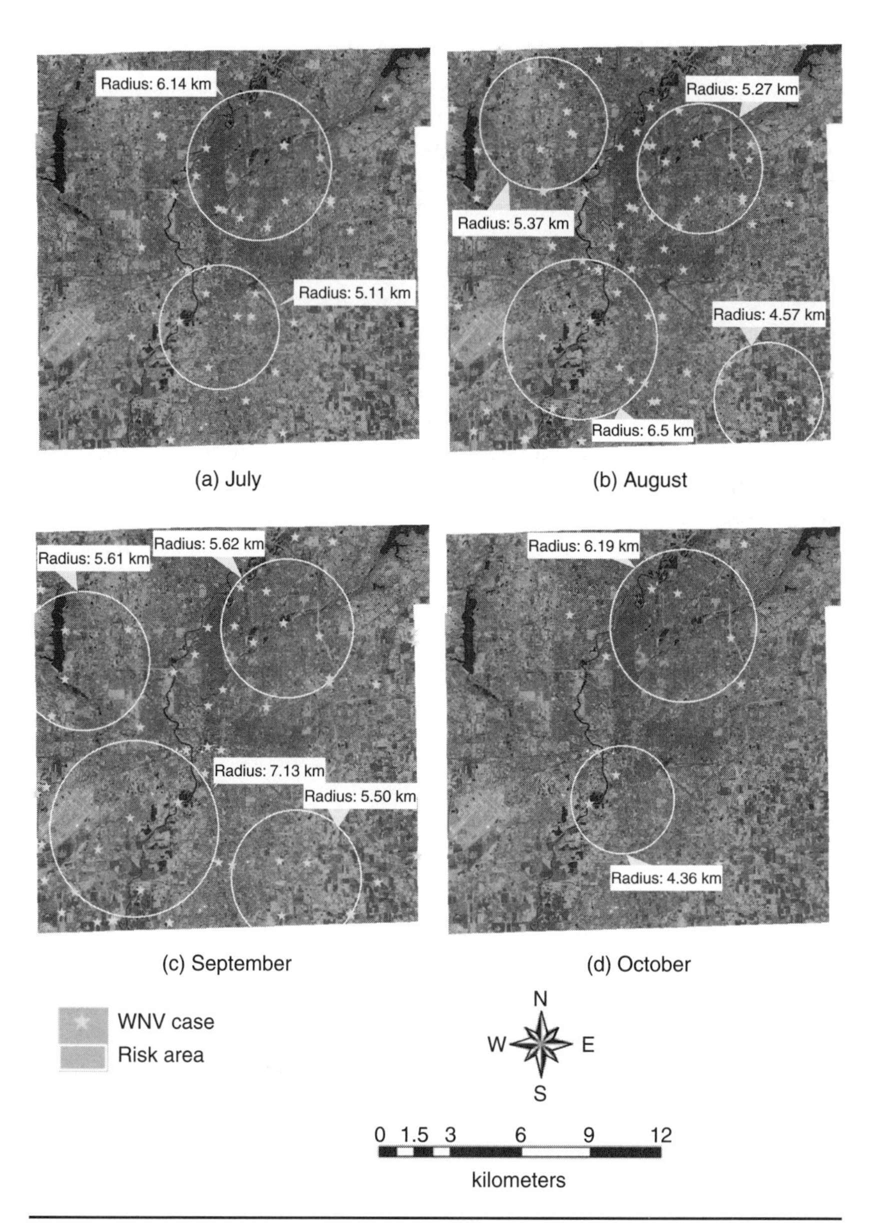

FIGURE 13.4 Risk areas in July, August, September, and October, in the years 2002 through 2007. See also color insert.

breeding combined with the warm temperatures in August and September. There was a cluster consisting of dense ponds and part of the White River in the southeastern city in each month. A possible explanation for this observation is that the ponds and the river provided plenty of moisture and still water for mosquito breeding.

Environmental Factors of WNV Dissemination

Table 13.2 shows the environmental variables that remained after the discriminant analysis; their coefficients, eigenvalues, and grouping accuracies; and the numbers of block groups with mosquito WNV records and without mosquito cases. According to the results of the stepwise Wilks' lambda discriminant analysis, some environmental variables seem to be more effective than others in differentiating block groups with mosquito WNV cases from those without WNV records. Area percentage of agriculture was one of the most effective variables in the discriminant analysis in all years except 2006. A higher proportion of agriculture land was associated with more WNV cases. This finding conforms to the risk-area analysis discussed earlier suggesting that agriculture lands in the southeastern city were always within high-risk clusters in different years. It could be explained that agriculture contributed to a cooler temperature in the surrounding areas in the summer (Liu and Weng, 2008; Weng et al., 2004) and that farming irrigation maintained a relatively constant moisture, which provided a favorable environment for mosquito breeding.

Total size of wetland was another important variable for WNV dissemination in the years 2002, 2005, and 2006. Larger wetlands were linked to more mosquito WNV records. According to regulations of U.S. Environmental Protection Agency (EPA), wetlands are those areas saturated by surface or ground water at a frequency and duration sufficient to support vegetation prevalence. A possible explanation of the observation is that high temperature and low to median precipitation in July and August in these three years created ideal conditions for the wetlands for mosquito breeding.

Total length of streams plays an important role in mosquito WNV dissemination in the years 2002, 2003, and 2007. Longer streams were associated with more WNV cases. This observation is in agreement with a WNV study in Mississippi that suggested that stream density contributed to WNV risk based on dead-bird occurrences (Cook et al., 2006). Both stream density and total length of streams were originally input in the discriminant analysis, but only the total length of streams remained after the analysis. This result suggests that curvy streams with still water provided a favorable environment for mosquitoes breeding. Area percentage of water remained after the discriminant analysis for the years 2004 and 2007. Water information derived from ASTER imagery included big rivers, lakes, and dams. Higher percentage of water sources certainly contributed to more mosquito WNV cases.

In addition to agriculture, streams, wetlands, and some other water sources, certain other variables also showed a positive contribution to WNV dissemination in various years, including mean slope and elevation change for 2004, human population density in 2006, and distance to the closest pollutant and distance to the closest waste

Year	Variable	Classification Function Coefficient	Eigenvalue	Block Group with WNV/ without WNV	Grouping Accuracy
2002	Percentage of agriculture	0.143	0.226	46/612	86.90%
	Total length of streams	0.001			
	Total size of wetlands	0.001			
	Constant	–2.124			
2003	Percentage of agriculture	0.126	0.202	35/623	84.80%
	Total length of streams	0.001			
	Constant	–1.336			
2004	Percentage of agriculture	0.154	0.201	25/633	84.80%
	Percentage of water	0.037			
	Mean slope	1.908			
	DEM variation	0.049			
	Constant	–4.261			
2005	Percentage of agriculture	0.232	0.24	36/622	90.30%
	Total size of wetlands	0.001			
	Constant	–2.499			

Table 13.2 Environmental Factors That Remained after Discriminant Analysis; Their Coefficients, Eigenvalues, and Numbers of Block Groups with and without WNV Cases; and Grouping Accuracies

Year	Variable	Classification Function Coefficient	Eigenvalue	Block Group with WNV/ without WNV	Grouping Accuracy
2006	Total size of wetlands	0.001	0.157	29/629	85.30%
	Human population density	0.001			
	Constant	–2.141			
2007	Percentage of agriculture	0.051	0.195	16/642	88.90%
	Total length of streams	0.001			
	Percentage of water	0.205			
	Distance to the closest pollutant	0.001			
	Distance to the closest waste industry	0.003			
	Constant	–4.709			

Table 13.2 Environmental Factors that Remained after Discriminant Analysis; Their Coefficients, Eigenvalues, and Numbers of Block Groups with and without WNV Cases; and Grouping Accuracies (*Continued*)

industry for 2007. However, these variables were significant only for a specific year, not for the whole time period. Based on the results of discriminant analysis, high ground slope would be expected to contribute to the spread of WNV, which contradicts the fact that when surface slope is high, it is less able to hold mosquito eggs and larva during heavy rainfall. Human population density showed a positive contribution to WNV dissemination in 2006. This indicates that human behavior might have affected the mosquito habitat in 2006. Distance to the closest pollutant and distance to the closest waste industry showed a positive contribution to the spread of WNV in 2007, which may indicate that there are well-built protection systems in the pollutant and waste-industry sites that eliminated mosquito breeding sites. These explanations are tentative. Further studies are warranted to understand the complexity of environmental impacts on WNV dissemination.

13.3 Discussion and Conclusions

This study has conducted a spatio-temporal analysis of the relationship between WNV dissemination and environmental variables using the integration of remote sensing, GIS, and statistical techniques. Results indicate that although epidemic curves peak in the same month (i.e., August) from 2002 through 2006, the number of WNV cases varied by year and by month. In order to explain the temporal variations, two factors that were closely related to mosquito life cycling, namely, temperature and precipitation, were investigated based on the weather records of NOAA's National Weather Service Weather Forecast Office. Figure 13.5*a* and *b* present the monthly mean temperature and mean precipitation in 2002 and 2007 from March through October. Figure 13.5*c* shows the relationships between temperature, precipitation, and WNV cases. Low precipitation and warm temperature were found to be associated with more WNV cases. Based on Fig. 13.5, a possible explanation for the highest WNV counts in July 2002 is that the year experienced the highest temperatures and lowest precipitation in that month compared with any other year that provided an ideal natural environment for mosquito breeding. Years 2005 and 2006 had relatively higher temperatures and lower precipitation in August compared with the years 2003 and 2004. This could help to explain why both years had many more WNV cases in August than in 2003 and 2004. An obvious increase in rainfall from July through August in 2007 may have contributed to lower WNV counts in August 2007 compared with those in July 2007. Other factors also need to be considered to completely understand the temporal variation of WNV cases, especially any mosquito control and abatement measures taken by local governments or organizations.

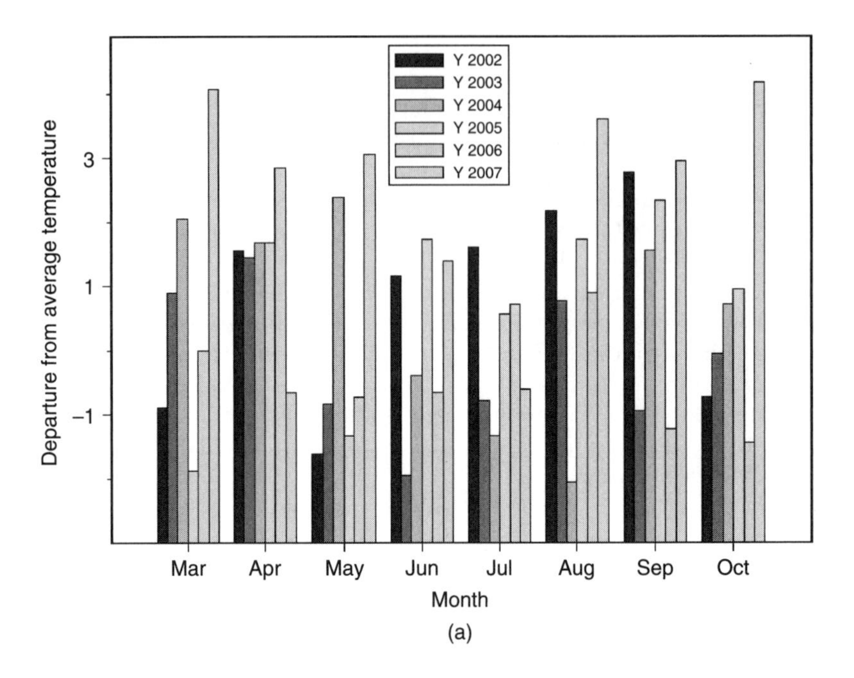

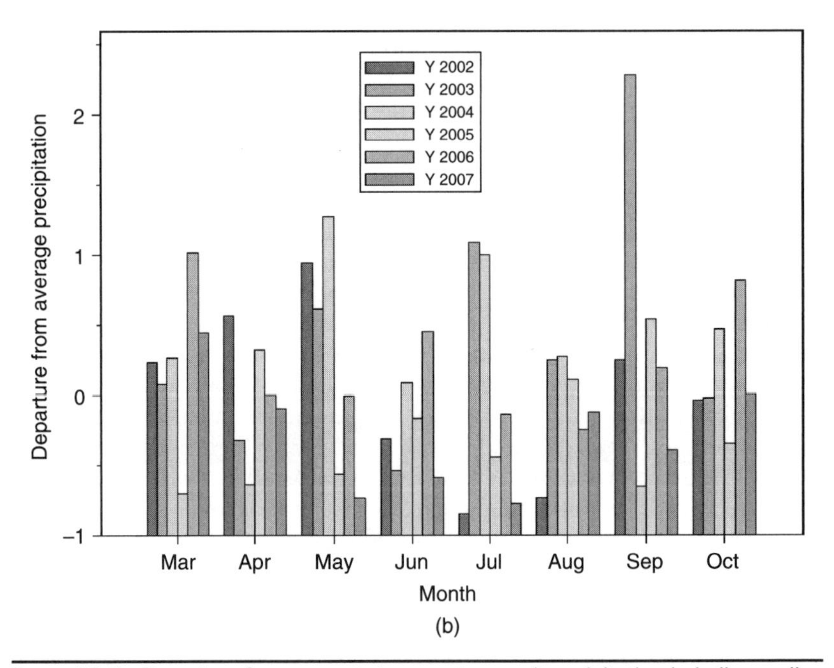

Figure 13.5 Departure from average temperature and precipitation in Indianapolis from March through October in 2002 through 2007. See also color insert.

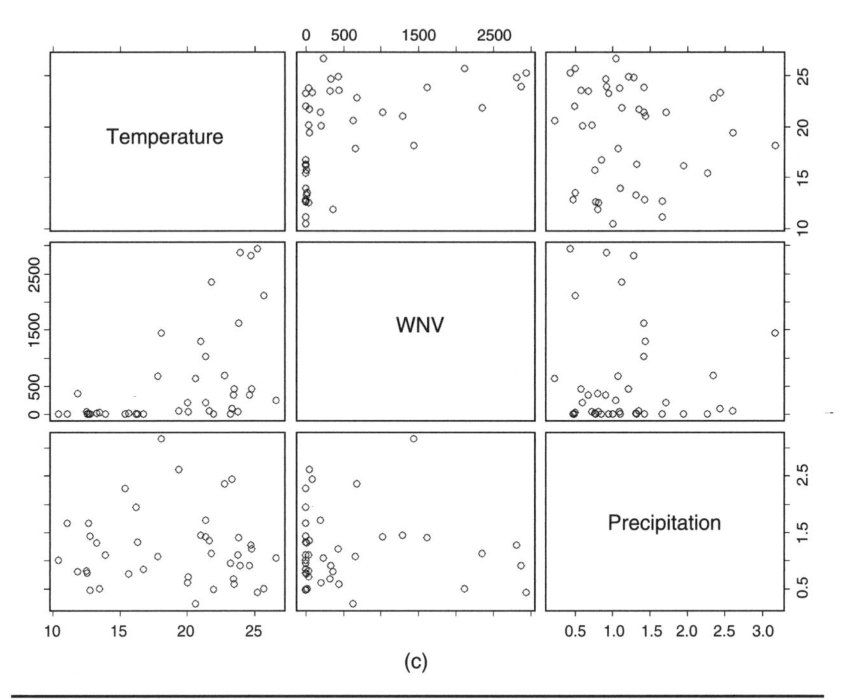

FIGURE 13.5 *(Continued)*

The research also found that WNV dissemination always started from the central longitudinal corridor and spread out to the west and east of the city. Different years and seasons had different high-risk areas. However, the southwestern and southeastern areas were believed to face the highest risk for WNV infection owing to their high percentage of agriculture and water surfaces. Major environmental factors that contributed to the outbreak of WNV in Indianapolis were percentages of agriculture and water, total lengths of streams, and total size of wetlands. Other environmental factors, such as elevation variations, mean slope, human population density, and distances to the closest pollutant and waste industry, need to be examined further in assessing their roles in the dissemination of WNV.

References

Anderson, J. F., Andreadis, T. G., Main, A. J., et al. 2006. West Nile virus from female and male mosquitoes (Diptera: Culicidae) in subterranean, ground, and canopy habitats in Connecticut. *Journal of Medical Entomology* **43,** 1010–1019.

Cooke, W. H., III, Grala, K., and Wallis, R. C. 2006. Avian GIS models signal human risk for West Nile virus in Mississippi. *International Journal of Health Geographics* **5,** 36; available at *www.pubmedcentral.nih.gov/articlerender.fcgi?artid=1618835.*

Dwass, M. 1957. Modified randomization tests for nonparametric hypotheses. *Annals of Mathematical Statistics* **8,** 181–187.

Dohm, D. J., and Turell, M. J. 2001. Effect of incubation at overwintering temperatures on the replication of West Nile virus in New York *Culex pipiens* (Diptera: Culicidae). *Journal of Medical Entomology* **38,** 462–464.

Dohm, D. J., O'Guinn, M. L., and Turell, M. J. 2002. Effect of environmental temperature on the ability of *Culex pipiens* (Diptera: Culicidae) to transmit West Nile virus. *Journal of Medical Entomology* **39,** 221–225.

Gingrich, J. B., Anderson, R. D., Williams, G. M., et al. 2006. Stormwater ponds, constructed wetlands, and other best management practices as potential breeding sites for West Nile virus vectors in Delaware during 2004. *Journal of the American Mosquito Control Association* **22,** 282–291.

Griffith, D. A. 2005. A comparison of six analytical disease mapping techniques as applied to West Nile Virus in the coterminous United States. *International Journal of Health Geographics* **4,** 18; available at *www.ncbi.nlm.nih.gov/pubmed/16076391.* Last accessed on June 15, 2008.

Hay, S. I., and Lennon, J. J. 1999. Deriving meteorological variables across Africa for the study and control of vector-borne disease: A comparison of remote sensing and spatial interpolation of climate. *Tropical Medicine and International Health* **4,** 58–71.

Herbreteau, V., Salem, G., Souris, M., et al. 2007. Thirty years of use and improvement of remote sensing applied to epidemiology: From early promises to lasting frustration. *Health & Place* **13,** 400-403.

Hodge, J. G., Jr., and O'Connell, J. P. 2005. West Nile Virus: Legal responses that further environmental health. *Journal of Environmental Health* **68,** 44–47.

Komar, N. 2003. West Nile virus: Epidemiology and ecology in North America. *Advances in Virus Research* **61,** 185–234.

Lian, M., Warner, R. D., Alexander, J. L., and Dixon, K. R. 2007. Using geographic information systems and spatial and space-time scan statistics for a population-based risk analysis of the 2002 equine West Nile epidemic in six contiguous regions of Texas. *International Journal of Health Geographics* **6,** 42; available at *www.ij-healthgeographics.com/content/6/1/42.*

Liu, H., and Weng, Q. 2008. Seasonal variations in the relationship between landscape pattern and land surface temperature in Indianapolis, U.S.A. *Environmental Monitoring and Assessment* **144,** 199–219.

Liu, H., and Weng, Q. 2009. An examination of the effect of landscape pattern, land surface temperature, and socioeconomic conditions on WNV dissemination in Chicago. *Environmental Monitoring and Assessment* (in press).

Liu, H., Weng, Q., and Gaines, D. 2008. Multi-temporal analysis of the relationship between WNV dissemination and environmental variables in Indianapolis, U.S.A. *International Journal of Health Geographics* **7(66),** 1–13.

Pierson, T. C. 2005. Recent findings illuminate research in West Nile virus. *Heart Disease Weekly,* 247.

Rainham, D. G. C. 2005. Ecological complexity and West Nile Virus: Perspectives on improving public health response. *Canadian Journal of Public Health* **96,** 34–40.

Reisen, W. K., Fang, Y., and Martinez, V. M. 2006. Effects of temperature on the transmission of West Nile virus by *Culex tarsalis* (Diptera: Culicidae). *Journal of Medical Entomology* **43,** 309–317.

Ruiz, M. O., Walker, E. D., Foster, E. S., Haramis, L. D., and Kitron, U. D. 2007. Association of West Nile virus illness and urban landscapes in Chicago and Detroit. *International Journal of Health Geographics* **6,** 10–20.

Ruiz, M. O., Tedesco, C., McTighe, T. J., et al. 2004. Environmental and social determinants of human risk during a West Nile virus outbreak in the greater Chicago area, 2002. *International Journal of Health Geographics* **3,** 8–18.

Sannier, C. A. D., Taylor, J. C., and Campbell, K. 1998. Compatibility of FAO-ARTEMIS and NASA Pathfinder AVHRR Land NDVI data archives for the African continent. *International Journal of Remote Sensing* **19,** 3441–3450.

Wang, Y., Zhang, X., Liu, H., and Ruthie, H. K. 1999. Landscape characterization of metropolitan Chicago region by Landsat TM. In *Proceeding of the ASPRS Annual Conference,* Portland, Oregon, May 19–21, pp. 238–247.

Weng, Q., Lu, D., and Schubring, J. 2004. Estimation of land surface temperature-vegetation abundance relationship for urban heat island studies. *Remote Sensing of Environment* **89**, 467–483.

William, K. R., Fang, Y., and Martinez, V. M. 2006. Effects of temperature on the transmission of West Nile virus by *Culex tarsalis* (Diptera: Culicidae). *Journal of Medical Entomology* **43,** 309–317.

Index

Note: Page numbers referencing figures are followed by an "*f*"; page numbers referencing tables are followed by a "*t*".

Q

R

V

W

X

Z